Models for Dependent Time Series

MONOGRAPHS ON STATISTICS AND APPLIED PROBABILITY

General Editors

F. Bunea, V. Isham, N. Keiding, T. Louis, R. L. Smith, and H. Tong

Monographs on Statistics and Applied Probability 142

Models for Dependent Time Series

Granville Tunnicliffe Wilson

Department of Mathematics and Statistics
Lancaster University
UK

Marco Reale

School of Mathematics and Statistics
University of Canterbury
New Zealand

John Haywood

School of Mathematics and Statistics
Victoria University of Wellington
New Zealand

CRC Press
Taylor & Francis Group
Boca Raton London New York

CRC Press is an imprint of the
Taylor & Francis Group, an **informa** business
A CHAPMAN & HALL BOOK

CRC Press
Taylor & Francis Group
6000 Broken Sound Parkway NW, Suite 300
Boca Raton, FL 33487-2742

First issued in paperback 2020

© 2016 by Taylor & Francis Group, LLC
CRC Press is an imprint of Taylor & Francis Group, an Informa business

No claim to original U.S. Government works

Version Date: 20150625

ISBN 13: 978-0-367-57052-1 (pbk)
ISBN 13: 978-1-58488-650-1 (hbk)

Library of Congress Cataloging-in-Publication Data

Tunnicliffe-Wilson, Granville.
    Models for dependent time series / Granville Tunnicliffe-Wilson, Marco Reale, John Haywood.
        pages cm. -- (Monographs on statistics and applied probability ; 142)
    "A CRC title."
    Includes bibliographical references and index.
    ISBN 978-1-58488-650-1 (hardcover : alk. paper) 1. Time-series analysis. 2. Autoregression (Statistics) 3. Mathematical statistics. I. Reale, Marco. II. Haywood, John (Mathematics professor) III. Title.

QA280.T86 2015
519.5'5--dc23   2015014849

Visit the Taylor & Francis Web site at
http://www.taylorandfrancis.com

and the CRC Press Web site at
http://www.crcpress.com

Contents

Preface

The aim of this book is to communicate an appreciation and understanding of the issues that arise, and the methodology that can be applied, when the dependence between time series is described and modeled. We intend that it will be of value to a wide range of those in the economic, physical and life sciences who have data in the form of multivariate (or vector) time series from which they wish to draw meaningful, applicable and statistically valid conclusions. The essential background is a good understanding of multiple linear regression. Some knowledge of univariate time series analysis would be an advantage.

The first four chapters, somewhat less than half the book, present the two main pillars of the subject that have been developed over the last sixty years or more. These are vector autoregressive modeling and multivariate spectral analysis. The first chapter is a gentle introduction that raises, in the context of several real examples, the statistical challenges that may be met in modeling the dependence between time series, because of the dependence within, as well as between, the series. In the second chapter we take care to clarify and distinguish the differing concepts of the autoregressive predictor, process and model, introducing issues such as multi-step prediction error as a criterion in the construction of a model. We also describe the properties of the autoregressive model that are useful in later chapters, such as the fourth chapter on the selection and estimation of autoregressive models. The third chapter, on spectral analysis, comes before this, as a preliminary to understanding the spectral properties of the autoregressive model. Methods associated with spectral analysis also contribute to improvements in the estimation of autoregressive models. Spectral analysis is very important in its own right, as a direct means of testing, for characterizing and estimating the dependence between time series. It is an aspect of time series analysis that has, for several decades, been eclipsed by the popularity of autoregressive modeling; it plays an important role in the later part of our book, which we hope will foster a much wider appreciation of the value of this subject.

This first part of the book provides the foundational material for the second part, which develops two different themes. Both of these are extensions to, and contribute to improvements in, the standard approach to autoregressive modeling. Chapter 5 presents methods for the construction of structural models by the identification of the significant links that describe the dependence of current variables upon each other and upon past variables. The examples we present illustrate how these models can describe the structure of this dependence using a relatively small number of coefficients. Autoregressive models represent the influence of the past upon current values of the series using a limited number of simple lagged values. In Chapter 6 we present models in which these are replaced by a number of prescribed combinations of lagged values, discounted into the far past. This is done in a way which preserves the rich structure of the autoregressive model, and leads also to the robust prediction of future values at longer lead times. Discounting can match the frequency at which the data is sampled, and the time lag is generally appropriate to the dependence within and between the series. This approach leads in a natural way to models for continuously recorded time series, which are the subject of Chapter 7. These models may also represent the processes underlying irregularly sampled series and Chapter 8 illustrates how they may be applied in this context. Chapter 9 provides a synthesis of our two extensions to the standard autoregressive model and related developments in the application of spectral analysis.

The style of the book is to avoid the formal approach to definitions and theorems that is standard in many similar texts. That formality has the advantage of clarity for those with a more mathematical background, but does not appeal, we believe, to many potential readers. There are still many lines of mathematical equations in the book, to specify the models that are used and present their properties, and some outline derivations of these properties. But in general the style is narrative and we use references to some of the excellent texts that provide formal proofs in order to provide for those who require them.

The novel material in the later chapters of this book has been developed by the authors, together with former research students at Lancaster University. Parts of the presentation are quite technical and some readers may wish to discover what these models can offer them by initially focusing on the examples. Much of the material has not been published elsewhere and some of the unpublished proofs of model properties have been omitted from the book, both to avoid disrupting the flow of the narrative and to restrain the length of the book. There are further technical details relating to the implementation of the models that are omitted. A website www.DependentTimeSeries.com® has therefore been established to provide access to this and other material related to the book. It will include the data sets used in the examples and some further details of how the models in the book were constructed. Files of the MATLAB® functions and other code used to analyze the examples and produce the illustrations used in the book will be included, together with documentation of this code. Technical documents will explain the estimation theory and methods used and the algorithms by which they are implemented. The site will also be updated with any necessary amendments to the book and further relevant material and references of which we are as yet unaware. There have been many recent developments in multivariate time series modeling which receive only a passing mention, if any, in this book. This does not imply any lack of appreciation, on our part, of the importance of these developments. It simply reflects the priority we have given to presenting our own recent work in the space available.

We are pleased to acknowledge the contribution to the development of the material in this book by former research students at Lancaster University. Alex Morton developed methodology and examples used in the continuous time multivariate models. Juan-Carlos Ibañez contributed to the use of state-space methodology and Milton Lo to the investigation of the properties of the models developed in the later chapters. The spectral analysis methods presented in the book were influenced by collaboration with Carl Scarrott, now senior lecturer in the School of Mathematics and Statistics at the University of Canterbury. We are also grateful to our Universities of Lancaster in England and of Canterbury and Victoria in New Zealand for granting periods of leave and for hosting and facilitating our various visits between these institutions during our collaboration on this book. Finally, we wish to thank our editor for his encouragement, support and patience and also the reviewers of the manuscript for their careful reading and constructive comments, which have been of great value for improving many aspects of the book.

Granville Tunnicliffe Wilson, Marco Reale and John Haywood

Chapter 1

Introduction and overview

1.1 Examples of time series

Time series are found in many places, very often presented as charts, in newspapers, business reports, scientific publications, and elsewhere. In *The Shorter Oxford English Dictionary*, Little et al. (1965), a series is defined as *a number of things of one kind, following one another in temporal succession*. In a time series the *things of one kind* are numerical records (observations or measurements) of successive, varying values of some given quantity.

This book is concerned with developing quantitative models for the relationships between two or more time series which might naturally be statistically associated or dependent on each other. In this chapter we present examples of such *multivariate time series* to illustrate their variety and the range of questions that may be asked about their structure and relationships. We will introduce the concept of lagged regression, which is the basis of autoregressive models, and give an informal account of the statistical challenges that arise in the construction and interpretation of these models, illustrating these points with reference to the examples.

Each of Figures 1.1 to 1.9 shows a set of time series. The number of samples in a series we will call its length. For all but one set of series illustrated in this chapter, these time points are *regularly sampled* or *equally spaced*, i.e., there is the same *sampling interval* or *sampling period* of time between any observation and the following one. The exception is the pair of series shown in Figure 1.1. These are the rates of extinctions and originations of genera compiled from marine fossil records subdivided into intervals of unequal length determined by stratigraphic boundaries. These series are analyzed in Kirchner and Weil (2000), who associate the extinction rate in any given interval with the latest time point of that interval, and the origination rate with the earliest time point of that interval. The sampling intervals then vary from 2.5 to 12 Myr (millions of years) and we describe the series as being *irregularly sampled* or *irregularly spaced*. In this and other examples we will typically capitalize the names of series when referring to them in abbreviated or stylized form in both text and figure titles; for example, we will write Extinctions for the times series of rate of extinction.

Most of the series illustrated in this chapter can be described as reasonably homogeneous in appearance over their recorded time span. We will generally assume that the relationships between the series in any set also remain unchanged throughout this time span. For longer time series this assumption may be checked in various ways. We make reference to these in Section 3.10.7, where we discuss further this important concern following our exposition of the autoregressive models and spectral analysis of time series.

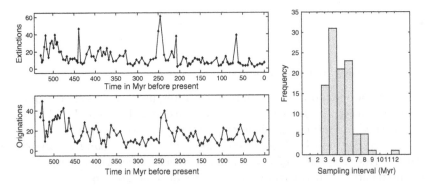

Figure 1.1 *Irregularly observed time series of rates of extinctions and originations of species, with the histogram of intervals between sampling time points of the series.*

It is desirable that time series are plotted against the true sample times but it is a convenient and quite common practice for time series with constant sampling intervals to be plotted against the series index, e.g., $1, 2, \ldots, 100$ for the series of length 100 of the monthly flour prices in three cities, shown in Figure 1.2. This is, after all, a simple re-scaling of the true sample times. The discipline of *discrete time series* analysis and modeling is associated with such series. The adjective discrete refers to the sample times and not the series values, which may be considered to be either discrete or continuous variables.

The processes that give rise to a time series sample may naturally occur over a continuum of time values, such as variations in blood oxygen saturation, pulse rate and respiration rate arising from monitoring a pre-term infant, illustrated in Figure 1.3. Background information on these records is provided by Lee et al. (2011). In fact the values plotted are separated by very short time intervals, being measurements made at the rate of 10 samples per second. Although, in a strict sense, they are discrete, it is nevertheless appropriate to model such high frequency sampled series using methods which represent them as arising over a continuum of time and to refer to them as *continuous time series*. We present such methods in Chapter 7. Again, we emphasize that the adjective continuous refers to the fact that time appears as a continuous variable in the representation of such a series and its model. The values of the time series plotted in Figure 1.3, particularly the series of pulse rate and respiration rate, also appear to vary as continuous functions of time. This is not necessary for a continuous time series, an example being the number of people in a queue, which naturally has occasional jumps in value. A continuous time series model is also appropriate for representing the underlying processes that give rise to irregularly sampled series such as those in Figure 1.1. We will apply such an approach to these series in Chapter 8.

Figure 1.2 *Monthly flour price indices in Buffalo, Minneapolis and Kansas City, over the period from August 1972 to November 1980. The first and last series are offset, respectively, lower and higher than the second series by 20 units.*

1.2 Dependence within and between time series

One of the first questions we might ask of a pair of time series is whether or not they are dependent, one upon the other. The treatment of this question is complicated by the fact that in most situations each time series is dependent *within itself*. Each observation of a single time series may be closely related in value to previous and subsequent observations of the same series. The standard test for the significance of the correlation between random samples of two variables is therefore not valid in this context. Indeed, it is easy to generate two artificial time series samples which are completely independent, one with the other, yet the sample correlation between them is very large. This is referred to as *spurious correlation* in Granger and Newbold (1974), although this term was used by Pearson (1896) for correlation induced by a common factor. It can arise between time series when there is no common factor; the correlation appears significant because it is referred to the standard test for correlation in random samples. That this is inappropriate in the time series context is explained by Yule (1926), who called the effect nonsense correlation.

A related point, for high frequency sampled series, is the effect of the sampling rate on the significance level nominally given by the standard test for correlation in random samples. We illustrate this in Table 1.1, in which each row shows the sample correlations $r_{i,j}$ between the first three variables shown in Figure 1.3. The values in the first row are formed from the full series of $n = 10{,}000$ points, those in the second row from the series of $n = 1000$ points of observations made just one per second, and those in the third row from a subsample of one observation every 10 seconds. The table also shows, in brackets, the nominal p-values for these correlations and the critical values r_α for the tests at the $\alpha = 5\%$ level.

Table 1.1 *Sample correlations between three Infant monitoring series and their nominal significance levels given by the standard test for correlation in random samples.*

n	$r_{1,2}$	$r_{1,3}$	$r_{2,3}$	r_α
10,000	0.061 (0.000)	0.346 (0.000)	−0.191 (0.000)	0.020
1000	0.061 (0.054)	0.348 (0.000)	−0.202 (0.000)	0.062
100	0.051 (0.614)	0.360 (0.000)	−0.239 (0.017)	0.197

The table shows how the standard test for significance of correlation between random samples is quite inappropriate in this context. The information content is very similar in the full series and those at lower sampling rates, whereas the p-values and critical values change considerably due to the dependence on the number of observations. For valid inference, methods are needed which take into account the dependence within as well as between the series.

This is an appropriate point to introduce the topic of spectral analysis of time series, which has long been used both to assess and to quantify the strength of relationships between time series. It is applicable to the discrete time series that we have illustrated above and is based simply on the linear transformation of each time series record to its discrete Fourier coefficients, which we prefer to call *harmonic contrasts*. These are linear combinations of observations with sinusoidal weighting. Each contrast is associated with a different frequency of sinusoidal weighting, a rapidly oscillating sinusoid having a high frequency and a slowly oscillating one a low frequency. If a series is rapidly varying, its harmonic contrasts will be larger at higher than lower frequencies, with the reverse for slowly varying series. For a single series there is very little correlation within its set of harmonic contrasts associated with different frequencies. However, if two series are dependent, their harmonic contrasts will be correlated at corresponding frequencies. The significance of the sample correlation between the harmonic contrasts of two series can therefore be readily assessed. This is, however, only one aspect of the spectral analysis of dependent time series which we will present in Chapter 3. The quantitative description of a set of dependent time series given by their spectral analysis, or the estimation of their *spectrum*, is of value in all subsequent chapters of this book.

Correlation coefficients are a standard means of quantifying relationships between variables. Also widely used are the regression coefficients of one variable upon another, in the context of either simple or multiple linear regression. In these contexts correlation and regression are equivalent, in that a full

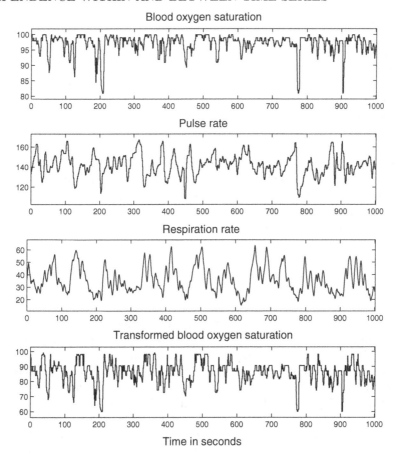

Figure 1.3 *Records from monitoring the blood oxygen saturation, pulse rate and res-piration rate of a pre-term infant. The lowest plot shows the blood oxygen saturation after an arcsine transformation.*

description of the correlations between a set of variables can also be repre-sented by a sufficient collection of regression relationships between them. Such a collection may be given by the regression of each variable upon the set of previous variables taken in some chosen order, though this is not necessarily unique. In the time series context, however, there is a natural representation using the time ordering. The *lagged regression* of the current value of *each* time series on *past* values of the *whole set* of series is then known as a mul-tivariate or *vector autoregression*. This autoregression is not only a predictor of the present values from the past, it can also be expressed as an *autore-gressive model* which represents the process by which successive values of the time series are generated. Including a sufficient number of past values in the regression also leads to valid statistical inferences regarding the dependence between the series. The *present* values of the other series are *not* included in the autoregression and consequently the regression errors for the different series may be correlated.

It is also natural to model the relationships between the harmonic contrasts of two or more time series using regression. This provides quantitative measures of the nature and strength of dependence between time series, which can be transformed to provide estimates of the coefficients in the lagged regression between the series. This approach is also described in Chapter 3.

Chapter 2 considers the theoretical nature of lagged dependence between time series, as a prelude to Chapter 4, which presents practical aspects of constructing autoregressive models of observed time series. These models include, as regressors, the set of all past values up to a finite lag known as the *order* of the model, and criteria for selecting this order are presented, such as Akaike (1973). Where necessary we will call these *canonical* autoregressive models to distinguish them from *structural* autoregressive models, which we present in Chapter 5. These are autoregressive models which *do* include regression terms to represent the dependence between the *present* values of the different series. The statistical procedure we describe for constructing these models suggests how, for each series, the terms in its dependence upon both present and past values may be selected. This can lead to a *sparse* model which represents the series using far fewer terms than the canonical autoregression. The statistical methodology we use for this purpose is that of *graphical modeling*, Whittaker (1990), Lauritzen (1996) and Edwards (2000), which generates diagrams representing the dependence between present and past series values.

Autoregressive models of time series are readily able to answer many of the questions that might be asked concerning their dependence. Many such questions can be expressed in terms of the prediction of the series, such as how well future values of one series might be predicted from past values of the whole set. But prediction can be used in a wider sense than just forecasting. Consider, for example, the series shown in Figure 1.4 of the numbers of moths caught in traps and the local environmental conditions. From an appropriate model one could predict what might have been the effect on the moth population, as measured by the Moth count series, if the rainfall had been lower in the spring.

Prediction is also useful in planning and control. Consider, for example, the record of weekly sales of a consumer product shown in Figure 1.5 with the related variations in price and promotional advertising over a four year period. A suitable model may be used to assess the effect on future sales of changes in the pattern of pricing and promotion. The dynamics of the inter-relationship between time series can be understood and characterized in terms of impulse and step responses, which are properties of the model. These responses describe, for example, the additional effect on the rate of originations of species of a very large extinction event, or the effect on the blood oxygen saturation of an infant of a sudden increase in their respiration

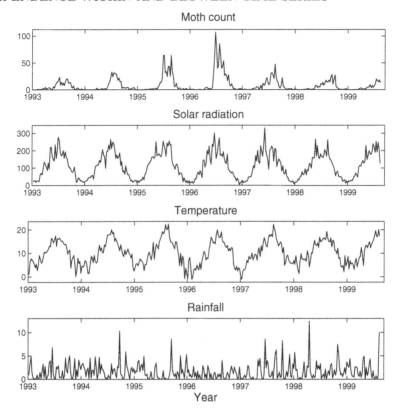

Figure 1.4 *Weekly averages of moth trappings and environmental factors which might explain their variation. These records were abstracted from the database of the UK Environmental Change Network and made available by the Centre for Ecology and Hydrology at Lancaster University.*

rate. There is also the question of whether the model can fully explain, or predict, the observed patterns of variation, such as cycles, in a group of series. Such patterns are evident in some of the plots shown in Figure 1.6 of quarterly indicators of the UK pig market. The indicators of the number of new animals introduced into the breeding herd and the profitability of the market to the producer both appear to have a cyclical pattern. It is important that the model can represent such patterns if it is used to construct simulated values of much greater length than the recorded series. For example, Figure 1.7 shows the daily interest rates of the USA dollar for a range of seven terms to maturity. Simulation of these series may be used to investigate the long term behavior of trading strategies that are based upon them. For this purpose it is important that the relationships between the series are correctly modeled.

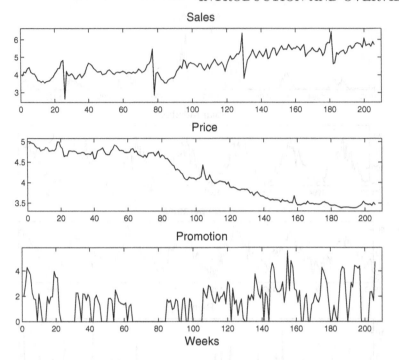

Figure 1.5 *Weekly sales of a consumer product with the retail price and a measure of advertising promotion of the product. These data have been scaled and their precise details withheld to protect commercial interests of the manufacturer.*

1.3   Some challenges of time series modeling

Both spectral and autoregressive modeling are empirical, data lead approaches to understanding the dependence between time series. These stand in contrast to the approach, also described by some as structural modeling, in which models are formulated on the basis of theoretical or hypothetical mechanisms, then fitted to the observed time series. In practice each of these approaches can complement the other. If complex economic models fail to give forecasts as accurate as empirical time series models, it may be necessary to re-think their structure. Empirical modeling also encounters many statistical challenges, and a good understanding of the context of the data can be very useful in suggesting, for example, the direction and time lag of dependencies which should be included in the model. It should always be checked that the implications of a fitted empirical model conform with the general understanding of the structure of the time series.

Such considerations are, of course, appropriate to all statistical modeling, and many of the challenges to multivariate time series modeling are familiar from the general context of regression modeling. Time series modeling does

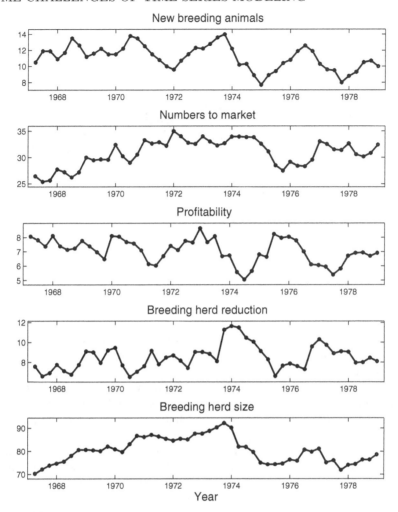

Figure 1.6 *Twelve years of quarterly indicators of the state of the UK pig market. A further description of the data and their source are given in Andrews and Herzberg (1985, p. 169). The data shown here have been scaled down respectively by factors of 10,100,1,1000 and 10 to make them more similar in size. Although not clearly evident to the eye, each of these series also exhibits some fixed seasonal quarterly pattern which is removed by simple regression before further modeling.*

have the advantage that it lends itself readily to the visual assessment of data, statistical summaries and model properties by graphical means. Sometimes lagged relationships between series may be quite visually evident, as in Figure 1.6, where the series of New breeding animals appears to follow Profitability. Sometimes there is a hint, as in the Extinctions and Originations series, and

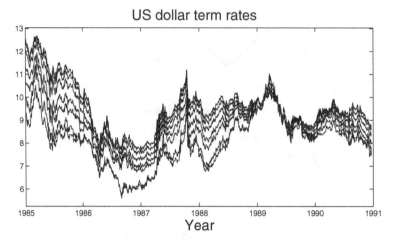

US dollar term rates

Figure 1.7 *The daily interest rates of the U.S. dollar for terms of 6 months and 1, 2, 3, 5, 7 and 10 years. At the start of the series these correspond in order to the lowest through the highest rate.*

sometimes it is not at all obvious, as in the dependence of Moth count on Rainfall shown in Figure 1.4, though, as we will see in Chapter 3, it is readily revealed by spectral analysis.

Strong dependence between the explanatory variables themselves, known as multicollinearity, can cause great difficulties in multiple linear regression, and the same is true in time series modeling, especially when the regression includes lagged values. The practical symptom is the difficulty of variable selection. Consider the series of Profitability and the New breeding animals of the Pig market series in Figure 1.6. We first note that the peak value in the sample cross-correlation between these series is 0.864, between New breeding number and Profitability two quarters previous. Now the present value of New breeding animals is very well predicted by regression on the values of both New breeding animals and Profitability in just the previous quarter, with 90% of the variance explained. This is represented in the diagram on the left in Figure 1.8 and conforms with what is to be expected, since the market can react quickly to an increase in Profitability by introducing new pigs into the breeding herd. The present value of Profitability also appears to be very well predicted by a regression on this same pair of previous values, as represented in the central diagram of Figure 1.8, with 63% of its variance explained. These two lagged regression equations can therefore be combined to give a bivariate first order autoregressive model for the two series. Furthermore, it can be checked that the properties of this model do reflect the observed properties of the two series, in particular their irregular cyclical pattern, which can be well reproduced by simulation of this model.

The dependence of Profitability on New breeding animals at the previ-

Present value

One quarter previous

Two quarters previous

Three quarters previous

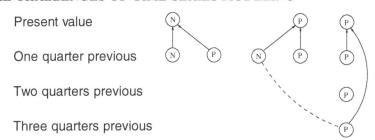

Figure 1.8 *Diagrams representing the prediction of the present values of New breeding animals (N) and Profitability (P) from values at previous quarters.*

ous quarter is not, however, consistent with the known structure of the pig market. We can understand why this dependence is statistically plausible by noting that the Profitability series can also be predicted by regression upon its own lagged values at both one and three previous quarters, with 64% of the variance explained. This is represented in the right hand diagram in Figure 1.8. The dashed line linking the Profitability at a lag of three quarters in this diagram with New breeding animals at a lag of one quarter in the central diagram is a reminder of the strong correlation previously noted between these variables. Thus the New breeding animals at lag one can act as a proxy for Profitability at lag three and take its place in the statistically plausible predictor for Profitability represented in the central diagram.

This example illustrates another of the modeling challenges, that of omitted variables. The Profitability series does in fact depend strongly, as expected, on the number of pigs brought to market (Numbers to market) and the Breeding herd size. When these variables are introduced into the model, the New breeding animals at the previous quarter no longer contributes to the model. Good empirical modeling requires that all the interrelated series should be included in the data set. However, economic systems are very large and it may be necessary to restrict the set of series to be included, whilst recognizing that they may be affected to some extent by excluded series. For example, the three Flour price series in Figure 1.2 are found to be closely interrelated, but follow a long term cycle which is not well explained by their interaction. This cycle is most likely the reflection of external economic factors affecting the general movement of all three series. Models for these series are presented by Tiao and Tsay (1989), Grubb (1992) and Athanasopoulos and Vahid (2008).

The cycle in the market for pig meat in the USA has been recognized and studied for many years as the *hog cycle*; see Quenouille (1957). It may be explained in part by the internal dynamics of series such as those shown in Figure 1.6, by the interplay of profitability and supply and the time scale of pig breeding and rearing. But there are many agricultural cycles recognized (see Granger and Hughes, 1971), and the price of cereals used in the feed and the price and production of other meats can also be expected to have some effect.

Closely related to the omission of variables is the aggregation of variables. For example, the breeding cycle of pigs extends to about two quarters of the year, so the total breeding herd effectively contains two or three cohorts at different stages of this cycle. We might expect improved model prediction if the numbers in these different cohorts were known, whereas only the total is known. A model in which the cohort numbers appear as latent variables has been formulated and fitted by Bodwick (1988). Such models may be developed at a second stage following empirical modeling, given appropriate insight into the structure of the system.

Another challenge is what are generally known as errors in variables, or, in the time series context, noisy observations, where the added noise is either random or has very little correlation within itself or with other observations. The errors may be due to limitations of measurement instruments or the size of data samples. If noise is superimposed on variables which are also collinear, it has the effect of drawing into the regression equation variables which otherwise might not be selected as explanatory. This improves the strength of the prediction, but can confound the inference as to which are the truly causal variables. To overcome this problem it is desirable sometimes to include a term for measurement errors or observation noise explicitly in time series models. Such models will not then be linear in the parameters and may be more difficult to estimate. Observation noise terms may be introduced at a second stage following the development of an empirical model. In Chapter 8 some of the models we present for the series in Figure 1.1 include observation noise terms.

1.4 Feedback and cycles

Time series which appear to have strong cycles generally exhibit multi-collinearity when fitting autoregressions. Such cycles are often the consequence of strong feedback between the series and may be seen in economic, biological and engineering systems. A classical example is the predator-prey cycle where the rapid increase in the number of predators leads to a severe depletion of prey. The subsequent collapse through starvation of the predator population allows an explosion in the number of prey and the renewal of the cycle. This can result in a persistent cycle with a fixed period dominating both series. From the viewpoint of spectral analysis, the harmonic contrasts of such series will typically be dominated by a few very large values associated with the cycle, the remainder being relatively small. The regression between the harmonic contrasts of two such series then has similarities with a simple regression in which a small number of very extreme or outlying points outweigh the remaining points of relatively small magnitude. Consequently, the information about the relationship between the series may be very limited. When attempting to model the dependence between the series, it can be difficult to select the correct variables and lags in an autoregression and to determine causality. Each series might just as well be predicted from its own past values as from a model including past values of the others.

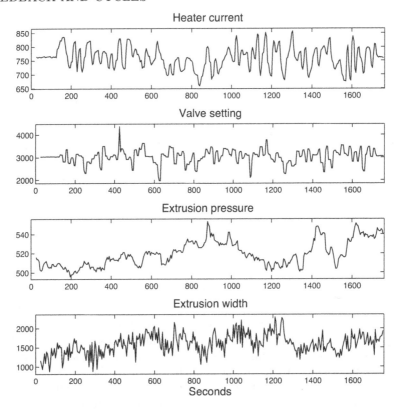

Figure 1.9 *Input series Heater current and Valve setting and output series Extrusion pressure and width, from an open loop experiment on a plastic Film extrusion process.*

In the context of engineering control of a continuous manufacturing system, the feedback from an output to an input variable is designed to return the output to a given set point whenever it departs appreciably from this value. The control algorithm is generally deterministic, which naturally leads to highly collinear dependence between the output and lagged values of the input. Poor design of a control system can, however, lead to large swings in the variables, from one extreme to another. The successful control of such a system using autoregressive modeling to determine the dependence between the variables is the theme of Akaike and Nakagawa (1988). Figure 1.9 shows the results of an open loop experiment on a system for continuous extrusion of plastic film which was subject to similar problems when operated under control, or closed loop conditions. The term open loop refers to the fact that the control system which would otherwise set the input is disconnected. Here, the inputs, the current to the heater which softened the plastic and the opening of the valve through which it was extruded, were manipulated so as to affect the pressure within the extrusion and its width, which were the outputs. The inputs for this experiment follow nearly random settings, though adjusted from time to time to ensure that the outputs remained in an acceptable range. From such records the dependence of the outputs on the inputs can be readily determined.

If the control system is very efficient in its task, there will be very little, apart from minor disturbances, remaining in the controlled output variable. It will then be almost impossible to identify from observations the system by which the input variable affects the output. In many systems, however, whether marketing, physical, biological, environmental or economic, there are sufficient internal disturbances to allow reasonable opportunity for identification.

1.5   Challenges of high frequency sampling

It is not easy to define precisely what is meant by high frequency sampled data, but one characteristic is that the harmonic contrasts at lower frequencies are much greater in magnitude than those at higher frequencies, which may be associated mainly with observation noise. In some situations the low magnitude at high frequency is the result of deliberate smoothing of the series by applying moving averages to reduce observation noise. Unsurprisingly, the smoothness of the series again leads to multicollinearity in lagged regression. A simple treatment of this problem might appear to be sub-sampling, taking, say, every third point of the series. This is not an uncommon step after a high frequency sampled series has been smoothed, but in some contexts it would lead to an unacceptable loss of information. We will show in Chapter 3 how spectral analysis can avoid sub-sampling and circumvent this problem by down-weighting the contribution from the higher frequencies. In Chapter 6 we introduce an extension of autoregressive models which is designed to fit the lower frequency information in the series without sub-sampling. The simple lagged regressors of the standard autoregressive model are replaced by regressors which are defined linear combinations of values stretching to much greater lags. These are similar in appearance to regressors at a regular subset of lags, but avoid the information loss which might occur from sub-sampling or the omission of lags not included in a subset regression. Moreover, these models transform naturally and simply into models for continuous time series, which are described in Chapter 7.

For high frequency sampled series which are also smooth in appearance, the present value may be very well predicted by the immediately previous value, from which it differs only by a small amount. Models which introduce higher lags, or past values of other series, as predictors may show little significant improvement in fit. This further challenge to model identification can be met by replacing the present value, as the variable to be predicted, by a combination of present and future values. We show how this approach can be simply incorporated into the extended autoregressive models presented in

Chapters 6 and 7. There is a long history to the idea of fitting time series models, including autoregressions, to predict not the present value, but one several steps into the future; see, for example, Findley (1985), Haywood and Tunnicliffe Wilson (1997) and Chevillon (2007). The justification is that any model will be imperfect and models are in practice commonly used for multi-step prediction, so the model will be more robust if the criterion of fitting is related to its application. The approach we advocate achieves such robustness and also implicitly places more weight on the lower frequencies in the data.

1.6 Causal modeling and structure

Causal structure is implicit in the design of control systems and in some contexts causality can be demonstrated by experimentation. Otherwise, causality is difficult to prove by empirical modeling. For many years the Granger test for causality, as it has come to be called (see Granger, 1969), has been widely applied in scientific as well as economic contexts. This is based on testing whether certain coefficients are zero in the autoregressive model. We have already pointed out that factors such as observation noise can confuse this picture. If there is a common causal variable outside of a system which affects different variables within a system at different time lags, it may then appear that one variable within the system causes another. Our approach is to construct empirical models with as great a simplicity of form and parameterization as possible, consistent with faithfully representing the observed statistical structure of the series. Such a model will allow comparison of different possible interpretations of causal relationships between the variables, but the conclusions may have to rest upon plausibility rather than proof.

In Chapter 6, where we define the extension of the autoregressive model, we say that a set of regressors, constructed as defined linear combinations of lagged values, are the *states* associated with the immediate past. The concept of state variables, or simply *states*, as we use it here, is simply that of the set of variables used in the model at any particular time to predict the next set of observations. These are empirical states, formed from the observed series. For the canonical autoregressive model, the states associated with the immediate past are simply a finite number of lagged values. The general concept of the state variables of a system at any particular time is that of the set of quantities necessary and sufficient to determine the future evolution according to the system model. For the linear autoregressive models that we will use, the empirical states are unique up to a linear transformation.

Linear models may be formulated in which the states are latent variables which are not directly observable. A simple example is the autoregressive model with observation noise, where the process following the model cannot be directly observed. More generally, such *state space* models typically employ latent variables to embody assumptions about the dynamic structure of the series, in contrast to empirical models, which are developed so as to reflect the observed structure in the data. See, for example, Harvey and Chung (2000) and the ensuing discussion. These models may nevertheless be fitted to observed series and used to produce predictions. The value of empirical models is that they can provide a benchmark against which such state space models, their properties and predictions, can be assessed.

1.7   Some practical considerations

As we will see from examples in the following chapters, the construction of models is not entirely automatic. There are statistical decisions and judgments to be made in selecting a final model, and part of this process is the careful checking of model properties against those of the sample, such as lagged correlations and the spectrum. A further valuable check is to withhold a fraction of the observations from the end of the series when fitting the model, then to use the model to predict these observations. The model can be assessed according to the accuracy of the predictions, and whether the prediction intervals are realistic. A long established practice is to check that the residuals after fitting the autoregressive model are uncorrelated. The portmanteau tests used for this purpose are important (see Hosking, 1980), but Haywood and Tunnicliffe Wilson (2009) show that they are not necessarily powerful against some model shortcomings related to high lead time prediction.

The method of fitting can also be important. In simple lagged regression the time index of the first observation which can be regressed upon past values is necessarily one plus the model order. For models of high order, or for short series, this leads to a loss of information that can be recovered using maximum likelihood estimation. As we will see in Chapter 8, when a simple trend term is included in the model, the estimation of the low frequency properties may be biased. This can be alleviated by using a modified form of likelihood; see Tunnicliffe Wilson (1989). For short time series, the balance between getting a good model fit and over-fitting can be a difficult issue, and the criteria for selecting the model order should be carefully considered.

The alignment of a set of time series, by adding or subtracting a fixed time shift to or from the times associated with one or more series, has an impact on a multivariate autoregressive model. Careful alignment can simplify the model, particularly in systems where there is a substantial delay between an input and output variable. The response of the output can then be seen to follow the input more closely and the series represented by an autoregressive model of lower order. We only consider linear models, but simple non-linear transformations, such as taking logarithms of the series, may improve the strength and accuracy of linear relationships that can be established between the series. In Figure 1.3 the last plot shows the effect of applying an arc-sine transformation to the blood oxygen saturation. This transforms a highly skewed distribution into one which is appreciably more symmetric and improves subsequent modeling. We will refer to this transformed variable as the Oxygen level, shortened to the Oxygen series where space is limited, as in figure titles (a practice which we also apply to other series names provided there is no ambiguity). This is just one of the considerations from the wider context of regression modeling that are also applicable to time series. Another is whether the data are homogeneous – is there any reason to believe that the relationships between the series may change part way through the data set? Has the quality, accuracy or definition of the series changed? Have measuring instruments been changed or re-calibrated part way through?

A related point is that one (or more) series in a multivariate set may be explanatory for the others, but may itself be determined or manipulated so that it does not appear particularly homogeneous through time. Examples are the price and promotion series in Figure 1.5. These may in part be determined in response to movements in sales, or even in anticipation of such movement, but may also be determined by the policies and budgets of marketing managers, which can change from year to year. Of course, a similar situation arises in econometric modeling, where interest rates and other variables may be determined by government officials.

A theory of estimation can be applied to autoregressive models which include such inhomogeneous explanatory variables if it can be assumed that their lagged sample correlations with past values of themselves and the other series will approach definite values as the series length increases. In practice, our treatment will parallel that of the wider regression context, so that inhomogeneous explanatory variables will generally be treated the same as the others. Care will be taken, though, to check on the sensitivity of estimation procedures to such a treatment. The Sales series in Figure 1.5 clearly shows the effect of a pre-Christmas peak and a post-Christmas trough. This can be modeled by introducing a new explanatory variable which is a simple indicator, being zero over the whole period except Christmas week, when it takes the value one. We will show in Chapter 3 how introducing this variable enables us to estimate the lagged effects of Christmas and answer the question of whether the total effect of the festival is a net increase in sales.

The examples of multivariate time series presented in this chapter are varied, but the largest number of series in any set is just seven. Sets which contain a much larger number of series are now common in areas as diverse as financial modeling and neuroscience. The difficulty is that, whilst the number of observations grows in direct proportion to the number of series, the potential number of parameters in a multivariate autoregressive model grows as the square of the number of series. Thus, for the Pig market series, a model of order 4 has potentially 115 parameters including variances, and this number is quite a high proportion of the total sample size of 240. However, the methods of analysis we describe in Chapter 5 result in a model which, though of order 4, includes only 35 parameters. This is because each series depends only on a subset of the other series, and even then only on selected lags. The methods we describe in Chapter 6 also enable the dependence on higher lags to be represented using fewer parameters. By these means it should be possible to model data sets containing much larger numbers of series.

In constructing these models our aim is to obtain a reliable description and representation of the structure of the dependence between the series. The main interest in these models may be the knowledge of the strength and nature of this dependence, but much more may be derived from the models, such as forecasts of future values or extraction of component cycles; see Morton and Tunnicliffe Wilson (2001). In the following chapters we have space to describe only a selection of these model applications. Our emphasis will be on the empirical construction of the models from observed time series.

Chapter 2

Lagged regression and autoregressive models

2.1 Stationary discrete time series and correlation

We start this chapter by considering the first and second order moments of time series – their means, variances, covariances and correlations. If the series are Gaussian, i.e., the joint distribution of any set of values is multivariate normal, these moments completely determine their distribution. We do not consider other forms of distribution, in large part because of the practical difficulties this raises, but also because assuming that the series are Gaussian can deliver solutions which are reasonably robust to departures from this assumption. In particular, the estimation of model parameters by maximum likelihood under the Gaussian assumption can be justified under much wider assumptions about the distribution. See, for example, Section 10.8 of Brockwell and Davis (1987) for relevant results in the context of univariate time series and Chapter 6 of Hannan (1970) in the multivariate case.

In the context of simple linear regression, given the correlation ρ between variables Y and X with respective means μ_Y, μ_X and standard deviations σ_Y, σ_X, the minimum mean square error linear predictor of Y given X is $\mu_Y + \rho(\sigma_Y/\sigma_X)(X - \mu_X)$, with mean square error $(1 - \rho^2)\sigma_Y^2$. This predictor is also called the projection of Y upon X. In the next section we will go on to develop similar minimum mean square error projections of present values of time series upon past values. These will depend on the means, variances and correlations of the series. An alternative would be to parallel the standard introduction to simple linear regression by fitting linear least squares regressions of present upon past values of observed time series. However, this would involve questions of the properties of estimates of correlations and regression parameters, the answers to which would require a careful specification of what model is being estimated. It is for this reason that we first motivate the idea of the autoregressive model in the context of minimum mean square error prediction, and postpone consideration of model estimation to Chapter 4.

Our notation is to use m for the number of discrete time series in an observed data set, and n for their length, which we assume is the same for all the series in a set, except where we explicitly state otherwise. We will typically use $x_{i,t}$ for the value of series i at time t and x_t for the column vector of values $x_{i,t}$, $i = 1, \ldots, m$. The component time series then form the rows of the matrix with elements $x_{i,t}$. Some authors, such as Brockwell and Davis (1987), use the reverse order of these indices, which corresponds more naturally to how a set of time series may be held as adjacent columns in a data file. This is not, however, a critically important convention.

Although it is formally correct only to use notation such as $x_{i,t}$ for the numerical value of the observed series, for convenience and simplicity of notation we will also use it in expressions such as $\mathrm{E}(x_{i,t})$ and $\mathrm{Var}(x_{i,t})$ for its expectation (or mean) and variance. Here, it would be conventional to use the capital letter X in place of x, to indicate that we are considering a random variable which could take a range of values from a probability distribution. The formal notation for the stochastic process consisting of the sequence of random variables over an infinite time span is then $\{X_t\}$. Our observed series is a realization from this process over a finite time range. Again, we will also use the notation x_t for this process; our usage will cause no confusion because the context will almost always make the meaning quite clear, and where it does not, we will make it explicit.

The most basic assumption we make relates to the idea that the general structure and appearance of the time series remains the same throughout the record, that it is the same in the past, and will be the same in the future. As with other hypotheses, this may in practice be falsifiable but cannot be proven. A simple check, for a sufficiently long series, is to divide it into sections and compare statistical descriptions and models fitted to each section. This homogeneity through time is usually expressed formally by the condition of stationarity. Even this is qualified, with the most useful form being stationarity of the first and second moments, known as second order or weak sense stationarity. This says that the series expected (or mean) values $\mathrm{E}(x_{i,t}) = \mu_i$ stay constant through time, and, most essentially, that the covariance between any two terms in the series depends only on the time lag between them. Thus

$$\mathrm{Cov}(x_{i,t}, x_{j,t-k}) = \Gamma_{i,j,k} \tag{2.1}$$

depends upon the series indices i and j besides the lag k, but is the same for all time t. The corresponding correlation we will denote by $\rho_{i,j,k}$. Note that the order of these indices and the sign of the lag are important because interchanging the indices reverses the sign of k. We agree here with Brockwell and Davis (1987), though some authors such as Hannan (1970) choose to use $t + k$ in place of $t - k$. Our choice reflects the fact that we intend to relate the first series $x_{i,t}$ to past values $x_{j,t-k}$ of the second series. In vector-matrix notation we write

$$\mathrm{Cov}(x_t, x_{t-k}) = \Gamma_k \tag{2.2}$$

where Γ_k has elements $\Gamma_{i,j,k}$ for $1 \leq i, j \leq m$. For lag zero we write

$$\mathrm{Var}(x_t) = \Gamma_0 = V \tag{2.3}$$

for the variance matrix of x_t. We call Γ_k the autocovariance function or just the covariance function of x_t. Where it is necessary to avoid ambiguity we will add the series identifier as a subscript, as in μ_x, V_x and $\Gamma_{x,k}$. When modeling real series, a mean level that is not constant may be represented by deterministic trend terms. These can be estimated by fixed regressors and subtracted from the series to ensure stationarity of the mean. In the following development, unless otherwise stated, we will assume that the process we are modeling has been corrected, if necessary, for any such trend terms, which usually include a constant. Even when there are no trend terms, we will also correct for the constant mean, so that we can take $\mu_x = 0$. The autocorrelation function corresponding to Γ_k is more useful in plots because it can be referred to the natural range of correlations from -1 to $+1$. For simplicity we will use the abbreviation acf for both this and the autocovariance function; the context will make clear which is meant.

Homogeneity through time is a property of a somewhat wider class of processes, known as integrated processes, obtained by cumulatively summing the terms in a stationary process. These processes are therefore characterized by the fact that their first, or possibly higher, order difference is a stationary process. They are not themselves stationary and their distribution can only be properly specified conditionally upon some initial values, so they do not have an autocovariance function. We will describe our approach to these processes at the end of Chapter 4.

The condition of stationarity can be usefully applied to higher order moments of a time series, particularly for deriving properties of estimators of parameters in models of the series. The condition can be applied to the whole distribution of the series, but this is a very strong assumption which cannot realistically be verified. Where the assumption of second order stationarity fails, it may be readily falsified, given sufficient data, by tests that focus on particular ways in which it may fail, such as through structural breaks or regime changes, and methods which detect and model such changes, as in Davis et al. (2006).

We define the autoregressive model in the next section of this chapter, but motivate it by first considering the construction of lagged predictors by the projection of present values of a time series upon past values. The coefficients of these predictors are determined by the covariance function of the series and for numerical illustration we will use the values shown in Figure 2.1 of the lagged correlations of five series presented as a matrix of plots of $\rho_{i,j,k}$ against the lag $k = 0, \ldots, 40$. This acf is, in fact, generated from a structural autoregressive model of the five Pig market time series shown in Figure 1.6, after they have each been corrected for fixed seasonal quarterly patterns of small magnitude. This model is a simplified version of one fitted in Section 5.11, but the source of this numerical acf is not directly relevant to its use as an illustration. We will treat it as a known quantity to illustrate how autoregressive models are motivated and constructed. Nevertheless, we comment briefly to provide context for the patterns seen in Figure 2.1.

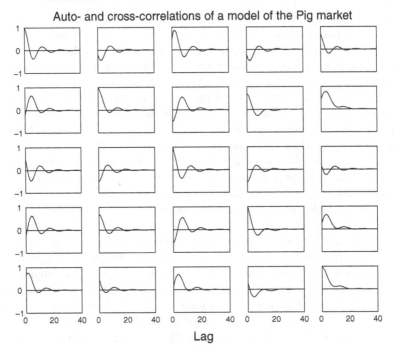

Figure 2.1 *The lagged correlations of the Pig market model as a matrix of plots with the correlations between series i and lagged values of series j shown in row i and column j.*

The plot in row i and column j of Figure 2.1 is of the correlations, derived from the model of the Pig market, between series i and lagged values of series j, with the order of the series being that shown in Figure 1.6. For a pair of distinct series this is their cross-correlation function which is abbreviated to ccf. For example, the first row of the figure corresponds to the series of New breeding animals and the third column to the Profitability series, and the corresponding plot in the third column of the first row reveals a peak around lag 2. This model property reflects the visually evident relationship in Figure 1.6 between New breeding animals and lagged values of Profitability on which we commented in Chapter 1. We also remarked there that the peak sample cross-correlation between these series was at lag 2, and in general the model cross-correlations reflect well the *main* features of the sample cross-correlations of the series. However, we will postpone to the next chapter the question of how to construct *sample values* of the acf from a set of observed time series. The sample cross-correlations exhibit considerable statistical variability, and, in particular, decay less rapidly at higher lags than is evident in the model cross-correlations of Figure 2.1. The set of values of any model acf is, of course, constrained, not simply by the fact that the cross-correlations are less than one in absolute value. The covariance matrix of any collection of series values $x_{i,t}$ for a set of pairs i, t, has elements taken from the acf. This matrix must always be positive definite. Generating the acf from a model process guarantees this property.

Numerical illustrations based upon the acf in Figure 2.1 arise throughout this chapter, the first seven sections of which cover the main properties of autoregressive models. The later sections cover more technical properties, in the sense that they are useful in the methodologies presented in later chapters on the estimation and checking of autoregressive models.

2.2   Autoregressive approximation of time series

We now proceed to build an autoregressive representation of the covariance structure of a general stationary process by determining the predictor $\hat{x}_t$ of the present value of x_t as a linear combination of all past values, or regressors, x_{t-k} up to some specified maximum lag p. Let the prediction error be e_t, so we may express

$$\begin{aligned} x_t &= \Phi_1 x_{t-1} + \Phi_2 x_{t-2} + \cdots + \Phi_p x_{t-p} + e_t \qquad (2.4) \\ &= \hat{x}_t + e_t \end{aligned}$$

using matrix coefficients $\Phi_1, \ldots, \Phi_p$. We determine these coefficients so that the variance of each component of the error e_t is a minimum. In fact, this also minimizes the variance of any prescribed linear combination of the components of e_t and the determinant of the variance matrix of e_t. The condition for this minimum is that the covariance $\text{Cov}(e_t, x_{t-k})$ between the error and each of the regression variables is zero. We also say that the predictor $\hat{x}_t$ is the *orthogonal projection* of x_t on the regression variables and the resulting equations are therefore also known as *normal equations*. Taking the covariance of each term in (2.4) with x_{t-k}, we obtain the equations for $\Phi_1, \ldots, \Phi_p$ in block matrix form:

$$\begin{pmatrix} \Gamma_1 & \Gamma_2 & \cdots & \Gamma_p \end{pmatrix} =$$

$$\begin{pmatrix} \Phi_1 & \Phi_2 & \cdots & \Phi_p \end{pmatrix} \begin{pmatrix} \Gamma_0 & \Gamma_1 & \cdots & \Gamma_{(p-1)} \\ \Gamma_{-1} & \Gamma_0 & \cdots & \Gamma_{(p-2)} \\ \vdots & \ddots & \ddots & \vdots \\ \Gamma_{-(p-1)} & \Gamma_{-(p-2)} & \cdots & \Gamma_0 \end{pmatrix}. \qquad (2.5)$$

We will call the block $p \times p$ matrix in these equations G_p. It is symmetric because $\Gamma_{-k} = \Gamma'_k$. The block entries are the same along any diagonal, and this structure is described as block Toeplitz. By taking the covariance of every element in (2.4) with e_t, we obtain $\text{Cov}(x_t, e_t) = \text{Var}(e_t)$; then by taking the covariance of every element in (2.4) with x_t, we obtain the expression for

$$\text{Var}(e_t) = V_e = \Gamma_0 - \begin{pmatrix} \Phi_1 & \Phi_2 & \cdots & \Phi_p \end{pmatrix} \begin{pmatrix} \Gamma_{-1} \\ \Gamma_{-2} \\ \vdots \\ \Gamma_{-p} \end{pmatrix}. \qquad (2.6)$$

Equations (2.5) and (2.6) are known as the multivariate Yule–Walker equations.

We now acknowledge that the coefficients $\Phi_1, \ldots, \Phi_p$ and the error variance V_e of this autoregressive predictor do in fact depend on the chosen order p. This should be indicated formally using a further subscript p to these symbols, which we have omitted for the sake of clarity. What is important is that we will use these coefficients to define an *approximating* model to the original process x_t. This model is identical in form to the prediction equation (2.4) but has a much wider interpretation which coincides only in the sense that it also determines the predictor of x_t from $x_{t-1}, \ldots, x_{t-p}$.

Although motivated and derived as a predictor, we therefore now interpret (2.4) as defining a process in which each successive value x_t is *generated* from previous values $x_{t-1}, \ldots, x_{t-p}$ and an *innovation* e_t which has variance V_e and is uncorrelated with *all* past innovations e_{t-k}. We say that e_t is *multivariate white noise*. We will call this process a VAR(p) model, where VAR stands for Vector AutoRegressive and p the model order. This is established nomenclature, though our preference elsewhere is to use the adjective multivariate rather than vector. To emphasize its distinct interpretation from here on as a model rather than a predictor, we write again the first line of (2.4) as the model equation:

$$x_t = \Phi_1 x_{t-1} + \Phi_2 x_{t-2} + \cdots + \Phi_p x_{t-p} + e_t. \qquad (2.7)$$

When this model is derived by construction of the predictor in (2.4), the process x_t which it defines is not, in general, identical to the original process x_t, but approximates its structure. However, we keep the same notation, because we will use it as a model to construct predictions and derive various properties of the original process. The main general property of the model is that the covariance properties and minimum mean square error linear predictions of any stationary process can be approximated with arbitrary accuracy by a VAR(p) model of sufficiently high order. Other important properties of the approximating VAR(p) are

1. It is unique (2.5) always has a unique solution.

2. It defines a unique stationary process x_t, with successive values generated by (2.7).

3. The innovation e_t is also uncorrelated with *all* past values $x_{t-1}, x_{t-2}, \ldots$

4. The approximating process has the same variance V_x and covariances $\Gamma_1, \ldots, \Gamma_p$ as those of the original process.

This last property reveals the manner in which the approximating process coincides with the original. We say that the approximating model *captures* these covariances of the original process. However, this property is lost if the model is modified to become a subset autoregression, i.e., some of the components of lagged values of $x_{t-1}, \ldots, x_{t-p}$ are omitted from the set used as predictors in (2.4). The coefficients in the projection of x_t onto the subset of predictors are then determined by simply removing the corresponding terms in the Yule–Walker equations (2.5). However, the process generated by the model derived from subset autoregression is not necessarily stationary. Even if it is stationary, the covariances of this approximating process may fail to coincide with any of the covariances of the original process. In Chapter 5 we introduce structural autoregressive models which include only a sparse subset of the predictors up to lag p. It is then important to apply the methods which we describe there for checking that these models are good approximations to the covariance structure of the series to which they are fitted.

The prediction of x_t in (2.4) improves as the specified order p increases, in the sense that V_e decreases, or at least does not increase. In the limit as $p \to \infty$, the prediction $\hat{x}_t$ and the error e_t converge, as do the coefficients Φ_k for any given k. We make only one, very reasonable, assumption, that the limiting variance V_e is strictly positive definite, i.e., no linear combination of the series is perfectly predictable from its past. In that case we can show that e_t is multivariate white noise and is known as the linear innovation series of x_t. This justifies the corresponding assumption for e_t in the approximating model. See Whittle (1953) and references therein.

However, although the expression $x_t = \hat{x}_t + e_t$ holds for the limiting prediction and innovation, it is not always possible to express the prediction as an infinite sum of past values:

$$
\begin{aligned}
\hat{x}_t &= \Phi_1 x_{t-1} + \Phi_2 x_{t-2} + \cdots \\
&= \sum_{k=1}^{\infty} \Phi_k x_{t-k}.
\end{aligned}
\tag{2.8}
$$

When this expression fails, a model of high order is characteristically required to obtain a good approximating autoregression. We illustrate this by two simple univariate examples for which we examine the pattern of autoregressive coefficients when the model order becomes large. For the first example we take $\Gamma_0 = 1$ and $\Gamma_k = (\cos k)/2$ for $k > 0$ and for the second we take $\Gamma_0 = 1$, $\Gamma_1 = -0.5$ and $\Gamma_k = 0$ for $k > 1$. Figure 2.2 shows, for each of these examples, a simulation of the series, the autocorrelation functions up to lag 20, a plot of V_e for increasing orders up to $p = 20$ and the coefficients Φ_k for the predictor of that order.

For both of these examples the limiting value of V_e is 0.5, and the model of order 20 comes close to this. However, for the first example the prediction error coefficients Φ_k all tend to zero in the limit, so the infinite sum (2.8) becomes zero. For the second example the coefficients Φ_k all tend to minus one in the limit, so the infinite sum (2.8) becomes $-\sum_{k=1}^{\infty} x_{t-k}$, which does not converge because the terms in the sum do not tend to zero. Conditions may be readily imposed on the process x_t which do ensure that the infinite sum for the predictor (2.8) is true, but we do not need them; they would clearly exclude these valid examples. We can always achieve a model that is acceptable for practical application by using an approximating autoregression with a sufficiently high order. We have to give consideration to problems of

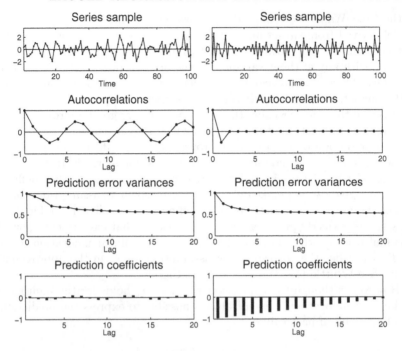

Figure 2.2 *Examples of two univariate processes and their prediction coefficients. On the left the autocorrelations are* $(1 + \cos k)/2$ *and on the right they are zero except for the value of* -0.5 *at lag 1.*

numerical accuracy in examples such as these, because the high order predictor in the first example is the sum of a large number of small quantities, and in the second example is the sum of quantities that partially cancel. However, modern computers, when used with well conditioned numerical procedures, are sufficiently accurate for fitting high order approximating models when these are necessary.

Consideration of the limiting prediction error or innovation e_t as the order of predictor increases also leads to an expression for the original process x_t as a combination of present and past values $e_t, e_{t-1}, e_{t-2}, \ldots$. This is known as the (multivariate) Wold decomposition, or Wold representation (see Whittle, 1953):

$$
\begin{aligned}
x_t &= e_t + \Psi_1 e_{t-1} + \Psi_2 e_{t-2} + \cdots + f(t) \\
&= e_t + \sum_{k=1}^{\infty} \Psi_k e_{t-k} + f(t).
\end{aligned}
\tag{2.9}
$$

Here, $f(t)$ is the *deterministic* component of the series in the sense that, given the model, it can in theory be determined exactly from the series and can itself be linearly predicted exactly from any stretch of its values in the far past, $f(t-k)$ for $k \geq K$, however large K may be. For the first example above, this decomposition reduces to

$$x_t = e_t + A \cos t + B \sin t \qquad (2.10)$$

so all the coefficients Ψ_k are zero and the deterministic part is a cyclical function. The coefficients A and B can be precisely determined from any stretch of values in the far past by simple regression. They are, technically, values of random variables, in that if we were given two different processes but with the identical autocovariances $\Gamma_k = (1 + \cos k)/2$, they would yield different values of the pair A, B, and these have the properties of being samples from uncorrelated distributions with mean 0 and variance 0.5. It can be checked from (2.10) that, indeed, $\mathrm{Cov}(x_t, x_{t-k}) = (\cos k)/2$ for $k > 0$. But for any single process we can treat this cyclical part as being deterministic.

For the second example the Wold representation is

$$x_t = e_t - e_{t-1} \qquad (2.11)$$

so there is no deterministic part, but $\Psi_1 = -1$. Again it is easily checked that x_t has the specified autocovariances.

The decomposition (2.9) is *always* valid for stationary multivariate series provided the limiting value of V_e is positive definite: the infinite sum always converges. The deterministic part is typically the sum of cyclical components, but may be more complex. In our example of the Moth count series in Figure 1.4 we will introduce deterministic components to represent and correct for the seasonal cycles and we will also remove quarterly cycles from the Pig market series in Figure 1.6, though they are not large in magnitude or very obvious to the eye. A series that has no deterministic term $f(t)$ in the Wold representation (2.9), or for which this term has been removed so that just the infinite sum of past innovations remains, is called a purely non-deterministic process; see Brockwell and Davis (1987, p. 189). We display this representation separately because of its importance in later development:

$$x_t = \sum_{k=0}^{\infty} \Psi_k e_{t-k}, \qquad (2.12)$$

where $\Psi_0 = I$. The autocovariances of the series x_t having the representation (2.12) are related to the coefficients in this expression by

$$\Gamma_k = \sum_{j=0}^{\infty} \Psi_{j+k} V_e \Psi_j'. \qquad (2.13)$$

It will also be of use later to remark that a series which has a Wold representation that extends only to a finite number q of past innovations, i.e.,

$$x_t = e_t + \Psi_1 e_{t-1} + \Psi_2 e_{t-2} + \cdots + \Psi_q e_{t-q}, \qquad (2.14)$$

is known as a vector moving average (VMA) process of order q. Such a process is characterized by the property that its autocovariances only extend to a finite lag: $\Gamma_k = 0$ for $k > q$. Any process with this property can always be represented as a VMA of the form (2.14).

2.3   Multi-step autoregressive model prediction

The VAR(p) model of a process x_t given by (2.7) can be used to construct a *multistep* prediction of the process at a *lead time* of h steps into the future. When we consider prediction we generally assume that observations are known or given up to time t, and that values at later times are unknown. We therefore re-write the model as

$$
\begin{aligned}
x_{t+1} \;=\;\; & \Phi_1 x_t + \Phi_2 x_{t-1} + \cdots + \Phi_p x_{t-p+1} \\
& + e_{t+1},
\end{aligned}
\tag{2.15}
$$

indicating that the right hand side (RHS) of the first line is the prediction of the first of the future values x_{t+1} expressed as a linear combination of the latest p observations $x_t, \ldots, x_{t-p+1}$. The second line is the prediction error e_{t+1}.

To construct the prediction of the next value x_{t+2} we first express this using the model equations as

$$
\begin{aligned}
x_{t+2} \;=\;\; & \Phi_1 x_{t+1} + \Phi_2 x_t + \cdots + \Phi_p x_{t-p+2} \\
& + e_{t+2},
\end{aligned}
\tag{2.16}
$$

and substitute for x_{t+1} on the RHS of the first line using the expression for this quantity in (2.15). This introduces a term in e_{t+1} in addition to e_{t+2}. In a similar manner we can take the model equation for x_{t+h} and substitute successively for $x_{t+h-1}, x_{t+h-2}, \ldots, x_{t+1}$ to derive in general:

$$
\begin{aligned}
x_{t+h} \;=\;\; & \Phi_1^h x_t + \Phi_2^h x_{t-1} + \cdots + \Phi_p^h x_{t-p+1} \\
& + e_{t+h} + \Psi_1 e_{t+h-1} + \cdots + \Psi_{h-2} e_{t+2} + \Psi_{h-1} e_{t+1}.
\end{aligned}
\tag{2.17}
$$

The RHS of the first line is an expression for the prediction of x_{t+h}, again as a linear combination of the latest p known observations, but with coefficients that also depend on the lead time h, which is indicated by a superscript to these coefficients. The second line is an expression for the prediction error as a linear combination of the unknown future innovations uncorrelated with all known observations. It coincides with the first h terms in the Wold representation of x_{t+h}. The variance of this error is

$$
V_e(h) = V_e + \Psi_1 V_e \Psi_1' + \cdots + \Psi_{h-1} V_e \Psi_{h-1},
\tag{2.18}
$$

which is useful for constructing error limits about the predictor.

The practical way of constructing the multi-step prediction derived from an approximating VAR(p) is to apply the following device. Using the model equation (2.7), generate successive values $x_{t+1}, x_{t+2}, \ldots, x_{t+h}$ with t updated to $t+1, t+2, \ldots, t+h$ but *replacing* the successive values of the innovations $e_{t+1}, e_{t+2}, \ldots, e_{t+h}$ all by *zero* on the RHS. The value of x_{t+h} so generated is just the required prediction $\hat{x}_{t,h}$, as may be confirmed by setting these innovations to zero in (2.17). The terms on the second line are then zero and x_{t+h} takes the value of the prediction on the RHS of the first line. This procedure can be summarized as generating predictions by forming

$$\hat{x}_{t,h} = \Phi_1\,\hat{x}_{t,h-1} + \Phi_2\,\hat{x}_{t,h-2} + \cdots + \Phi_p\,\hat{x}_{t,h-p} \qquad (2.19)$$

for $h = 1, 2, \ldots$ where, however, any term $\hat{x}_{t,h-k}$ on the RHS with $h \leq k$ is interpreted as the known present or past value x_{t+h-k}. For $h = 1$ all the terms on the RHS are such known quantities. For $h > p$ all terms on the RHS are predictions of future values.

By a similar consideration the coefficients Ψ_k can be obtained from the recursion

$$\Psi_k = \Phi_1\,\Psi_{k-1} + \Phi_2\,\Psi_{k-2} + \cdots + \Phi_p\,\Psi_{k-p} \qquad (2.20)$$

which is applied for $k = 1, 2, \ldots$ starting with $\Psi_0 = I$ and setting $\Psi_{k-j} = 0$ for $j > k$ on the RHS. The coefficients $\Phi_1^h, \ldots, \Phi_p^h$ of the multi-step prediction in (2.17) are then given by

$$\Phi_k^h = \Phi_k\,\Psi_{h-1} + \Phi_{k+1}\,\Psi_{h-2} + \cdots + \Phi_p\,\Psi_{h-1+k-p}. \qquad (2.21)$$

2.4   Examples of autoregressive model approximation

When the Yule–Walker equations are solved for increasing order p to determine approximating VAR(p) models for a process x_t, successive values of the prediction error variance V_e decrease to a limit. This limit may be attained by a finite value of p, in which case the approximating model represents a process with exactly the same acf as the original process. More generally, V_e will decrease indefinitely to its limit, but a model which provides a very good approximation will typically be found for some relatively low value of the order. We illustrate this point in this section. It provides a background to our description, in Chapter 4, of methods for determining this order when a model is fitted to an observed time series.

As an illustration, we will once again use the covariances, shown in Figure 2.1, derived from a model of the Pig market. Figure 2.3 illustrates the approximation by a VAR(p) model of the full five series. Because these covariances were derived from a model which can be represented as an exact VAR(2), this model is recovered by the approximating VAR(p) model for order $p = 2$. The error variance plots are shown for the determinant of the error variance matrix and three of the series, on a logarithmic scale. These are all constant for orders $p \geq 2$. The plot for the series of New breeding animals is constant

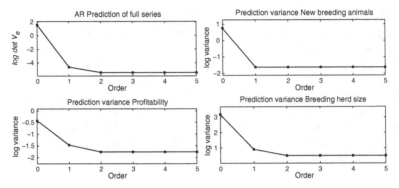

Figure 2.3 *The prediction error variances of autoregressive models of increasing order for the full set of Pig market series.*

from order one, but the determinant of the error variance matrix gives the overall indication of the order of the generating model. The approximating VAR(p) models for $p > 2$ are identical to the VAR(2) model except for extra coefficients $\Phi_3, \ldots, \Phi_p$, which are all zero.

To provide an illustration of the case when V_e reduces indefinitely, we restrict our consideration to a pair of the series, the Numbers to market and Profitability. Their auto- and cross-correlations appear in the second and third rows and columns in Figure 2.1 and are displayed again in Figure 2.4. The cross-correlations of Profitability on Numbers to market at positive lags are here displayed as cross-correlations of Numbers to market on Profitability at negative lags.

Figure 2.5 shows the determinant of the error variance matrix V_e and error variance plots for both series of this pair, again on a logarithmic scale. From the plots we see that the approximation by a VAR(p) model is improving up to order 4 or 5. In fact, the reduction in V_e continues indefinitely, but a model of order 5 would appear to give an accurate approximation to the limiting value.

2.5   The multivariate autoregressive model

In this chapter autoregressive models have only emerged as approximations derived from autocovariance functions. We have explained how (2.7) can define a process that is followed by such a model. We will occasionally refer to (2.7) as the canonical form of VAR model to distinguish it from the structural autoregressive model that we introduce in Chapter 5 and the extended form of model that we introduce in Chapter 6. In Chapter 4 we will describe how to fit canonical autoregressive models to observed time series and therefore need to define them in somewhat greater generality. The process definition we have

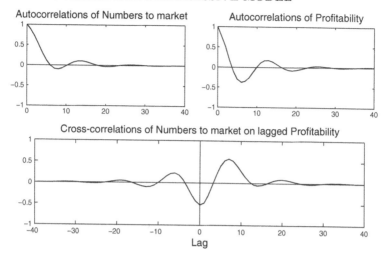

Figure 2.4 *The autocorrelations of the Numbers to market and Profitability series, with their cross-correlations shown for negative and positive lags.*

already given still stands. The only additional statement is of conditions that must be satisfied by an otherwise general set of coefficients $\Phi_1, \ldots, \Phi_p$ and the error variance V_e, in order that (2.7) determines a second order stationary process. We require that V_e is positive definite, but the main condition, which is necessary and sufficient and known as the *stationarity condition*, is that the coefficients Ψ_k generated from $\Phi_1, \ldots, \Phi_p$ by the recursion (2.20) converge to zero. The generated process then has the convergent Wold representation (2.9) but includes no deterministic component.

This stationarity condition can also be expressed in terms of the matrix polynomial

$$\Phi(z) = I - \Phi_1 z - \Phi_2 z^2 - \cdots - \Phi_p z^p. \tag{2.22}$$

By defining also the (infinite) power series

$$\Psi(z) = I + \Psi_1 z + \Psi_2 z^2 + \cdots, \tag{2.23}$$

the recursion (2.20) can be translated as the relationship

$$\Phi(z)\Psi(z) = I. \tag{2.24}$$

The stationarity condition is that $\det\{\Phi(z)\}$, a polynomial of degree mp, is non-zero for all values of $|z| \leq 1$. This is the same as requiring that all the zeros of $\det\{\Phi(z)\}$ have modulus greater than 1, a condition that can be directly checked. Note that, given $\Phi(z)$, there is generally more than one power series $\Psi(z)$ that satisfies (2.24). Even in the most simple univariate case with $\Phi(z) = 1 - \phi z$, there are solutions $\Psi(z) = 1 + \phi z + \phi^2 z^2 + \cdots$ and $\Psi(z) = -\phi^{-1} z^{-1} - \phi^{-2} z^{-2} - \cdots$. We require a solution in positive powers

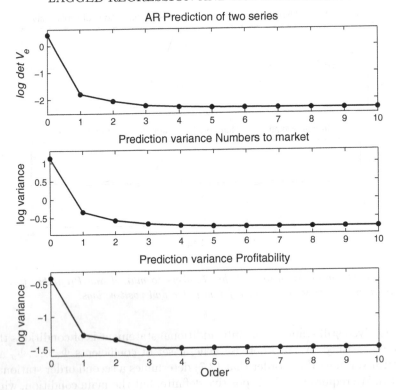

Figure 2.5 *The prediction error variances of autoregressive models of increasing order for a pair of the Pig market series.*

of z with coefficients that converge to zero, and the stationarity condition is necessary and sufficient for this.

If an autoregressive model is proposed with coefficients satisfying the stationarity condition, its covariance function can be found by re-arranging the Yule–Walker equations (2.5) and (2.6). Given $\Phi_1, \ldots, \Phi_p$ and V_e, these can be written as linear equations in the elements of $\Gamma_0, \ldots, \Gamma_p$. For higher lags we can calculate for $k = p+1, p+2, \ldots$

$$\Gamma_k = \begin{pmatrix} \Phi_1 & \Phi_2 & \cdots & \Phi_p \end{pmatrix} \begin{pmatrix} \Gamma_{k-1} \\ \Gamma_{k-2} \\ \vdots \\ \Gamma_{k-p} \end{pmatrix}, \tag{2.25}$$

which is obtained by taking the covariance of every term in the model equation (2.7) with x_{t-k}. For an approximating VAR(p), (2.25) can be considered as generating approximations to the covariances of a process beyond lag p.

2.6   Autoregressions for high lead time prediction

The prediction of time series at higher lead times, $h > 1$, has received a great deal of attention, as referenced in Section 1.5. To understand the issues involved we set out here four different methods for the construction of such predictors and discuss their advantages and drawbacks. For each approach, the predictor of x_{t+h} is of autoregressive form, being a linear combination of the p most recent values $x_t, \ldots, x_{t-p+1}$. Each uses the principle of minimum mean square error to derive coefficients for the predictor. One of the methods requires the solution of non-linear equations for the coefficients; the other three use linear projection and we will illustrate these by applying them, again, to the series with the acf shown in Figure 2.4. As the order p increases, these three methods all approach the same optimal predictor, which is the projection of x_{t+h} upon all past values $x_t, x_{t-1}, x_{t-2}, \ldots$. However, for finite order p these approaches have different properties, which are illustrated in Figure 2.6. The methods are as follows:

(i) Construct the VAR(p) model that minimizes the error variance of the prediction at lead time $h = 1$, and use (2.19) to construct $\hat{x}_{t,h}$ for lead time $h > 1$. The drawback is that the prediction error variance will not in general be a minimum unless the model exactly approximates the process, because $\hat{x}_{t,h}$ will not be the minimum variance projection of x_{t+h} upon $x_t, \ldots, x_{t-p+1}$. To take the simple example of a univariate series x_t, the minimum variance prediction of x_{t+2} from x_t is $\hat{x}_{t,2} = \rho_2 x_t$, where ρ_2 is the lag 2 autocorrelation of x_t. But the prediction from a minimum variance approximating AR(1) model is $\hat{x}_{t,2} = \rho_1^2 x_t$, which necessarily has a greater error variance unless $\rho_2 = \rho_1^2$. This approach does have the advantage that the same VAR(p) model can be used to construct predictors at any lead time.

(ii) Construct the predictor by projection of x_{t+h} upon $x_t, \ldots, x_{t-p+1}$. This has the minimum error variance of the finite order predictors. It has the drawback that it does *not* provide a process model from which other properties may be derived. Also, a different predictor is required for each choice of lead time h. Even so, this approach has received much attention because of its optimality for a given order and lead time without the need for constructing a general process model; see Bhansali (2007), Marcellino et al. (2006) and Chevillon (2007). An inadequately specified VAR(p) model, with too low an order, can give poor predictions at higher lead times using the approach of (i), in which case this predictor can have a clear advantage.

(iii) Use a VAR(p) model but determine its coefficients to minimize the prediction error variance at the specified lead time $h > 1$. This has the drawback of requiring the solution of non-linear equations. Taking the univariate AR(1) model $x_t = \phi x_{t-1} + e_t$, for which the predictor of x_{t+2} is $\phi^2 x_t$, the required coefficient is determined by solving $\phi^2 = \rho_2$. If $\rho_2 \geq 0$, this predictor coincides with the projection predictor of approach (ii) above, but if $\rho_2 < 0$, the error variance, which is then minimized by taking $\phi = 0$, is much greater. However, Haywood and Tunnicliffe Wilson (1997) provide a reliable iterative scheme for determining this predictor for univariate ARMA models. Applied to the finite order autoregressive predictor, it has the advantage over approach (ii) of giving a single process model which can be used to construct predictions at all lead times. Its predictor at lead time h does, typically, coincide with that using the approach in (ii), and it is similarly robust to an inadequately specified VAR(p) model, with too low an order.

(iv) We introduce a fourth approach because it is included in a natural manner in the extensions to autoregressive models which we introduce in Chapters 6 and 7. It is particularly relevant to high frequency sampled series when the lead times at which forecasts may be required are much higher than the sampling interval. This approach is to minimize the prediction error variance of an exponentially discounted *combination* of future values:

$$x_{t,\rho} = x_{t+1} + \rho x_{t+2} + \rho^2 x_{t+3} + \cdots = \sum_{h=1}^{\infty} \rho^{h-1} x_{t+h} \qquad (2.26)$$

for some discount factor ρ. Higher values of ρ will give more weight to predicting higher lead times. The advantage of this approach is that coefficients obtained by projecting $x_{t,\rho}$ onto $x_t, x_{t-1}, \ldots, x_{t-p+1}$ can be simply transformed to the coefficients of a VAR(p) model which can therefore be used to predict at *all* lead times. Weighting these predictions in the same manner as in (2.26) yields a prediction of $x_{t,\rho}$ which is consistent, i.e., exactly the same as its projection. This VAR(p) model is also stationary and tends to produce predictions at higher lead times that are closer to the optimal projections than those given by the VAR(p) model that minimizes the error variance of the prediction at lead time $h = 1$.

Figure 2.6 illustrates the points made in this comparison for the approaches described in (i), (ii) and (iv). The model acf is used for the series of Numbers to market and Profitability previously considered. The three plots show prediction error variances of x_{t+5} based upon the bivariate terms $x_t, x_{t-1}, \ldots,$ x_{t-p+1}, for increasing order $p = 1, \ldots, 10$. The first plot shows the generalized variance for predictions of the pair of series. The second plot shows the variance for the first series of the pair, the Numbers to market, and the third shows the variance for the second series, the Profitability. Each plot contains three lines corresponding to the three different predictors. As the order increases, they all approach the optimum prediction based upon the whole set of past series values. The black line (lowest) shows the logarithmic variances for the *projection* predictor of (ii), which by definition has the lowest variance of any linear predictor. The light gray line (highest) shows the logarithmic variances of the *autoregressive* predictor of (i), derived from the approximating VAR(p) model that minimizes the prediction error variance at lead time 1. Note that the first order predictor is worse even than that of order zero, which simply predicts the series by its mean value. The second order predictor is also rather poor compared with the projection predictor.

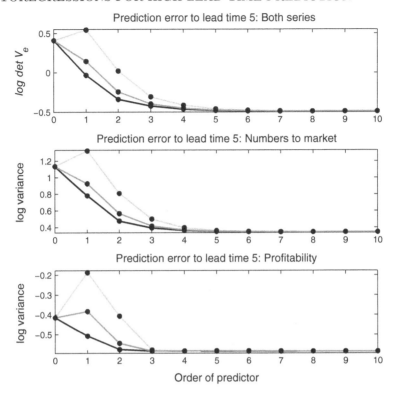

Figure 2.6 *The prediction error variances at lead time 5 of autoregressive predictors of increasing order p for a pair of the Pig market series. The black line (lowest) shows the logarithmic variances for the projection from lead time 5 onto the most recent p observations; the light gray line (highest) shows the logarithmic variances from the approximating AR(p) model that minimizes the prediction error variance at lead time 1; the dark gray line (intermediate) shows the logarithmic variances from the approximating AR(p) model that minimizes the prediction error variance of a linear combination of values at increasing lead times with exponentially decaying weights.*

The dark gray (intermediate) lines in the plots of Figure 2.6 show the prediction error variances using the approach in (iv), with a value of $\rho = 0.75$, also applied to construct the predictor at lead time $h = 5$. It is much closer to the projection predictor than the previous autoregressive predictor, especially for orders of two or more. There is, of course, some penalty in the form of an increase in prediction error variance at the lead time of one, compared with the standard autoregressive model. For the model of order 1 in our example this is respectively 9% and 4% for the two series, and for the model of order 4 it is 3% and 0.1%.

Method (i) can be considered a special case of method (iv) taking $\rho = 0$, and it corresponds to the use of ordinary least squares regression when fitting the VAR(p) model to observations of a time series, as described in Chapter 4. In Chapters 6 and 7 we use a criterion corresponding to method (iv) to fit models to observed time series and show how use of a value of $\rho > 0$ can improve the approximation of the model to the data.

2.7   Model impulse response functions

A useful property of a VAR model is revealed by considering the effect of an innovation term $e_{j,t}$ in series $x_{j,t}$ at time t upon the value of the same or another series $x_{i,t+h}$ at a later time $t + h$. This effect is linear and is therefore described by a single coefficient. To aid our understanding of this and of many other properties of the VAR model (2.7), we now write it in a form which makes use of the backward shift operator B. This is a linear operator which acts on a time series to replace $x_{i,t}$ by $x_{i,t-1}$. It can be applied repeatedly so that, in general,

$$B^k x_{i,t} = x_{i,t-k}. \tag{2.27}$$

Using this operator, the VAR(p) model (2.7) may be re-arranged as

$$
\begin{aligned}
x_t - \Phi_1 B x_t - \Phi_2 B^2 x_t - \cdots - \Phi_p B^p x_t & \\
= \quad (I - \Phi_1 B - \Phi_2 B^2 - \cdots - \Phi_p B^p) x_t & \tag{2.28} \\
= \quad \Phi(B) x_t & = \quad e_t
\end{aligned}
$$

which implicitly defines the matrix polynomial operator

$$\Phi(B) = I - \Phi_1 B - \Phi_2 B^2 - \cdots - \Phi_p B^p. \tag{2.29}$$

This is exactly the same matrix polynomial as defined in (2.22) but with z replaced by B. We can also define the operator

$$\Psi(B) = I + \Psi_1 B + \Psi_2 B^2 + \cdots \tag{2.30}$$

which corresponds in a similar manner to (2.23). Also from (2.24) we have that these two operators are mutual inverses:

$$\Phi(B)\Psi(B) = I. \tag{2.31}$$

From (2.28) we therefore derive the Wold representation for the VAR(p) model as

$$
\begin{aligned}
x_t & = \quad \Phi(B)^{-1} e_t \tag{2.32} \\
& = \quad \Psi(B) e_t \\
& = \quad e_t + \Psi_1 e_{t-1} + \Psi_2 e_{t-2} + \cdots .
\end{aligned}
$$

The coefficient of $e_{j,t}$ in $x_{i,t+k}$ is the same that of $e_{j,t-k}$ in $x_{i,t}$, which is $\Psi_{i,j,k}$. As a sequence in k, this is known in the econometric literature as the impulse response (function) or IRF from an innovation or *shock* in series $x_{j,t}$ to subsequent values of series $x_{i,t}$. It is one way to indicate the impact of one series upon another. Figure 2.7 shows impulse response sequences for the effect of a shock in the Profitability series upon the New breeding animals and the Breeding herd size, as implied by the VAR(2) model of the five Pig market series. As the shock works its way through the different variables, market forces will eventually restore the system to equilibrium according to the pattern of these responses, which are here seen to follow a damped cycle. The response of the New breeding animals is fairly swift. The build up in the Breeding herd size takes longer and lasts longer.

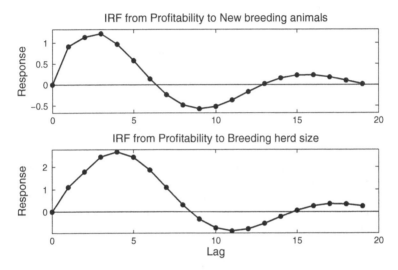

Figure 2.7 *The impulse response sequences for the effect of a shock in the Profitability series upon the New breeding animals and the Breeding herd size, as implied by the VAR(2) model of the five Pig market series.*

Note that we have not used the innovation variance matrix in this derivation of the impulse response. This is an omission because innovations in other series will in general be correlated with, and therefore influenced by, the innovation or shock $e_{j,t}$. However, this influence depends on the structural interpretation of the dependence between the innovations, and this may not be unique. For example, with two series we might choose $e_{2,t}$ to be represented by a regression upon $e_{1,t}$. An impulse in $e_{1,t}$ would then lead to an additional response via its predicted effect upon $e_{2,t}$. Or we might interpret this dependence the other way around—but we could not hold both interpretations together. In Chapter 5 we present our approach to the structural interpretation of the innovation variance using graphical modeling. One of our examples is the modeling of the Pig market. The impulse responses shown in Figure 2.7

have been chosen because they are consistent with the structural model we determine for these series.

The idea of using impulse responses to characterize the relationship between series has a somewhat different (and more longstanding) interpretation in the context of modeling linear systems for the purpose of control. Let us reduce the model to just two variables, an input x_t which is used to control an output y_t, and write their VAR representation in terms of its components as

$$
\begin{pmatrix} \Phi_{xx}(B) & \Phi_{xy}(B) \\ \Phi_{yx}(B) & \Phi_{yy}(B) \end{pmatrix} \begin{pmatrix} x_t \\ y_t \end{pmatrix} = \begin{pmatrix} f_t \\ g_t \end{pmatrix}. \tag{2.33}
$$

We separate these out as one equation which describes the dependence of the output on the input as

$$
y_t = \frac{-\Phi_{yx}(B)}{\Phi_{yy}(B)} x_t + \frac{1}{\Phi_{yy}(B)} g_t \tag{2.34}
$$

and a second equation that describes the feedback from the output to the input

$$
x_t = \frac{-\Phi_{xy}(B)}{\Phi_{xx}(B)} y_t + \frac{1}{\Phi_{xx}(B)} f_t. \tag{2.35}
$$

We will assume for now that there is no correlation between the innovation or shock series f_t and g_t, though this dependence could be accommodated by a structural representation. We are now interested in the *open loop* response of y_t to x_t, described by (2.34), when the feedback represented by (2.35), which closes the loop, is absent. This response is represented by the sequence ν_k given by the expansion

$$
\frac{-\Phi_{yx}(B)}{\Phi_{yy}(B)} = \nu(B) = \nu_0 + \nu_1 B + \nu_2 B^2 + \cdots. \tag{2.36}
$$

There is, in general, no guarantee that this response is stable, in the sense that the coefficients ν_k converge to zero. They may diverge even though the bivariate VAR model satisfies the stationarity condition. Indeed, the lack of stability of this response is a major reason for applying feedback control. To distinguish this from the IRF used in the econometric context, we will call it the open loop impulse response function or OLIRF.

In the absence of the noise terms, if the input series is controlled to be an impulse, set so that $x_t = 0$ for $t < 0$, $x_t = 1$, then $x_t = 0$ for $t > 0$, we obtain the output $y_t = \nu_t$. This is a natural explanation of the term impulse response function. The open loop step response function (OLSRF) is obtained if the input is a unit step with $x_t = 1$ for $t \geq 0$. It is simply the cumulative sum of the OLIRF.

For our example of the Pig market model, Figure 2.8 shows the OLIRF and OLSRF from the Profitability to the New breeding animals. These are derived simply by once more generating the Ψ_k sequence but removing any feedback by setting to zero all the elements of $\Phi(B)$ except $\Phi_{1,1}(B)$ and $\Phi_{1,3}(B)$. The

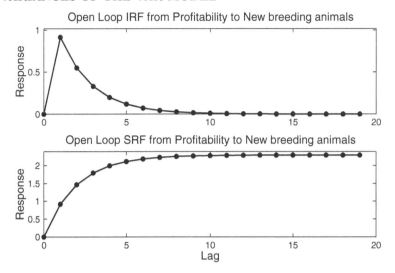

Figure 2.8 *The open loop impulse and step response of the New breeding animals to the Profitability series as implied by the VAR(2) model of the five Pig market series with feedback terms suppressed.*

response to a step increase in profitability is a gradual increase in the numbers of new animals taken into the breeding herd every quarter. By being more selective in the choice of feedback terms that are set to zero, one could use the impulse or step responses to investigate indirect responses from one series to another via intermediate series.

Our final point is a comparison of the econometric and control systems interpretation of the impulse response function in an open loop situation. Returning to the bivariate system (2.33), suppose that the feedback term $\Phi_{xy}(B) = 0$. Then by substituting for $x_t = \{1/\Phi_{xx}(B)\}f_t$ in (2.34), the econometric interpretation of the IRF from the shock f_t in x_t, to y_t, is given by the operator

$$\Psi_{yx}(B) = \frac{-\Phi_{yx}(B)}{\Phi_{yy}(B)\Phi_{xx}(B)} = \frac{\nu(B)}{\Phi_{xx}(B)}. \tag{2.37}$$

This is identical to $\nu(B)$ when we set $\Phi_{xx}(B) = 1$, but otherwise may be quite different.

2.8 Covariances of the VAR model

In this section we develop representations of the structure of the variance matrix of the finite span of series values $x_1, \ldots, x_n$. Some of the results we present apply to a series with a general covariance structure, but we also focus on the case where x_t follows a VAR(p) model. These results are of use in some of the computational methods employed for time series modeling later in the book. Consider the variance matrix of the *stacked* vector of this span of values:

$$\text{Var} X_n = \text{Var} \begin{pmatrix} x_1 \\ x_2 \\ \vdots \\ x_n \end{pmatrix} = \begin{pmatrix} \Gamma_0 & \Gamma_{-1} & \cdots & \Gamma_{-(n-1)} \\ \Gamma_1 & \Gamma_0 & \cdots & \Gamma_{-(n-2)} \\ \vdots & & \ddots & \vdots \\ \Gamma_{(n-1)} & \Gamma_{(n-2)} & \cdots & \Gamma_0 \end{pmatrix} = \tilde{G}_n. \qquad (2.38)$$

This is of the same form as the matrix G_p which occurs in the Yule–Walker equations (2.5) but with the reverse ordering of its block rows and columns because the series sequence is the reverse of the lagged sequence in (2.5). The notation recognizes their relationship but uses the tilde to distinguish them.

The multivariate Yule–Walker equations (2.5) may be solved by a recursive method given by Whittle (1963). For a specified order p this method provides solutions of the equations for all orders $k = 1, 2, \ldots, p$. It also exploits the patterned structure of the matrix in (2.5) in which the entries are constant down all block diagonals. By this means the number of numerical computations is reduced from the order of p^3 to the order of p^2. This has many applications, one of which is to generate a set of autoregressive relationships which fully characterize the covariance structure of a finite set of n successive values of a general stationary series. For this purpose we need to add a further subscript k to the autoregressive coefficients, before the lag index, as $\Phi_{k,1}, \ldots, \Phi_{k,k}$. This is to indicate the order k of the VAR(k) model with these coefficients, which are determined recursively for $k = 1, 2, \ldots, n - 1$. The prediction errors $e_{k,t}$ and their variances $V_{k,e}$ are similarly indexed. The relationships can then be represented as

$$x_t = \Phi_{k,1} x_{t-1} + \Phi_{k,2} x_{t-2} + \cdots + \Phi_{k,k} x_{t-k} + e_{k,t} \qquad (2.39)$$

with

$$\text{Var}\, e_{k,t} = V_{k,e}. \qquad (2.40)$$

Note that $e_{k,t}$ is the innovation in x_t orthogonal to the finite sequence x_{t-k}, $\ldots, x_{t-1}$.

The variance structure of the series over a finite time span can be represented using a succession of these equations illustrated here in matrix form for $n = 5$ after transferring all terms in x_t to the left hand side (LHS) of (2.39):

$$\begin{pmatrix} I & 0 & 0 & 0 & 0 \\ -\Phi_{1,1} & I & 0 & 0 & 0 \\ -\Phi_{2,2} & -\Phi_{2,1} & I & 0 & 0 \\ -\Phi_{3,3} & -\Phi_{3,2} & -\Phi_{3,1} & I & 0 \\ -\Phi_{4,4} & -\Phi_{4,3} & -\Phi_{4,2} & -\Phi_{4,1} & I \end{pmatrix} \begin{pmatrix} x_1 \\ x_2 \\ x_3 \\ x_4 \\ x_5 \end{pmatrix} = \begin{pmatrix} e_{0,1} \\ e_{1,2} \\ e_{2,3} \\ e_{3,4} \\ e_{4,5} \end{pmatrix}. \qquad (2.41)$$

We can write these in general as

$$L_n X_n = E_n \qquad (2.42)$$

where X_n is the stacked series values. The vector E_n is similarly the stacked innovation elements which are uncorrelated one with another so that its variance matrix D_n is block diagonal. In our example:

$$D_5 = \text{Var} \begin{pmatrix} e_{0,1} \\ e_{1,2} \\ e_{2,3} \\ e_{3,4} \\ e_{4,5} \end{pmatrix} = \begin{pmatrix} V_{0,e} & 0 & 0 & 0 & 0 \\ 0 & V_{1,e} & 0 & 0 & 0 \\ 0 & 0 & V_{2,e} & 0 & 0 \\ 0 & 0 & 0 & V_{3,e} & 0 \\ 0 & 0 & 0 & 0 & V_{4,e} \end{pmatrix}. \tag{2.43}$$

Note that we take the zero order prediction error of x_t to be simply $e_{0,t} = x_t$, so $V_{0,e} = V_x$.

This structure can also be usefully exploited to simulate a sample of the series with the given autocovariances. The first step is to generate the uncorrelated samples of $e_{0,1}, e_{1,2}, e_{2,3} \ldots$. We then use the direct relationship (2.39) to construct a sample of $x_1 = e_{0,1}$, $x_2 = \Phi_{1,2}x_1 + e_{1,2}$, $x_3 = \Phi_{2,1}x_2 + \Phi_{2,2}x_1 + e_{2,3}$, $\ldots$ and in general for $k = 1, 2, \ldots$

$$x_{k+1} = \Phi_{k,1}x_k + \cdots + \Phi_{k,k}x_1 + e_{k,k+1}. \tag{2.44}$$

We can also use (2.42) to characterize the structure of the variance matrix $\tilde{G}_n$ of the vector X_n of stacked values of $x_1, \ldots, x_n$ by its relationship to the innovation variance matrix D_n:

$$L_n \tilde{G}_n L_n' = D_n \tag{2.45}$$

from which

$$\tilde{G}_n = L_n^{-1} D_n (L_n')^{-1} \tag{2.46}$$

and

$$\tilde{G}_n^{-1} = L_n' D_n^{-1} L_n. \tag{2.47}$$

All of this is valid for any stationary multivariate time series from which VAR approximations of successively higher order can be generated. A special case arises if the series exactly follows a VAR(p) model. This will be of interest in our further development of VAR modeling. For $n > p + 1$ there is some simplification. Thus, for our example when $p = 2$, we have the simpler forms of (2.41) and (2.43):

$$\begin{pmatrix} I & 0 & 0 & 0 & 0 \\ -\Phi_{1,1} & I & 0 & 0 & 0 \\ -\Phi_2 & -\Phi_1 & I & 0 & 0 \\ 0 & -\Phi_2 & -\Phi_1 & I & 0 \\ 0 & 0 & -\Phi_2 & -\Phi_1 & I \end{pmatrix} \begin{pmatrix} x_1 \\ x_2 \\ x_3 \\ x_4 \\ x_5 \end{pmatrix} = \begin{pmatrix} e_{0,1} \\ e_{1,2} \\ e_2 \\ e_3 \\ e_4 \end{pmatrix} \tag{2.48}$$

and

$$D_5 = \text{Var} \begin{pmatrix} e_{0,1} \\ e_{1,2} \\ e_3 \\ e_4 \\ e_5 \end{pmatrix} = \begin{pmatrix} V_{0,e} & 0 & 0 & 0 & 0 \\ 0 & V_{1,e} & 0 & 0 & 0 \\ 0 & 0 & V_e & 0 & 0 \\ 0 & 0 & 0 & V_e & 0 \\ 0 & 0 & 0 & 0 & V_e \end{pmatrix}. \tag{2.49}$$

Thus, beyond the first two rows, and in general beyond the first p rows, the coefficients, the errors and their variances are just those of the VAR(p) model. Similarly for $k \geq p$, the simulation in (2.44) reduces to generation of values from the VAR(p) model. Also, the matrix of coefficients L_n as in (2.42) becomes a band matrix with non-zero elements only on the diagonal and the p block sub-diagonals. This carries through to the (symmetric) inverse covariance matrix $\tilde{G}_n^{-1}$, which has non-zero elements only on the diagonal, the p sub-diagonals and p super-diagonals. Moreover, apart from the upper left $p \times p$ block sub-matrix of $\tilde{G}_n^{-1}$, all the other elements are easily derived from the VAR(p) model coefficients using (2.47). We return to this topic after the next section.

2.9   Partial correlations of the VAR model

For a group of variables that have a joint multivariate normal distribution, the partial correlation between two variables in the group, say x and y, is the correlation between them conditional upon the remaining variables, say z, in the group. This may be found by correcting each of x and y for their linear regressions upon z, to give residuals $e = x - \alpha z$ and $f = y - \beta z$, say, with respective residual standard deviation σ_e and σ_f. Then the correlation between e and f is the partial correlation between x and y.

When the variables in the group are not necessarily Gaussian but their covariance matrix is known, we will define the partial correlation in the same way: we *correct for* linear regressions on the remaining variables, even though this may not be the same as *conditioning upon* these variables. Further, we will use the notation $\pi = \text{Corr}(x, y|z)$ for the partial correlation.

A fundamental property of the partial correlation is that a non-zero value indicates that including x as a linear predictor of y *in addition to* z will improve upon the prediction of y given z. Doing so will reduce the prediction error variance by a factor $(1 - \pi^2)$. Moreover, the revised predictor of y is of the form $\beta z + \pi(\sigma_f/\sigma_e)(x - \alpha z)$. For a set of variables with covariance matrix V and inverse $W = V^{-1}$, the variance matrix of a subset of variables, say y, x, conditional upon the remaining set z (or more generally, corrected for the remaining set if Gaussianity cannot be assumed), can be simply expressed as

$$V_{y,x|z} = W_{y,x}^{-1}, \tag{2.50}$$

where $W_{y,x}$ is the sub-matrix of W corresponding to the variables y, x. When y, x consists of just a pair of variables with indices i and j, their partial correlation given all the remaining variables is then (see Whittaker, 1990, p. 143)

$$\pi_{i,j} = -W_{i,j}/\sqrt{W_{i,i}W_{j,j}}. \tag{2.51}$$

In the context of univariate time series modeling, the partial auto-correlation function (pacf) has been widely used, following its advocacy by Box and Jenkins (1970). The partial correlation at lag k is $\pi_k =$

$\text{Corr}(x_t, x_{t-k} | x_{t-1}, x_{t-2}, \ldots, x_{t-k+1})$, i.e., the variables x_t and x_{t-k} are corrected by their regression on all intermediate terms in the series. Moreover, $\pi_k = \phi_{k,k}$, the coefficient of x_{t-k} in the approximating autoregressive predictor of order k of x_t. The characteristic property of an AR(p) model is that $\pi_k = 0$ for $k > p$.

For a univariate stationary time series, the partial autocorrelation coefficient arises naturally in the recursive solution to the Yule–Walker equations, but the case of multivariate series is more complex. The recursive method of Whittle (1963) generates, in the successive steps $k = 1, 2, \ldots$ a forward regression (of order $k - 1$), of x_t upon $x_{t-1}, \ldots, x_{t-k+1}$ with *forward* coefficients $\Phi_{k-1,1}, \ldots, \Phi_{k-1,k-1}$. It also generates a backward regression of x_{t-k} upon the same regressors, but *in reverse order*, with *backward* coefficients $\tilde{\Phi}_{k-1,1}, \ldots, \tilde{\Phi}_{k-1,k-1}$. Let the errors in these be respectively $e_{k-1,t}$ and $\tilde{e}_{k-1,t-k}$. Then the coefficient $\Phi_{k,k}$ in the approximating forward predictor of order k is given by the regression of $e_{k-1,t}$ upon $\tilde{e}_{k-1,t-k}$ and the coefficient $\tilde{\Phi}_{k,k}$ in the approximating backward predictor of order k is given by the regression of $\tilde{e}_{k-1,t-k}$ upon $e_{k-1,t}$. See Dégerine (1990) for various correlations that can be extracted between $e_{k-1,t}$ and $\tilde{e}_{k-1,t-k}$. There is, however, a generalization in terms of determinants: $\det \text{Var}(e_{k-1,t})$ and $\det \text{Var}(\tilde{e}_{k-1,t})$ are identical and are reduced by the factor $\det(I - \Phi_{k,k}\tilde{\Phi}_{k,k})$ in the step from $k - 1$ to k. Figure 2.9 shows a plot of one minus this factor, which represents the improvement in predictability in the step from $k - 1$ to k, and which we will call the partial R^2. This plot is computed using the model acf shown in Figure 2.1, but for the pair of series 3 and 4. We select these because it makes the point that the plot may not be monotonically decreasing.

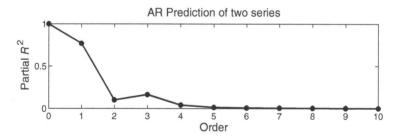

Figure 2.9 *The partial R^2 of autoregressive models of increasing order for a pair of the Pig market series.*

2.10   Inverse covariance of the VAR model

We return to the topic of the inverse covariance matrix of stationary series over a finite time span introduced in Section 2.8. This is relevant here, because (2.51) shows how this matrix is easily transformed to provide the partial correlations between any two variables in this set, conditional upon the remainder.

We will use this when we carry out structural modeling in Chapter 5. In Section 2.12 we present the basic role of the inverse covariance matrix in constructing the projection of one arbitrary set of series values upon any other set. It is also relevant to the calculation of the likelihood of a VAR(p) model given a sample $x_1, \ldots, x_n$ of the series.

If the series exactly follows a VAR(p) model $\Phi(B)x_t = e_t$, it may also be represented by a backward model of the same order,

$$x_t = \tilde{\Phi}_1 x_{t+1} + \tilde{\Phi}_2 x_{t+2} + \cdots + \tilde{\Phi}_p x_{t+p} + \tilde{e}_t. \tag{2.52}$$

Each series value is generated in reverse succession as a combination of p *future* values and a backward innovation $\tilde{e}_t$, which is white noise with variance $\tilde{V}_e$. We also express this model as

$$\tilde{\Phi}(B^{-1})x_t = \tilde{e}_t \tag{2.53}$$

where B^{-1} is the forward shift operator: $B^{-1}x_t = x_{t+1}$.

The forward and backward VAR models have identical coefficients for a univariate process, but this is not in general true for a multivariate process. Given a forward VAR(p) model, there is no very simple method of deriving the coefficients of the backward model. One way is to derive the covariances $\Gamma_0, \Gamma_1, \ldots, \Gamma_p$ of the process using the re-arranged Yule–Walker equations, then solve a backward set of Yule–Walker equations for the backward VAR(p) model. This just requires replacing each term Γ_k by $\Gamma_{-k} = \Gamma'_k$ in (2.5) and (2.6). The Whittle (1963) recursive method of solving the Yule–Walker equations will also generate the backward model besides recovering the forward model.

The relationship between a VAR(p) model and its covariances may be expressed concisely in operator terms by introducing the *covariance generating function* of the process:

$$\Gamma(z) = \sum_{-\infty}^{\infty} \Gamma_k z^k = \cdots + \Gamma_{-1}z^{-1} + \Gamma_0 + \Gamma_1 z + \Gamma_2 z^2 + \cdots. \tag{2.54}$$

The covariance between $\Phi(B)x_t = e_t$ and $\Phi(B)x_{t-k} = e_{t-k}$ then leads to

$$\Phi(z)\Gamma(z)\Phi(z^{-1})' = V_e \tag{2.55}$$

and hence

$$\Gamma(z) = \Phi(z)^{-1}V_e\left\{\Phi(z^{-1})'\right\}^{-1}. \tag{2.56}$$

The RHS may be uniquely expanded as a series in z^{-1} and z with coefficients that converge to zero, in which the coefficient of z^k is Γ_k. Carrying out this expansion numerically is very closely related to solving the Yule–Walker equations (2.5), (2.6) and (2.25). We will refer to the expression (2.56) when we consider the spectrum of the VAR(p) model in Chapter 4. Its inverse also has a useful application, as we note below. There is a corresponding expression to (2.56) in terms of the backward model:

$$\Gamma(z) = \tilde{\Phi}(z^{-1})^{-1}\tilde{V}_e\left\{\tilde{\Phi}(z)'\right\}^{-1}, \tag{2.57}$$

so that the *inverse covariance generating function* may be expressed using either as

$$\{\Gamma(z)\}^{-1} = \Phi(z^{-1})'V_e^{-1}\Phi(z) = \tilde{\Phi}(z)'\tilde{V}_e^{-1}\tilde{\Phi}(z^{-1}). \tag{2.58}$$

This involves only a finite set of powers of z from $-p$ to p and has coefficients known as *inverse autocovariances* given by

$$\Gamma_k^i = \sum_{j=0}^{p-k} \Phi_j'V_e^{-1}\Phi_{j+k} = \sum_{j=0}^{p-k} \tilde{\Phi}_{j+k}'\tilde{V}_e^{-1}\tilde{\Phi}_j \quad \text{for} \quad k = 0,\dots,p, \tag{2.59}$$

where we use the superscript i to denote the inverse autocovariance matrix, and set $\Phi_0 = \tilde{\Phi}_0 = -I$. Inverse autocorrelations for univariate series were introduced by Cleveland (1972).

The inverse of the matrix $\tilde{G}_n$ of $x_1, \dots, x_n$ defined in (2.38) has all its elements except those in the first (upper left) and last (lower right) $p \times p$ block sub-matrices given, for $r > s$, by

$$\left(\tilde{G}_n^{-1}\right)_{r,s} = \Gamma_{r-s}^i. \tag{2.60}$$

By equating and comparing the Gaussian probability densities of the forward and backward models, the complete matrix $\tilde{G}_n^{-1}$ can be found for any $n \geq p$ as

$$\tilde{G}_n^{-1} = M'WM - N'\tilde{W}N \tag{2.61}$$

where the $n \times n$ block matrix M and $p \times n$ block matrix N are

$$M = \begin{pmatrix} \Phi_0 & 0 & 0 & \cdots & 0 & 0 & 0 \\ \Phi_1 & \Phi_0 & 0 & 0 & \cdots & 0 & 0 \\ \vdots & \ddots & \ddots & \ddots & \ddots & \ddots & \vdots \\ \Phi_p & \cdots & \Phi_1 & \Phi_0 & 0 & \cdots & 0 \\ 0 & \Phi_p & \cdots & \Phi_1 & \Phi_0 & 0 & \vdots \\ \vdots & \ddots & \ddots & \ddots & \ddots & \ddots & 0 \\ 0 & 0 & 0 & \Phi_p & \cdots & \Phi_1 & \Phi_0 \end{pmatrix} \tag{2.62}$$

and

$$N = \begin{pmatrix} \tilde{\Phi}_p & 0 & 0 & \cdots & 0 & 0 & 0 \\ \tilde{\Phi}_{p-1} & \tilde{\Phi}_p & 0 & 0 & \cdots & 0 & 0 \\ \vdots & \ddots & \ddots & \ddots & \ddots & \ddots & \vdots \\ \tilde{\Phi}_1 & \cdots & \tilde{\Phi}_{p-1} & \tilde{\Phi}_p & 0 & \cdots & 0 \end{pmatrix}. \tag{2.63}$$

The matrices W and $\tilde{W}$ in (2.61) are conformable block diagonal with entries V_e^{-1} and $\tilde{V}_e^{-1}$ along their respective diagonals. See Dym and Young (1990) for a wider background to the expression (2.61) in the context of linear systems theory. There is a similar expression in which the roles of forward and backward models are interchanged. By using the identities in (2.59) we show here a form of $\tilde{G}_5^{-1}$ in which these roles are symmetric for the example of $n = 5$ and $p = 2$:

$$
\begin{pmatrix}
\tilde{V}_e^{-1} & -\tilde{V}_e^{-1}\tilde{\Phi}_1 & -\tilde{V}_e^{-1}\tilde{\Phi}_2 & 0 & 0 \\
-\tilde{\Phi}_1'\tilde{V}_e^{-1} & \left(\Gamma_0^i - \tilde{\Phi}_2'\tilde{V}_e^{-1}\tilde{\Phi}_2\right) & \Gamma_1^{i\,'} & \Gamma_2^{i\,'} & 0 \\
\Gamma_2^i & \Gamma_1^i & \Gamma_0^i & \Gamma_1^{i\,'} & \Gamma_2^{i\,'} \\
0 & \Gamma_2^i & \Gamma_1^i & \left(\Gamma_0^i - \Phi_2'V_e^{-1}\Phi_2\right) & -\Phi_1'V_e^{-1} \\
0 & 0 & -V_e^{-1}\Phi_2 & -V_e^{-1}\Phi_1 & V_e^{-1}
\end{pmatrix}.
\qquad (2.64)
$$

This example also illustrates a general feature of the matrix structure for $n > p$, that the coefficients of the VAR(p) model can be directly derived from the non-zero elements of the last block row and those of the backward model from the first block row. To make this clear we have set $\Gamma_2^{i\,'} = -\tilde{V}^{-1}\tilde{\Phi}_2$ and $\Gamma_2^i = -V^{-1}\Phi_2$ in the center of the first and last rows, respectively.

The inverse covariance matrix $\tilde{G}_p^{-1}$, derived directly via the model covariances or from either of the expressions (2.47) or (2.61), is useful for calculating the likelihood of the VAR(p) model. We will use sample values of the last block row of the inverse covariance matrix with $n = p+1$ for identifying structural models in Chapter 5 and will compare the inverse covariance matrix of the fitted model with these sample values as a means of model checking.

2.11   Autoregressive moving average models

Vector autoregressive moving average (VARMA) models have been studied for many years, with Quenouille (1957) among the earliest texts and Hannan (1970) providing an extensive treatment. A comprehensive theoretical exposition of the models is provided by Hannan and Deistler (1988) but in practice they have been difficult to apply. One approach is advocated by Tiao and Tsay (1989). The VARMA(p,q) model for a stationary multivariate time series x_t of dimension m is of the form

$$
\Phi(B)x_t = \Theta(B)e_t
\qquad (2.65)
$$

where $\Phi(B)$ has exactly the same structure and constraints as the autoregressive operator in the VAR model (2.28). The operator $\Theta(B)$ is of order q and given by

$$
\Theta(B) = I - \Theta_1 B - \Theta_2 B^2 - \cdots - \Theta_q B^q.
\qquad (2.66)
$$

It satisfies the constraint known as the *invertibility condition*, that $\det\{\Theta(z)\}$, a polynomial of degree $m\,q$, is non-zero for all values of $|z| < 1$. This is the same as requiring that all the zeros of $\det\{\Theta(z)\}$ have modulus greater than or equal to 1. Note that this permits zeros with modulus equal to 1, which are excluded in the stationarity condition upon $\Phi(B)$. The process e_t has exactly the same interpretation as in the VAR(p) model. It is multivariate white noise and is the innovation series of x_t in the Wold representation of x_t given by

$$\begin{aligned} x_t &= \Phi(B)^{-1}\Theta(B)e_t && (2.67) \\ &= \Psi(B)e_t. \end{aligned}$$

Unlike the VAR(p) model, the VARMA(p,q) representation of a process is not necessarily unique, in that models with different orders or coefficients may correspond to processes with exactly the same covariance structure and Wold representation. Conditions on the orders and coefficients for a unique representation of the VARMA model are presented in Hannan and Deistler (1988).

If $\det\{\Theta(z)\}$ has no zeros with modulus equal to 1, we can also determine e_t as

$$\begin{aligned} e_t &= \Theta(B)^{-1}\Phi(B)x_t && (2.68) \\ &= \Pi(B)x_t, \end{aligned}$$

so that we have an autoregressive representation of $\Pi(B)x_t = e_t$ which is necessarily of infinite order. If $\det\{\Theta(z)\}$ does have zeros with modulus equal to 1, the innovations e_t can be determined by taking the limit as $r \to 1$ of the RHS of (2.68) in which $\Theta(B)$ is replaced by $\Theta(rB)$, for $0 < r < 1$.

One motivation for the use of VARMA models is that they can arise naturally when a subset of r time series y_t is selected from a larger set of m time series x_t that follow a VAR(p) model. In general, y_t will not follow a VAR model of any finite order, but may be represented by a VARMA model.

To see this, suppose z_t are the $s = m - r$ series not included in the subset and that we can partition the model for x_t as

$$\left(\begin{array}{cc} \Phi_{yy}(B) & \Phi_{yz}(B) \\ \Phi_{zy}(B) & \Phi_{zz}(B) \end{array} \right) \left(\begin{array}{c} y_t \\ z_t \end{array} \right) = \left(\begin{array}{c} f_t \\ g_t \end{array} \right). \tag{2.69}$$

By eliminating z_t from these partitioned equations, the remaining series can be represented by

$$\Upsilon(B)y_t = \Lambda(B)e_t \tag{2.70}$$

where $\Upsilon(B)$ is, in general, an $r \times r$ matrix operator of order $(s+1)p$ and $\Lambda(B)$ is an $r \times m$ matrix operator of order sp. Finally, we note that $w_t = \Lambda(B)e_t$ is a stationary multivariate time series of dimension r which has a covariance function extending to a finite lag sp. As we remarked at the end of Section 2.2, w_t can then be represented as a VMA $w_t = \Omega(B)h_t$ where $\Omega(B)$ is an $r \times r$ matrix operator of order sp and h_t is the innovation series of w_t. Finally, we can write

$$\Upsilon(B)y_t = \Omega(B)h_t \tag{2.71}$$

which is a VARMA representation of y_t of the form (2.65).

The pair of series that we considered in Section 2.4 is a subset of 5 series following a VAR(2) model, so they will theoretically follow a VARMA model with an AR operator of order 8 and an MA operator of order 6. They did appear, however, to be very well approximated by a VAR(5) model. In practice, VAR models have proved to be capable of providing excellent approximations to many real multivariate time series. There are some contexts, however, when the order required of the approximating model may be unacceptably large. The extensions to VAR models that we present in Chapter 6 can overcome this limitation in many examples.

2.12   State space representation of VAR models

As we remarked in Chapter 1, we define the states of a linear model to be the set of variables required, at any given time t, to predict the next observation x_{t+1}. For a VAR(p) model, the natural set consists of $x_t, x_{t-1}, \ldots, x_{t-p+1}$, though we point out that these are not unique, because any fixed non-singular transformation of this set can also be taken as the state variables. The value of this concept, of a state vector, is that it enables us to represent the model as a first order Markovian transition from the state at one time point to the next. To be consistent with the notation used in (2.7) for our VAR(p) model, we express this transition as from time $t-1$ to time t (though it is conventional in systems theory to use the transition from t to $t+1$):

$$\begin{pmatrix} x_t \\ x_{t-1} \\ \vdots \\ x_{t-p+1} \end{pmatrix} = \begin{pmatrix} \Phi_1 & \Phi_2 & \cdots & \Phi_p \\ I & 0 & \cdots & 0 \\ & \ddots & \ddots & \vdots \\ \cdots & 0 & I & 0 \end{pmatrix} \begin{pmatrix} x_{t-1} \\ x_{t-2} \\ \vdots \\ x_{t-p} \end{pmatrix} + \begin{pmatrix} e_t \\ 0 \\ \vdots \\ 0 \end{pmatrix}. \qquad (2.72)$$

We define the state vector S_t, transition matrix T and disturbance vector E_t by writing (2.72) as the state transition equation

$$S_t = TS_{t-1} + E_t. \qquad (2.73)$$

The first line of (2.72) is just the VAR(p) model and the remaining lines update the state vector by copying over the most recent $(p-1)$ observations from the previous state. The eigenvalues of T can be shown to be the reciprocal zeros of $\det \Phi(z)$, which by the stationarity condition are all less than one in modulus.

We can view (2.73) as a VAR(1) process of dimension mp in the state vector S_t with autoregressive coefficient T, so the Yule–Walker equations (2.5) and (2.6) give

$$\Gamma_{S,1} = TV_S \quad \text{and} \quad V_E = V_S - T\Gamma'_{S,1}, \qquad (2.74)$$

which can be combined into $V_E = V_S - TV_ST'$. This is a discrete Lyapunov equation; see Anderson and Moore (1979, p. 64). Given V_E and T, it can be solved for V_S in many ways, including the solution of a set of $mp(mp+1)/2$ linear equations or the efficient numerical algorithm of Hammerling (1990).

The first block row of V_S holds the values of $\Gamma_0, \ldots, \Gamma_{p-1}$, and, in general, V_S is equal to G_p, the matrix arising in the Yule–Walker equations (2.5). In this example the value of x_t is derived from the state vector very simply as

$$x_t = \begin{pmatrix} I & 0 & \cdots & 0 \end{pmatrix} \begin{pmatrix} x_t \\ x_{t-1} \\ \vdots \\ x_{t-p+1} \end{pmatrix}. \tag{2.75}$$

We define the observation matrix H by writing (2.75) as the observation equation

$$x_t = HS_t. \tag{2.76}$$

Taken together, the transition equation (2.72) and observation equation (2.75) define the state space representation for the VAR(p) process x_t. State space representations of the general form shown in (2.73) and (2.76) are used to model a wide range of processes. More typically, the states S_t are *not* directly observed, only the observations x_t. The representation of the VAR(p) process is a restrictive example in several ways, because the disturbance vector E_t in (2.72) is degenerate with only m non-zero elements which are also equated to the process innovations. In the general model it may be any multivariate white noise. In a widely used time series model presented by Harvey (1989), the state consists of independent unobserved components to represent stochastic levels, trends and seasonality, each with an independent disturbance term. The observation equation for the univariate process x_t sums these components. Although the innovation in x_t depends upon the independent disturbances, it is not related to them in any simple manner, such as their sum. This *unobserved components model* also includes an observation error or noise term which is also commonly included in the general observation equation as $x_t = HS_t + A_t$, where A_t is also multivariate white noise and may be interpreted as measurement error or sampling variability. We will allow this possibility when we consider models for continuous time series, but remark that the observation noise A_t can formally be incorporated into the state variables with the transition matrix extended by corresponding zero diagonals. Our formulation omitting this term is not therefore restrictive in theory. The relationship between state space representations, VAR and VARMA models is presented in Hannan and Deistler (1988, Chap. 1).

The virtue of the state space model is that it has a Markovian structure under the Gaussian assumption. More generally, given the state S_t at time t, the linear predictor of any future observation x_{t+k} or any future state S_{t+k}, for $k > 0$, depends only on S_t and not on any other past state S_{t-k} or observation x_{t-k} for $k > 0$. Given a finite span of observations $x_1, \ldots, x_n$, this leads to a recursive procedure, called the Kalman filter (Kalman, 1960), for calculating the projection $\hat{x}_t$ of x_t upon $x_{t-1}, \ldots, x_1$ and, as an essential step in this procedure, the projection $\hat{S}_t$ of S_t upon $x_t, x_{t-1}, \ldots, x_1$. The recursive step is to calculate $\hat{x}_t$ from $\hat{S}_{t-1}$ then $\hat{S}_t$ from $\hat{S}_{t-1}$ and x_t, a step that can then be

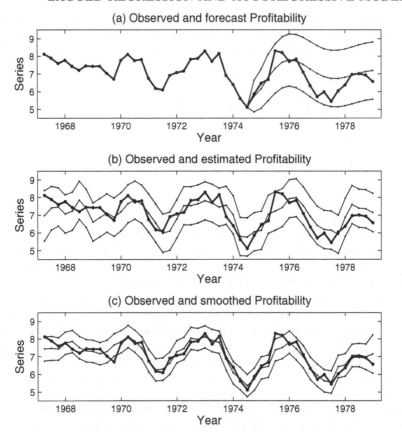

Figure 2.10 *(a) Forecasts of the Profitability series constructed using observations of all values of just three series, the Profitability series itself, the Numbers to market and the New breeding animals, up to the second quarter of 1974. (b) Estimates of the Profitability series constructed at each quarter using only observations of the New breeding animals up to the same quarter. (c) Estimates of the Profitability series constructed at each quarter using the whole series of observations of the New breeding animals. The observed series is shown by the heavy line and the forecasts and their two standard error limits are shown by the light lines.*

repeated with t increased successively from $t = 1$ to $t = n$. All that is needed to start the recursions is the covariance matrix.

There is a further recursive procedure, known as smoothing, which proceeds in the reverse direction from $t = n$ to $t = 1$ and generates the projection of the states S_t upon the whole set of observations $x_1, \ldots, x_n$. The covariance matrices of the filtered or smoothed state vector projections, which are also called state estimates, are generated by the recursions.

These procedures apply equally well to state space representations in which the transition matrix T, the variance matrix of the disturbance vector E_t and the observation matrix H all vary with time t. For an illustration of filtering and smoothing, we allow H to vary in a manner that restricts which components of x_t are observed at a given time. For the state space model of the VAR(p) process, this allows us conveniently to calculate the projection of any subset of series values in the span $x_1, \ldots, x_n$, upon any other subset of observed values. Figure 2.10 shows three illustrations of this. The upper plot shows estimates, or forecasts, of the Profitability series constructed using observations of all values of just three series up to the second quarter of 1974, the Profitability series itself, the Numbers to market and the New breeding animals. We emphasize that the model used was the VAR(2) of all five series which we have used throughout the illustrations in this chapter. The central plot shows the estimates of the Profitability series constructed at *each quarter* using only observations of the New breeding animals up to the *same quarter*. The Kalman filter was used to generate these. The lower plot shows estimates of the Profitability series constructed at *each quarter* using the *whole series* of observations of the New breeding animals by applying the smoother. These last two figures provide a rigorous assessment of the dependence between Profitability and the other series that is implied by the model. They show estimates of the values of Profitability *without* direct reference to any of its own past (or future) values, and using only the values of other series. The Profitability series is only used along with the others in constructing the process model. This is in contrast to predictions which are commonly presented based upon immediate past values of *all* the series, including that which is predicted. The extent to which one series depends on the other series rather than on its own past is then not clear.

The structure of the model is that Profitability depends only indirectly, and after some delay via the other series, upon the New breeding animals. The filtered estimates of Profitability in the central plot are not, therefore, particularly accurate. In contrast, the number of New breeding animals depends quite strongly upon the recent Profitability, so that in the lower plot of the filtered series, which uses information from *subsequent*, strongly related values of New breeding animals, the estimates of Profitability are much more accurate.

2.13 Projection using the covariance matrix

The recursive procedure of the Kalman filter is particularly efficient for projecting the future upon an increasing set of known series values as time progresses. However, the principles of projection that are applied within this recursive procedure can be applied in a *batch mode* to the whole series using only the model acf. For relatively short time series, this can be competitive with the Kalman filter in numerical efficiency, especially if the dimension of the state vector is high.

In general, given the covariance matrix V of a set of (zero mean) variables

X, the projection $\hat{X}_a$ of one subset, X_a, of the variables upon the comple-
mentary subset, X_b, can be achieved using either of two methods. The first
method is to determine from V the coefficients L in the projection $\hat{X}_a = L\,X_b$.
In our context of stationary multivariate time series, we used this method when
solving Yule–Walker equations for autoregressive coefficients. The equations
determining L are derived by solving

$$LV_{b,b} = V_{a,b} \tag{2.77}$$

where the subscripts of V indicate the selection of rows and columns appropriate to the two sets of variables. The conditional variance of $\hat{X}_a$ is then given
by $V_{a,a} - V_{a,b}L' = V_{a,a} - V_{a,b}V_{b,b}^{-1}V_{b,a}$.

The second method is to solve a set of equations directly for the numerical
values of $\hat{X}_a$ given the numerically observed values of X_b. This method can be
considered as dual to the first. The equations are determined by the inverse
covariance matrix $V^{-1} = W$ of the whole set of variables and are simply

$$W_{a,a}\hat{X}_a = -W_{a,b}X_b \tag{2.78}$$

where the subscripts of W again indicate the selection of rows and columns
appropriate to the two sets of variables. For Gaussian variables, this result
derives directly from the term $Q = X'V^{-1}X$ in the exponential part of the
joint density. Conditioning on X_b is achieved simply by fixing the values of X_b
in this joint density, which gives $\hat{X}_a$ as the conditional mean and $W_{a,a}^{-1}$ as the
conditional variance. The result may also be derived from the first method by
matrix algebra.

For the VAR(p) model, the inverse covariance matrix G_n^{-1}, as given in
(2.47) and (2.61), is a relatively simple band matrix, so the second method is
particularly efficient. The projections obtained using this second method are
necessarily identical with those derived using the Kalman filter and smoothing
when the set of known variables is fixed, as in the upper and lower plots of
Figure 2.10. For the central plot in this figure, at any given time the past
observations constitute the set of known variables. This set is increasing as
time progresses through the plot, and the recursive nature of the Kalman filter
is well suited to this context.

2.14   Lagged response functions of the VAR model

This chapter introduced autoregressive models by considering the projection
of current values of a set of series upon a finite set of their past values. Another
projection of interest is of the current value of one series, say y_t, upon a finite
span of past and future values of another series x_t, expressed as

$$y_t = \sum_{k=-p}^{p} v_k x_{t-k} + n_t. \tag{2.79}$$

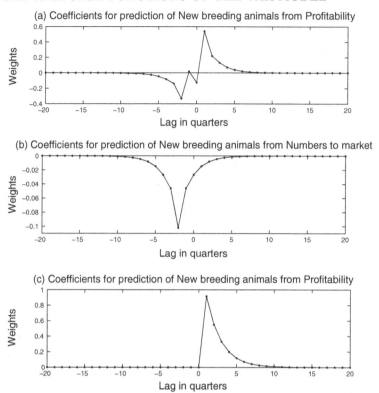

Figure 2.11 *Plots (a) and (b) show the lagged coefficients upon, respectively, Profitability and the Numbers to market, in the projection of the New breeding animals upon past and future values of the four other series. Plot (c) shows the lagged coefficients from the projection upon Profitability alone after the removal of selected model terms so that there is no feedback loop from New breeding animals to Profitability.*

As the order p increases, the error term n_t will become uncorrelated with all past and future values of x_t and the coefficients v_k in this projection will typically converge for each fixed lag k. These methods can be readily extended to the projection of one series $x_{i,t}$ in a set upon all the remaining series:

$$x_{i,t} = \sum_{j \neq i} \sum_{k=-p}^{p} v_{i,j,k} x_{j,t-k} + n_t. \qquad (2.80)$$

As p increases, we will call the limiting values of these coefficients $v_{i,j,k}$ the lagged response coefficients, or simply the lagged response, of the one series, to the series upon which it is projected. In an open loop system, the lagged response of the output to the input, determined by this projection, will be the causal open loop impulse response, and all the coefficients at negative lags will be zero. However, in general, the coefficients at negative lags will not be zero, and they cannot represent causal dependence of one series upon the others; they merely have a predictive role in the statistical sense of the projection (2.80). Nevertheless, we will call them lagged responses, with coefficients at negative lags described as non-causal.

To illustrate the pattern of coefficients, we use the first projection method described in the previous section to determine, for the Pig market series, the coefficients in the projection of New breeding animals at the first quarter of 1972 upon past and future values of the four other series over the period of the records. Although this is a finite span of past and future values, the coefficients have effectively reached the limit over the infinite span. The first two plots in Figure 2.11 show the coefficients in this projection of the series of New breeding animals upon the lagged values of, respectively, the Profitability and the Numbers to market. Note that the coefficients at *positive* lags apply to *past* values of the predicting series, which conforms with the convention for the coefficients in autoregressive models. In these plots we see that weights are applied to future as well as past values — the lagged response is non-causal. Figure 2.11(b) shows that there is useful information to determine the New breeding animals from the Numbers to market at lag -2 quarters, i.e., six months *in the future*.

Causal or *one-sided* dependence cannot be illustrated for the Pig market model using projections of the form (2.80), except by modifying the model used to generate the covariance structure of our example by removing terms by which Profitability depends directly or indirectly upon present and past values of New breeding animals. This pruning of the model actually removes the cyclical pattern in the model autocorrelations, but does serve to illustrate that the coefficients now determined by the projection of New breeding animals upon the single series of Profitability, shown in Figure 2.11(c), indicate dependence only upon the past. They are in fact identical to the open loop impulse response from Profitability to New breeding numbers derived directly from the Pig market model and shown in Figure 2.8.

We will see that the coefficients in projections of the form (2.80) can be conveniently estimated from observed time series using the spectral methods of the next chapter. These methods are, however, most valuable in open loop situations, where the causal dependence of an output series on one or more input series is to be determined and where there is complete absence of any feedback whereby the output series influences future values of the input series. The spectral methods will then indicate dependence of the output upon the inputs by the presence of significant coefficients at positive lags alone. The requirement of an open loop context for the estimation of these response functions does limit the applicability of standard spectral methods, but there are several examples in the next chapter that demonstrate how useful they can be. An extension of spectral methods to the estimation of causal responses in the closed loop context is presented in Chapter 9.

Chapter 3

Spectral analysis of dependent series

3.1 Harmonic components of time series

The spectral analysis of time series refers generally to methods based upon decomposing the series into sinusoidal cycles of different frequencies. Classical Fourier analysis is most appropriately used for representing deterministic, or *fixed*, periodic functions as the sum of cycles, and we will also find this useful, for example, in the context of time series with regular seasonal patterns. However, the spectral analysis of time series, although motivated initially by the search for such hidden periodic patterns, derives directly from the quite distinct property of (second order) stationarity. From this it follows that a time series can be represented as the sum of *random* cyclical, or *harmonic*, components that are uncorrelated one with another. In the foregoing we have emphasized the words *fixed* and *random* because they provide a meaningful and relevant distinction between classical Fourier analysis and time series spectral analysis for those familiar with linear mixed effects models which may contain both fixed and random effects. The terminology of harmonic and spectral analysis derives from the notion that musical sound and colored light can each be broken down into components consisting of, respectively, pure tones and primary colors of the spectrum, which are associated with different frequencies of oscillation.

We start our presentation by showing how such components are computed from a finite sample of a time series. We will present a somewhat informal introduction to the subject, and the distributional properties that we state are approximations applicable to finite length samples of a time series. They should properly be qualified by a measure of their accuracy, which increases with the sample length, but in practice provide a useful and acceptable working basis. We refer to Bloomfield (1976), Priestley (1981), Brillinger (1981), Brockwell and Davis (1987) and Shumway and Stoffer (2000) for both details of the theory and valuable practical procedures. In the last section of this chapter, we give a short introduction to the representation of a time series of infinite length in terms of harmonic components.

We will introduce the subject in the context of a pair of univariate series x_t and y_t, describing their individual and pairwise spectral properties and the spectral dependence of y_t upon x_t. These ideas generalize naturally to the context of a multivariate time series x_t where we wish to investigate the spectral dependence of one component series upon the remainder.

Given, therefore, values $x_1, \ldots, x_n$ of a zero mean univariate time series, our approach to spectral analysis is based upon the construction of two sets of quantities, which we will call (cosine and sine) *harmonic contrasts*.

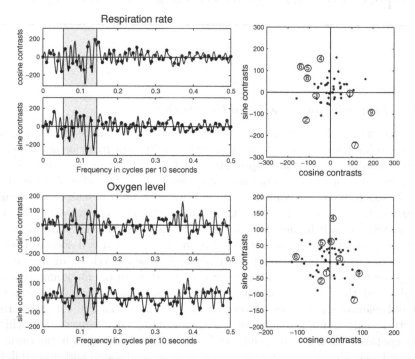

Figure 3.1 *On the left is the pair of cosine and sine contrasts plotted against frequency for each of two series, with the values at the harmonic frequencies shown as points on the curves. To the right, for each of the series, is a scatter diagram of the cosine contrasts against the sine contrasts at the harmonic frequencies. The shaded area on each plot defines a frequency band, and the points on the scatter diagrams corresponding to the contrasts within these bands are replaced with numbered circles centered on the same points.*

These are continuous functions of frequency f, defined as

$$a(f) = \sum_{t=1}^{n} x_t \cos(2\pi ft) \quad \text{and} \quad b(f) = \sum_{t=1}^{n} x_t \sin(2\pi ft). \qquad (3.1)$$

In practice, they are calculated at a discrete set of frequencies $f_k = k/N$ for some integer $N \geq n$, and we set

$$a_k = a(f_k) \quad \text{and} \quad b_k = b(f_k). \qquad (3.2)$$

The choice of N as some round number close to $4n$ is generally convenient and ensures the smooth appearance of the harmonic contrasts when plotted against frequency. Here we take the frequency f to be the number of cycles per sampling interval, but this can be scaled to the number of cycles per unit time. Note that the contrasts are periodic functions of frequency: $a(f + k) = a(f)$ and $b(f + k) = b(f)$ for $k = 1, 2, \ldots$. Moreover, $a(-f) = a(f)$ is an even function of frequency and $b(-f) = -b(f)$ is an odd function, so plotting them in the range $0 \leq f \leq 0.5$ is sufficient to give the complete picture of their behavior. Figure 3.1 shows the cosine and sine harmonic contrasts for both the Respiration rate and Oxygen level series of Figure 1.3, but both sampled at intervals of 10 seconds to give 100 points in each series.

Our use of the term *contrasts* rather than the more conventional term *Fourier coefficients* for the quantities in (3.1) and (3.2) is borrowed from the language of Analysis of Variance, where contrasts are linear combinations of the responses at different levels of a factor. When these responses might follow a polynomial form across the levels, they are transformed to linear combinations which are typically the coefficients of orthogonal polynomials. The harmonic contrasts that we define are exact orthogonal combinations only for the particular frequencies $f_k = k/n$ for integer k and n equal to the series length. We will restrict our use of the term *harmonic frequencies* to this specific set of frequencies, although we use the term harmonic contrasts for any range of frequencies f.

The series values can be represented as the sum of component cycles with coefficients proportional to the discrete set of harmonic contrasts as

$$x_t = \frac{1}{N} \sum_{k=0}^{N-1} \{a_k \cos(2\pi f_k t) + b_k \sin(2\pi f_k t)\}, \tag{3.3}$$

or from the continuous set as

$$x_t = \int_{-0.5}^{0.5} \{a(f) \cos(2\pi f t) + b(f) \sin(2\pi f t)\} \, df. \tag{3.4}$$

We will call the terms inside the summation of (3.3) or the integral of (3.4) the harmonic components of x_t. In practice, when integrals over frequency such as those in (3.4) are to be a evaluated numerically, expressions of the form (3.3) are used with a sufficiently high value of N. The correspondence between the two expressions would appear more direct if the limits of the integral were taken from zero to one. The limits from minus one half to plus one half are more conventional and give the same value because the arguments are periodic.

Statistical analysis of the harmonic contrasts is the basis of spectral modeling for one or several time series. The essential property upon which this rests is that the contrasts within any given *band* of frequencies, such as that shown by the shaded section of the plots in Figure 3.1, are uncorrelated with the contrasts in any other non-overlapping band. Within a single plot the contrasts are effectively uncorrelated if separated in frequency by an interval of at least $\delta = 1/n$, and we have marked by points on the plots the contrasts which lie on the harmonic frequencies with this separation. The dependence between two series can then be characterized by the covariances and regression relationships between their respective harmonic contrasts, in bands which cover the whole frequency range. However, these relationships will not in general be homogeneous across the whole frequency range. The frequency bands must then be chosen both sufficiently wide to cover several harmonic frequencies and sufficiently narrow for the relationships to be effectively homogeneous within the band.

There are nine marked harmonic frequencies within the shaded bands in Figure 3.1. So, for example, to build a regression relationship within this band we can take, at each of these frequencies, the pair of cosine and sine contrasts for the Respiration rate as the responses and the corresponding pair for Oxygen level as the regressors, giving a total of 18 regression equations. We explain the structure of these equations shortly. Each of these pairs is plotted as a numbered circle centered upon their position, in the plot of the whole set of pairs of harmonic contrasts in Figure 3.1. The numbers correspond for the two series to the sequence of frequencies in the band. There is a fairly obvious similarity between the pattern of these points for the two series which is suggestive of a significant relationship.

Spectral methods provide an invaluable tool for the practical statistical procedures of testing and modeling the relationships between time series. They were developed in the two decades or so following 1945 with an emphasis upon engineering applications. With advances in computing power, they were soon applied in statistics and econometrics; see Bartlett (1955), Blackman and Tukey (1958) and Granger and Hatanaka (1964). These procedures were termed *frequency domain methods* to contrast with *time domain methods* based upon analyzing relationships between lagged values of the series. However, conclusions drawn by using one of these methodological domains usually translate (or can be transformed) simply into the other domain, and the use of both can give greater insight into the analysis than using one alone. Spectral analysis may more readily reveal important features of the dynamic behavior of observed time series, which it is essential that time domain parametric models adequately represent. *Frequency domain regression*, between the harmonic contrasts, can also give a direct answer to the question of whether there is any linear dependence between two or more observed series. By transforming the results back into the time domain, estimates of the lagged dependence between series can be readily obtained.

3.2 Cycles and lags

The unique role played by the circular functions, sine and cosine, in the representation of stationary time series rests on a particular connection between them in relation to time shifts or lags. Figure 3.2 shows the pair of cosine and sine functions of frequency $f = 0.13$ cycles per sampling interval, plotted

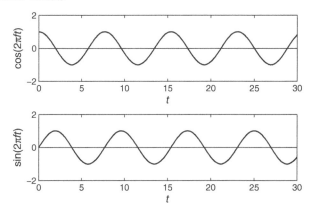

Figure 3.2 *The functions* $\cos(2\pi f t)$ *and* $\sin(2\pi f t)$ *for* $f = 0.13$.

as continuous functions of time t, though we will only be concerned with the values at (discrete) integer t. The connection between these functions that we will shortly find useful for understanding the pattern of relationships between harmonic contrasts is that the sine function is just the cosine function lagged by a quarter of a cycle, *and* the cosine function can equally well be described as *minus* the sine function lagged by a quarter of a cycle. Thus cosine and sine can generally be interchanged provided that this sign reversal is respected.

The next point is that a time lag in a sinusoidal function can be usefully viewed as corresponding to a rotation. This is illustrated by Figure 3.3, which shows (above) the side view of a corkscrew of similar appearance to the cosine function in Figure 3.2 and (below) the view after a rotation of the corkscrew to the left by one quarter turn. This now appears to be lagged by one quarter cycle and is similar to the sine function.

Rotating the corkscrew to the left by a general fraction ϕ of a turn leads to the appearance of a lag by the same fraction of a cycle. This is seen in the plot in Figure 3.4 of a general sinusoidal cycle of *amplitude* R and phase lag ϕ, which can always be expressed as a linear combination of the two basic sinusoidal functions:

$$x(t) = R\cos\left\{2\pi(ft - \phi)\right\} = A\cos(2\pi f t) + B\sin(2\pi f t). \qquad (3.5)$$

The coefficients $A = R\cos 2\pi\phi$ and $B = R\sin 2\pi\phi$ are represented geometrically in the same figure as the Cartesian coordinates of a point expressed in polar form as (R, ϕ). We point out here that our convention for the sign of the phase is the reverse of that used in several other texts. That is why we refer to it here as the phase lag, though for brevity we will usually just refer to it as the phase. Our usage means that when we consider the causal dependence of one series upon another, the phase will generally be positive in the same manner as the lag.

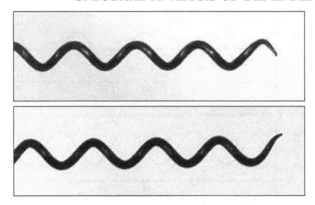

Figure 3.3 *Above is a side view of a corkscrew and below the view when it is rotated to the left by one quarter turn. This corresponds to the appearance of a shift to the right of the corkscrew by one quarter cycle.*

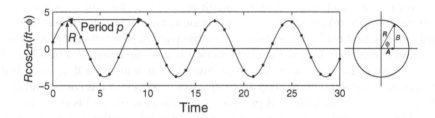

Figure 3.4 *A sinusoidal cycle with the values at the integer sampling times emphasized, and a circular representation of its coefficients.*

Any time lag applied to $x(t)$ will increase the phase lag ϕ and transform the coefficients A and B by a corresponding rotation on the circle in Figure 3.4. More generally, any linear combination of lagged values of $x(t)$ can be similarly represented by a rotation and scaling of these coefficients, but this transformation depends upon the frequency.

Before demonstrating this, note that, although the cycle in Figure 3.4 is plotted as a continuous function of time, it will be observed only at the discrete integer sampling intervals shown. These observed points are marked, and though they appear fairly regular, they will not be the same in successive cycles unless the period is an integer. An example is an annual cycle observed weekly which has a period of 52.18. There is also no reason why the maximum of the cycle should be observed at a sampling point. From here on we will restrict time to integer values and write x_t in place of $x(t)$.

Consider now a series y_t formed by applying a linear lagged operation to

$x_t = \cos(2\pi ft)$ as in

$$y_t = \sum_{k=-\infty}^{\infty} v_k x_{t-k}, \tag{3.6}$$

with $\sum_k |v_k|$ finite. Then y_t is also a cycle of the same frequency:

$$y_t = G(f)\cos(2\pi ft - \Psi(f)) = \tau(f)\cos(2\pi ft) + \nu(f)\sin(2\pi ft) \tag{3.7}$$

where

$$\tau(f) = \sum_{k=-\infty}^{\infty} v_k \cos(2\pi fk) \quad \text{and} \quad \nu(f) = \sum_{k=-\infty}^{\infty} v_k \sin(2\pi fk). \tag{3.8}$$

The functions of frequency $G(f)$ and $\Psi(f)$ are called the *gain* and *phase* of the *transfer function* relating y_t to x_t. If, more generally,

$$x_t = R\cos\{2\pi(ft - \phi)\} = A\cos(2\pi ft) + B\sin(2\pi ft), \tag{3.9}$$

the transfer function gives

$$y_t = G(f)R\cos\{2\pi(ft - \phi - \Psi(f)\} = C\cos(2\pi ft) + D\sin(2\pi ft). \tag{3.10}$$

Figure 3.5 illustrates how the transfer function multiplies the amplitude of the cycle by $G(f)$ and adds $\Psi(f)$ to the phase. The coefficients C and D in (3.10) of the cycle in y_t are given by the orthogonal matrix transformation of the coefficients of the cycle in x_t:

$$\begin{pmatrix} C \\ D \end{pmatrix} = \begin{pmatrix} \tau(f) & -\nu(f) \\ \nu(f) & \tau(f) \end{pmatrix} \begin{pmatrix} A \\ B \end{pmatrix}. \tag{3.11}$$

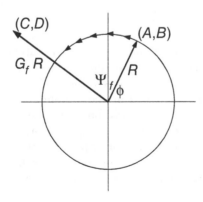

Figure 3.5 *The effect of a transfer function on the amplitude and phase of a cycle. The dependency of the gain $G(f)$ and phase $\Psi(f)$ upon frequency f is, for clarity, indicated by a subscript in the figure.*

Note that the pattern of the matrix in (3.11) reflects our previously stated connection between the cosine and sine functions. The dependence $\tau(f)$ of the cosine coefficient of y_t upon the cosine coefficient of x_t is the same as that of the sine coefficient of y_t upon the sine coefficient of x_t. The dependence $\nu(f)$ of the sine coefficient of y_t upon the cosine coefficient of x_t is minus that of the cosine coefficient of y_t upon the sine coefficient of x_t.

The transfer function uniquely determines the lagged coefficients v_k in the relationship (3.6) as

$$v_k = \int_{-0.5}^{0.5} \tau(f) \cos(2\pi k f) + \nu(f) \sin(2\pi k f) df. \tag{3.12}$$

When the operation (3.6) is applied to a general *input* series x_t to produce an *output* y_t, it is called a *linear filter* and is characterized by both the coefficients v_k and the transfer function. Note that the linear lagged relationship between time series appearing in the projection (2.79) is of this form, and we later show how the transfer function of this relationship can be estimated by frequency domain methods.

The properties of the gain and phase close to frequency zero are of particular interest. At frequency zero $\nu(f)$ is zero, so the phase will be either zero or one half, corresponding, respectively, to a positive or negative low frequency dependence between the series. The gain at frequency zero, with the sign so determined by the phase, is just $\tau(0) = \sum_k v_k$. It represents the magnitude of the long term dependence between the series, which is of particular interest in many applications, as we will point out in later examples. For a dependence which is restricted to a single lag, i.e., all v_k are zero except for some v_K, the phase will be a linear function of frequency with slope equal to the lag K. More generally, if the phase is approximately linear at low frequencies, its slope indicates an average lag in the relationship between the series. The phase is only unique up to the addition of an arbitrary integer, and it is usual to choose this at each frequency to ensure that the plot of the phase is continuous, starting from its value at frequency zero. It may therefore range outside the limits of ± 0.5 to which a phase is more conventionally restricted.

We explained earlier why the frequency range extends from zero to one half in plots of the harmonic contrasts in Figure 3.1. The frequency represents the fraction of a cycle completed in one sampling interval of the series. The maximum frequency of $f = 0.5$ corresponds to one cycle every two sampling intervals and is known as the *Nyquist* frequency. Any sinusoidal curve with a frequency greater than this (or a shorter period) can be shown to be identical, at the sampling time points, to a sinusoid with a frequency lower than the Nyquist frequency. We say that these frequencies are *aliased* at the sampling times.

This is illustrated in Figure 3.6, where the thicker solid line shows a shorter stretch of the cycle of frequency $f = 0.13$ shown in Figure 3.4. Superimposed on this are two other cycles, one plotted with a thin line, the other with a dotted line. These are both aliased with the solid line, having the same values

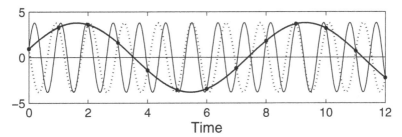

Figure 3.6 *A cycle with the values at the integer sampling times emphasized and two aliased cycles superimposed.*

at the sampling intervals, but frequencies above the Nyquist limit. The thin line has frequency $f = 1 + 0.13$ and therefore has one extra whole cycle included within each sampling interval. The dotted line has frequency $f = 1 - 0.13$ and completes just less than one whole cycle in a sampling interval. In general, the frequencies that are aliased with f are $K \pm f$ for integer K, with K extra cycles completed within the sampling interval. Reducing the sampling interval by a suitable factor would resolve the ambiguity of these aliased cycles. The theory of spectral analysis assumes that the sampling interval has been chosen sufficiently small that there are no aliased cycles, i.e., all cycles have a frequency below the Nyquist limit. In practice one should be aware that some unusual feature of the spectrum might be explained by aliasing of a higher frequency cycle.

3.3 Cycles and stationarity

The unique role of the circular functions in the analysis of a stationary time series can be illustrated by considering a series consisting of just one component cycle, such as those in the representation (3.3):

$$x_t = A\cos(2\pi ft) + B\sin(2\pi ft) \qquad (3.13)$$

where A and B are random variables. Then the joint distribution of A and B is only consistent with x_t being a weakly stationary time series if $\mathrm{Var}(A) = \mathrm{Var}(B)$ and $\mathrm{Cov}(A, B) = 0$. This may be confirmed by the direct calculation of the autocovariance function of x_t, which is then $\gamma_k = \mathrm{Var}(A)\cos(2\pi fk)$. If (3.13) is extended by a similar term $C\cos(2\pi gt) + D\sin(2\pi gt)$ at a distinct frequency g, then besides a similar constraint on the joint distribution of C and D, to be consistent with stationarity this pair must also be completely uncorrelated with A and B. This result is accorded geometric intuition by considering the scatter diagram of the harmonic contrasts in Figure 3.1. Applying a time lag to the sample by replacing t by $t - k$ for some k would rotate the point corresponding to frequency f to the left by $2\pi fk$, so that points at frequencies f and g would be rotated relative to each other by the angle

$2\pi(f - g)k$. Stationarity of the series requires that the general appearance of these points should not be affected by such rotations arising from the time shift. This is confirmed by their approximate circular symmetry and random scatter, which is consistent with the aforementioned variance and covariance constraints. Similar constraints apply to the joint distribution of the harmonic contrasts of two jointly stationary series. Recalling the connection between the cosine and sine functions, we find that, at a given frequency, the covariance between the cosine contrasts of two series should be the same as that between their sine contrasts, and the covariance between the cosine contrast of one series and the sine contrast of the other should be minus that between the sine and cosine contrast. For a single series, similar considerations imply that sine and cosine contrasts should have the same variance and zero covariance. The resulting pattern of the covariances for a jointly stationary pair of series will now be presented.

3.4   The spectrum and cross-spectra of time series

We now consider pairwise second moment properties at a single frequency f of the harmonic contrasts (3.1) defined for jointly weakly stationary time series x_t and y_t. These are required to describe the properties of frequency domain regression between the series. We will attach a single subscript of x or y for quantities involving a single series, and both subscripts for jointly defined quantities, for example, $\gamma_{yx,k}$ for the lagged covariance between the series. For the following results we assume that the autocovariances of the series are absolutely summable, e.g., $\sum_k |\gamma_{x,k}| < \infty$. Then for large n we have

$$n^{-1}\text{Var}\{a_x(f)\} = n^{-1}\text{Var}\{b_x(f)\} \quad = \quad 0.5 S_x(f) \qquad (3.14)$$

$$n^{-1}\text{Cov}\{a_x(f), b_x(f)\} \quad = \quad 0$$

where $S_x(f)$ is the *spectral density*, also called the power spectrum or more simply the *spectrum*, of x_t defined by

$$S_x(f) = \sum_{k=-\infty}^{\infty} \gamma_{x,k} \cos(2\pi k f). \qquad (3.15)$$

The harmonic contrasts of y_t have corresponding properties to these.

This is an appropriate point to describe some of the background to spectral analysis. This lies in the search for hidden cycles by fitting the regression

$$x_t = A\cos(2\pi f t) + B\sin(2\pi f t) + e_t \qquad (3.16)$$

to a single observed series. The two regressors are very close to orthogonal and exactly so at the harmonic frequencies $f_k = k/n$. The estimated coefficients are then

$$\hat{A} = (2/n)a(f) \quad \text{and} \quad \hat{B} = (2/n)b(f) \qquad (3.17)$$

and the sum of squares due to the regression is

$$(2/n)\{a(f)^2 + b(f)^2\}. \tag{3.18}$$

For x_t stationary with acf $\gamma_{x,k}$ and containing no fixed cycles, the expected sum of squares due to the regression is readily evaluated as

$$2 \sum_{k=-(n-1)}^{(n-1)} (1 - |k|/n)\, \gamma_{x,k} \cos(2\pi k f) \simeq 2S_x(f) \quad (n \text{ large}), \tag{3.19}$$

from which the definition (3.15) is motivated. These ideas originate with Schuster (1898), who introduced the *periodogram* as the plot of the squared amplitude of the cycle, $\hat{A}^2 + \hat{B}^2$, against the discrete set of periods n/k corresponding to the harmonic frequencies.

Turning now to the *joint* properties of the harmonic contrasts in x_t and y_t, these are given by

$$n^{-1}\mathrm{Cov}\{a_y(f), a_x(f)\} = n^{-1}\mathrm{Cov}\{b_y(f), b_x(f)\} = 0.5P_{yx}(f) \tag{3.20}$$

$$n^{-1}\mathrm{Cov}\{a_y(f), b_x(f)\} = -n^{-1}\mathrm{Cov}\{b_y(f), a_x(f)\} = 0.5Q_{yx}(f),$$

where $P_{yx}(f)$ is the *co-spectrum* between y_t and x_t and $Q_{yx}(f)$ is the *quadrature-spectrum*, given by

$$P_{yx}(f) = \sum_{k=-\infty}^{\infty} \gamma_{yx,k} \cos 2\pi k f \tag{3.21}$$

$$Q_{yx}(f) = \sum_{k=-\infty}^{\infty} \gamma_{yx,k} \sin 2\pi k f.$$

Together these constitute the *cross-spectrum*. Note that if the order of y and x is interchanged, the sign of the quadrature-spectrum is reversed, which is why we are now careful to use subscripts in these variables.

The variance matrix of the two pairs of harmonic contrasts may then be set out as

$$\frac{2}{n}\mathrm{Var}\begin{pmatrix} a_y(f) \\ b_y(f) \\ \cdots \\ a_x(f) \\ b_x(f) \end{pmatrix} = \begin{pmatrix} S_y(f) & 0 & \vdots & P_{yx}(f) & -Q_{yx}(f) \\ 0 & S_y(f) & \vdots & Q_{yx}(f) & P_{yx}(f) \\ \cdots & \cdots & & \cdots & \cdots \\ P_{yx}(f) & Q_{yx}(f) & \vdots & S_x(f) & 0 \\ -Q_{yx}(f) & P_{yx}(f) & \vdots & 0 & S_x(f) \end{pmatrix}. \tag{3.22}$$

3.5   Dependence between harmonic components

We are now able to show how the regression of the harmonic contrasts in y_t upon the contrasts in x_t can be used to determine the coefficients in the projection of y_t upon lagged values of x_t, which, following (2.79), we now write as

$$y_t = \sum_{k=-\infty}^{\infty} v_{y\,x,k} x_{t-k} + n_t. \tag{3.23}$$

Using (3.11), the relationships between the harmonic contrasts in y_t and x_t can be expressed

$$\begin{pmatrix} a_y(f) \\ b_y(f) \end{pmatrix} = \begin{pmatrix} \tau_{y\,x}(f) & -\nu_{y\,x}(f) \\ \nu_{y\,x}(f) & \tau_{y\,x}(f) \end{pmatrix} \begin{pmatrix} a_x(f) \\ b_x(f) \end{pmatrix} + \begin{pmatrix} a_n(f) \\ b_n(f) \end{pmatrix}, \tag{3.24}$$

where, from (3.8), $\tau_{y\,x}(f)$ and $\nu_{y\,x}(f)$ are the transfer functions of the linear filter coefficients $v_{y\,x,k}$ in (3.23) and $a_n(f)$ and $b_n(f)$ are the harmonic contrasts of the noise n_t. Our interpretation of (3.24) is that, at each frequency, the cyclical component of y_t is explained as the cyclical component of x_t with its amplitude and phase modified by the transfer function, together with a cyclical component due to noise. It is derived by substitution in (3.23) for x_t in terms of its harmonic contrasts (3.4). However, values of y_t over the finite time range, $t = 1, \ldots, n$, will typically depend upon some lagged values of x_t outside this range. Their omission leads to *end-effect* approximations in the relationship (3.24), but these become negligible in large samples. Estimating the regression (3.24), which we describe in the next section, then provides us with estimates of the transfer function gain and phase, the lagged response coefficients $v_{y\,x,k}$ and the noise autocovariances.

Using the covariance properties specified in (3.22), we find from (3.24) that

$$\frac{2}{n} \mathrm{Cov} \left\{ \begin{pmatrix} a_y(f) \\ b_y(f) \end{pmatrix}, \begin{pmatrix} a_x(f) \\ b_x(f) \end{pmatrix} \right\} =$$

$$\begin{pmatrix} P_{y\,x}(f) & -Q_{y\,x}(f) \\ Q_{y\,x}(f) & P_{y\,x}(f) \end{pmatrix} = \begin{pmatrix} \tau_{y\,x}(f) & -\nu_{y\,x}(f) \\ \nu_{y\,x}(f) & \tau_{y\,x}(f) \end{pmatrix} \begin{pmatrix} S_x(f) & 0 \\ 0 & S_x(f) \end{pmatrix}. \tag{3.25}$$

The transfer function coefficients are then simply determined as

$$\begin{aligned} \tau_{y\,x}(f) &= P_{y\,x}(f)/S_x(f) \tag{3.26} \\ \nu_{y\,x}(f) &= Q_{y\,x}(f)/S_x(f). \end{aligned}$$

Similarly, by considering the variance between the harmonic contrasts in y_t and simplifying, we find from (3.24) that

$$\begin{aligned} S_y(f) &= S_x(f) \left\{ \tau_{y\,x}(f)^2 + \nu_{y\,x}(f)^2 \right\} + S_n(f) \tag{3.27} \\ &= S_x(f) G_{y\,x}(f)^2 + S_n(f), \end{aligned}$$

where $G_{yx}(f)$ is the transfer function gain.

A measure of the strength of the relationship is given by the multiple squared correlation. In the context of spectral analysis this is known as the *squared coherency* and is given by

$$R_{yx}(f)^2 = \frac{P_{yx}(f)^2 + Q_{yx}(f)^2}{S_y(f)\,S_x(f)}. \tag{3.28}$$

The numerator on the right of (3.28) is the *squared cross-amplitude spectrum* $A_{yx}(f)^2$. Of most practical interest, however, is the squared coherency and the phase of the transfer function. As (3.26) shows, the transfer function is proportional to the cross-spectra, so its phase $\Psi_{yx}(f)$, as given by (3.7), is also known as the phase of the cross-spectrum. From this we may express the co-spectrum and quadrature-spectrum as

$$P_{yx}(f) = A_{yx}(f)\cos\{2\pi\Psi_{yx}(f)\} \tag{3.29}$$

$$Q_{yx}(f) = A_{yx}(f)\sin\{2\pi\Psi_{yx}(f)\}.$$

The squared coherency also measures the proportion of the spectrum $S_y(f)$ explained by the regression (3.24); note that both contrasts have the same variance. The spectrum of the noise term in this regression is then given by

$$S_n(f) = S_y(f)\left\{1 - R_{yx}(f)^2\right\}. \tag{3.30}$$

3.6   Bivariate and multivariate spectral properties

All of the quantities defined in the previous section are illustrated in Figure 3.7 using the individual and pairwise spectral properties derived from the autocovariances of the Pig market model.

The first row shows the spectra of the series of New breeding animals and the Profitability. These both peak at a frequency of about 0.08 cycles per quarter, corresponding to a period of about 12 quarters. This is the length of the irregular cycle seen in the series and the decaying cyclical patterns seen throughout the model autocorrelations of Figure 2.1. The second row shows the cross-amplitude spectrum, which has a similar peak, and the squared coherency between the New breeding animals and the Profitability. This has its highest value close to one, indicating strong dependence, around the peak frequency of the spectra.

The third row shows the gain and phase of the transfer function in the dependence of New breeding animals upon the Profitability. The gain of approximately 1.8 around the peak frequency indicates the relative amplitude of the cycles in the two series. The slope of the phase plot is approximately 2.0 at low frequencies, which conforms with the understanding that Profitability is predictive for New breeding animals. This is confirmed by the coefficients of their lagged dependence in the last row, derived from the transfer function using (3.12). These are predominately associated with positive lags.

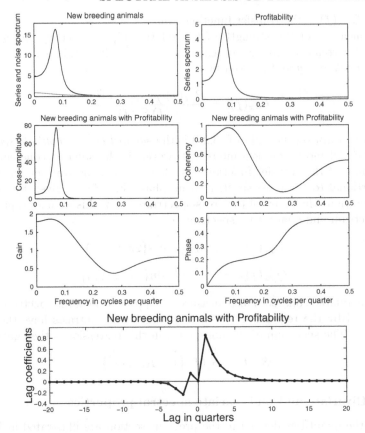

Figure 3.7 *In the first row is shown, on the left, the spectrum of the New breeding animals series, together with the spectrum (in gray) of the noise process remaining when the dependence upon Profitability has been explained. On the right is the spectrum of the Profitability series. In the second row is shown the cross-amplitude spectrum of the two series and their squared coherency. The third row shows the gain and phase of the dependence of the New breeding animals upon Profitability, and the final row shows the coefficients in their lagged dependence.*

The gain and phase may also be displayed in a single polar plot, as shown in Figure 3.8. The points marked on this plot indicate the frequency at intervals of 0.05. The radial distance is the gain, and the angle, in fractions of the circle, shows the phase.

All the plots in Figure 3.7 are of pairwise properties that depend only on the autocovariances and cross-covariance of the two series. Of general interest are the properties associated with the dependence of one series upon *two or more* other explanatory series. This dependence may be represented by extending the covariance matrix (3.22) and regression equation (3.24) to include contrasts of these further series. The covariance matrix is extended with 2×2

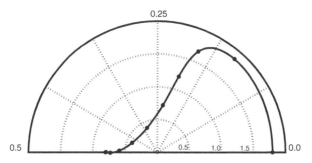

Figure 3.8 *The points marked on this plot indicate the frequency at intervals of 0.05. The radial distance is the gain, and the angle, in fractions of the circle, shows the phase.*

blocks for each pair of series and the regression equation by further pairs of explanatory contrasts.

The properties estimated in this context of multivariate dependence are plotted in Figure 3.9. We refer to all of these as partial quantities to distinguish them from the pairwise regression quantities in Figure 3.7, from which they will generally differ.

There are several methods of computing this multivariate dependence, and one which is helpful to the understanding is to use the expression (2.50) for the conditional variance. At each frequency the covariance matrix V, extended to include all the series, is inverted. Then the four by four sub-matrix of the inverse W, corresponding to a selected pair of dependent and explanatory series, is extracted. On inversion, this is the variance matrix of the selected pair conditional upon the remaining series. It has exactly the same form as (3.22) and can be used as before to construct the squared coherency, gain and phase and lagged coefficients. The squared coherency is now termed the *partial squared coherency*, though for brevity of expression we will call it simply the partial coherency. This measures the *further* fraction of the spectrum $S_y(f)$ explained after allowing for the dependence upon all other series. The gain, phase and lagged coefficients are just those appropriate to the selected explanatory variable in the multiple regression of the dependent series on *all* the explanatory series. The noise spectrum in the multiple regression is given by the noise spectrum of any of these partial regressions and can be used to compute the *multiple squared coherency*, or more simply the multiple coherency, of y_t with the whole set of explanatory series x_t as

$$R_{yx}(f)^2 = 1 - S_n(f)/S_y(f). \qquad (3.31)$$

Note that in Figure 3.9, the partial coherency is lower than the squared coherency in Figure 3.7, because the other series have a strong explanatory effect: the multiple coherency is very high. The lagged coefficients are also

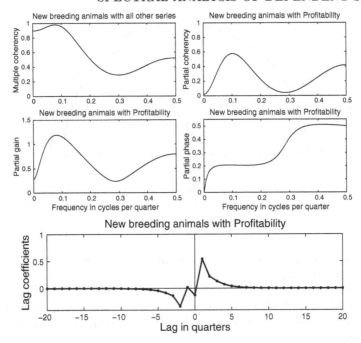

Figure 3.9 *All these plots relate to the spectral analysis of the New breeding animals series, taking all four of the remaining Pig market series as explanatory variables. The first row shows, on the left, the multiple coherency in this analysis and, on the right, the partial coherency of Profitability. The second row shows the gain and phase of the dependence upon Profitability, and the third row shows the coefficients in the lagged dependence.*

distinctly different and are, in fact, identical to those derived directly from the series covariances and displayed in Figure 2.11(a).

3.7   Estimation of spectral properties

We first present a fairly direct method of estimating, from observations of the time series, the various quantities displayed in Figures 3.7 and 3.9. Essentially, all we need are estimates of the spectra of single series and cross-spectra between pairs of series, such as shown in (3.22). However, we motivate these estimates by rewriting the regression equation (3.24) between contrasts in a form suitable for least squares estimation of the transfer function coefficients $\tau_{yx}(f)$ and $\nu_{yx}(f)$:

$$\begin{pmatrix} a_y(f) \\ b_y(f) \end{pmatrix} = \begin{pmatrix} a_x(f) & -b_x(f) \\ b_x(f) & a_x(f) \end{pmatrix} \begin{pmatrix} \tau_{yx}(f) \\ \nu_{yx}(f) \end{pmatrix} + \begin{pmatrix} a_n(f) \\ b_n(f) \end{pmatrix}. \qquad (3.32)$$

Now return to Figure 3.1 and take the nine harmonic frequencies within the shaded bands. Take each pair of marked contrasts $a_{y,k}$ and $b_{y,k}$ of Respiration rate as a pair of responses in (3.32), with the corresponding pair of marked contrasts $a_{x,k}$ and $a_{y,k}$ of Oxygen level as the regressors in the design matrix. Assuming that the spectral properties remain constant throughout the shaded band, we can collect these pairs to provide $m = 18$ regression equations for $d = 2$ coefficients τ_{yx} and ν_{yx}. The variables in any one equation are uncorrelated with those in another, giving m degrees of freedom in the response and $m - d = 16$ for the residual. The estimates are $\hat{\tau} = 1.584$ and $\hat{\nu} = 0.020$, both with standard error 0.257. If the series are Gaussian, the contrasts will be also, and therefore the estimates. As in standard linear regression, there are much wider conditions under which the estimates may be shown to be approximately normal (see for example Hannan (1970, p. 288) and Brillinger (1981, p. 26)), and we will assume that such conditions apply. The regression is then found to be highly significant, as is seen by referring the F statistic of 19.02 to its critical 95% value of $F_{2,16} = 3.634$. This can be transformed to a critical value of 0.3123 for the squared coherency of $R^2 = 0.7039$, using the transformation

$$R^2 = 1 - \frac{m - d}{m - d + dF}. \tag{3.33}$$

The estimated gain and phase are $\hat{G} = 1.584$ and $\hat{\Psi} = 0.002$. These estimates are generally associated with the central point in the frequency band, which is $f = 0.1$ in this case. The procedure can be replicated for frequency bands of the same width centered upon each harmonic frequency to construct estimates of spectral properties across the whole frequency range. Care must be taken at frequencies close to the endpoints $f = 0$ and $f = 0.5$ of this range. The frequency band is narrowed by these limits; the degrees of freedom is reduced and the variance of the estimates is increased.

The estimates so constructed by the regression (3.32) can alternatively be derived by substituting into (3.26) the sample estimates of the spectra and cross-spectra in (3.22) constructed from the harmonic contrasts in the frequency band $0.05 < f_k < 0.15$. These are

$$\hat{S}_y = \frac{2}{n} \sum_k \left(a_{y,k}^2 + b_{y,k}^2 \right) / m \tag{3.34}$$

$$\hat{S}_x = \frac{2}{n} \sum_k \left(a_{x,k}^2 + b_{x,k}^2 \right) / m$$

$$\hat{P}_{yx} = \frac{2}{n} \sum_k \left(a_{y,k} a_{x,k} + b_{y,k} b_{x,k} \right) / m$$

$$\hat{Q}_{yx} = \frac{2}{n} \sum_k \left(b_{y,k} a_{x,k} - a_{y,k} b_{x,k} \right) / m.$$

Spectral analysis is readily extended to several series using estimates of the form given in (3.34) for the spectrum of each individual series and cross-spectra between all pairs of series. Confidence intervals for the multiple coherency and gain, phase and partial coherency from all explanatory series, and critical values for all the coherencies can then by derived by standard methods of multiple regression. Perhaps the most important practical question regards the choice of bandwidth. Two residual degrees of freedom in the regression are lost for each explanatory series, which is of particular concern if there are many of these. A wider band gains more degrees of freedom and therefore lower variance of the estimates, but if the spectral properties vary substantially across the band, their estimates will necessarily be some form of average of these, and therefore biased. As part of an extensive discussion of the choice of bandwidth, Priestley (1981, p. 537) presents an argument that this should be chosen to be proportional to $n^{-1/5}$, but the precise value depends on the variability of the spectrum, particularly the presence of any narrow peak which would be smoothed out by too wide a band. The estimates of the lagged response are, however, less sensitive to the bandwidth, except that a wider band will lead to greater smoothing and response values which only extend to low lags before decaying away. A narrow band can therefore be used to estimate this response, provided the residual degrees of freedom in the spectral regression remains positive.

This section has introduced the basic principle by which estimates of spectral properties can be constructed. The standard method of spectral analysis was developed from these principles, with the addition of two particular refinements which we will explain in the next section. Figure 3.10 shows the dependence of Respiration rate on Oxygen level derived using these standard methods. Note the significance limit set for the squared coherency and also the confidence limits on the estimate. Rather than showing confidence limits around the estimates of the lagged response, we have placed them around zero, so that the estimated values lying outside these limits can be taken to be significant. All the limits become wider close to the end points of the frequency range, except for the phase, which necessarily has a precise half integer value at these points. Evidence for a highly significant relationship between the series is given by the coherency gain and lagged responses. The significant responses at both positive and negative low lags indicate what is to be expected in a healthy respiratory system, that the Respiration rate is not purely causally dependent on the Oxygen level, but that there is some feedback from the Respiration rate to Oxygen level. The analysis shown here cannot disentangle these inter-dependencies, but an extension of spectral analysis presented in Chapter 9 shows how this can be done.

3.8   Sample covariances and smoothed spectrum

This is the natural point in our book to define the sample values $C_{i,j,k}$ of the auto- and cross-covariances $\Gamma_{i,j,k}$ constructed from observations $x_1, \ldots, x_n$ of

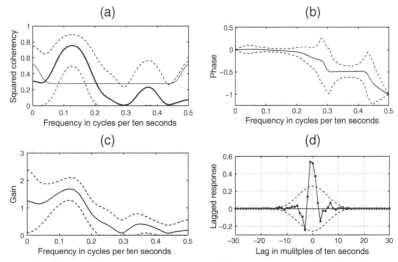

Figure 3.10 *Estimated spectral properties of the dependence of infant Respiration rate on Oxygen level using a bandwidth of 0.1: (a) The squared coherency between the series, with significance limit and upper and lower confidence limits, (b) the phase of the dependence, with confidence limits, (c) the gain of the dependence, with confidence limits, (d) the lagged response of the dependence, with significance limits.*

a zero-mean (or mean-corrected) multivariate time series. These are

$$C_{i,j,k} = (1/n) \sum_{t=1+k}^{n} (x_{i,t} x_{j,t-k}) \quad \text{for} \quad k = 0, \ldots, n-1. \tag{3.35}$$

For $k \geq n$ we take the values as zero. In bivariate illustrations we will replace $x_{i,t}$ by y_t and $x_{j,t}$ by x_t and use sub-indices y and x in place of i and j. Figure 3.11 shows the auto- and cross-correlations obtained by taking the Respiration rate for y_t and Oxygen levels for x_t, as in the previous section. The mean-corrected series of 100 values at 10 second sampling intervals are also shown in the figure.

Note that for $k > 0$ the sample covariances are biased in the sense that $E(C_{i,j,k}) = (1 - k/n)\Gamma_{i,j,k}$. As an extreme example, in Figure 3.11 the sample cross-correlation at the maximum lag of 99 is calculated from just the single product of the first and last observations in the respective series, and the factor $(1 - k/n)$ is $1/100$, a very large bias towards zero. However, for any fixed lag k the bias vanishes as this factor approaches 1.0 in large samples. For example, if we used a record of the series over a period ten times longer, this factor at the same lag of 99 would be 0.9. The bias could be removed by replacing the divisor n, before the summation in (3.35), by $(n - k)$. Priestley (1981, p. 323) discusses this issue, concluding that the divisor n gives the more satisfactory estimate. An important reason is that this estimate generally has a lower mean square error. It also ensures that the auto- and cross-correlation

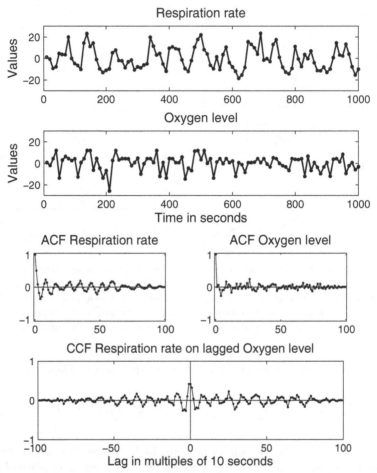

Figure 3.11 *Plots of Respiration rate and Oxygen level sampled every 10 seconds and corrected for their mean, together with their sample autocorrelation and cross-correlation functions.*

estimates lie in the range from -1 to 1, which is not guaranteed if the divisor $n - k$ were to be used. We will use the divisor n as in (3.35).

We now introduce the *sample spectra*, sometimes called the raw spectra and cross-spectra, of a pair of series y_t and x_t. These are constructed from pairs of harmonic contrasts at a *single* frequency:

$$S_y^*(f) = \left\{a_y(f)^2 + b_y(f)^2\right\}/n \qquad (3.36)$$
$$S_x^*(f) = \left\{a_x(f)^2 + b_x(f)^2\right\}/n$$
$$P_{yx}^*(f) = \left\{a_y(f)a_x(f) + b_y(f)b_x(f)\right\}/n$$
$$Q_{yx}^*(f) = \left\{b_y(f)a_x(f) - a_y(f)b_x(f)\right\}/n.$$

We use the $*$ in this notation to distinguish the sample spectra from the estimated spectra obtained in (3.34) by averaging harmonic contrasts over a frequency band.

On substituting for the harmonic contrasts (3.1) and simplifying, these sample spectra are expressed in terms of the sample covariances as

$$S_y^*(f) = \sum_{k=-(n-1)}^{(n-1)} C_{yy,k} \cos(2\pi k f) \qquad (3.37)$$

$$S_x^*(f) = \sum_{k=-(n-1)}^{(n-1)} C_{xx,k} \cos(2\pi k f)$$

$$P_{yx}^*(f) = \sum_{k=-(n-1)}^{(n-1)} C_{yx,k} \cos(2\pi k f)$$

$$Q_{yx}^*(f) = \sum_{k=-(n-1)}^{(n-1)} C_{yx,k} \sin(2\pi k f).$$

These are the natural sample values corresponding to the definitions (3.15) and (3.21). Consider now any *smoothed spectral estimate* obtained by symmetrically weighted averaging of the sample spectra about each frequency. *Any* such estimate can be expressed as the sums in (3.37) but with the covariances scaled by a term w_k at lag k, for example,

$$\hat{S}_y(f) = \sum_{k=-(n-1)}^{(n-1)} w_k C_{yy,k} \cos(2\pi k f). \qquad (3.38)$$

This formula enables us to conveniently calculate smoothed spectrum estimates at all frequencies. The weighting applied to average the spectra is called a *spectral window*, and the corresponding sequence w_k is called a *lag window*. Many considerations relating to the estimation of spectra using this approach are presented in Jenkins (1961) and Parzen (1961). Figure 3.12 shows three examples of spectral and lag windows. The spectral windows are centered on the frequency $f = 0.1$, which was used in our initial example (3.34) of averaging over adjacent harmonic frequencies. The top row corresponds to this averaging, with the uniform weights shown on the right. The corresponding lag window shows a cyclical pattern of decay at lower lags, but rises in a symmetric fashion. This is intuitively unappealing because we would expect higher lag covariances to be less important.

Forming the average of the sample spectrum over the same band but at half the frequency intervals is illustrated in the second row. The lag window does now decay away, but is still cyclical with some negative values, which also seem counterintuitive. Using a uniform continuous weighting over the frequency band would give an almost identical lag window. The third row shows

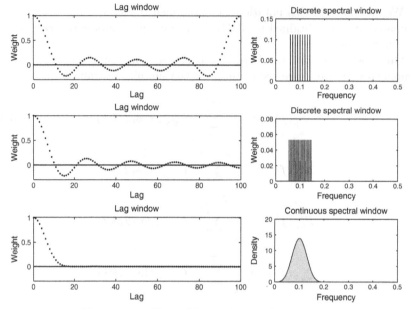

Figure 3.12 *Plots of lag and spectral windows.*

a spectral window with continuous weighting of apparently normal shape, for which the lag window dies away smoothly. This is the window used to form the estimates shown in Figure 3.10. The shape of the spectral window is that of the distribution of the sum of four uniform random variables, which is why it is so close to normal. Its width is chosen so that it gives approximately the same reduction in variance when applied to a uniform spectrum, as in the two previous examples.

Figure 3.13 shows, on the left, the raw spectrum of the Respiration rate as a gray line, together with the spectrum smoothed using the discrete spectral window in the top row of Figure 3.12 as a black line. On the right are shown two smoothed spectra, the black line from using the normal spectral window in the third row of Figure 3.12. The gray line is the smoothed spectrum obtained using the same shape of window, but of one quarter the width. The difference between the two black line spectra is mostly cosmetic, rather than of statistical significance, with the one on the right being smoother in appearance. The variability of the sample spectrum is described by the exponential distribution about the true spectrum because it is proportional to the sum of squares of two independent normal variables with common variance. An exponential random variable ranges between 0.05 and 3 times its expected value with probability 0.9, which explains the variability of the raw spectrum. The smoothed spectrum estimate formed from the averaging of 9 pairs of independent harmonic contrasts will have a scaled chi-squared distribution on 18 degrees of freedom provided the true spectrum remains constant over

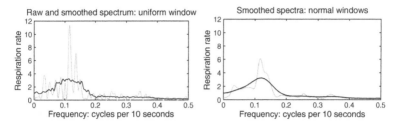

Figure 3.13 *Plots of raw and smoothed spectra of the Respiration rate.*

the bandwidth. Such a random variable ranges between 0.44 and 1.46 times its expected value with probability 0.9, so 90% confidence limits on the true spectrum may be found by dividing the smoothed spectrum by these factors. When a uniform spectral window is used, the degrees of freedom associated with this scaled chi-squared distribution is generally given by $2 B n$ where n is the series length and B the bandwidth of the window. When the spectral window is not uniform, the distribution of the smoothed spectrum may be approximated by a scaled chi-squared distribution with the same mean and variance using an appropriate choice of the scaling factor and degrees of freedom. A notional bandwidth B may be placed upon a non-uniform window by setting $2 B n$ equal to this degrees of freedom.

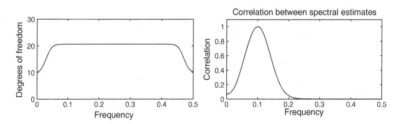

Figure 3.14 *Plots of the degrees of freedom of the smoothed spectrum estimates and the correlation between the estimate at frequency $f = 0.1$ and other frequencies.*

Figure 3.14 shows, on the left, the degrees of freedom as a function of frequency for the spectrum shown on the right in Figure 3.13 and which is estimated using the normal spectral window in the third row of Figure 3.12. As we remarked earlier, fewer independent values of the raw spectrum are available at the endpoints of the frequency range, so there is a reduction in the degrees of freedom. Away from the endpoints the degrees of freedom is close to 20, which is that specified by setting the notional bandwidth equal to 0.1. On the right of Figure 3.14 is shown the correlation between the spectrum estimate at frequency $f = 0.1$ and the estimates at other frequencies. This illustrates a general point that the correlation is negligible at frequencies that are separated by a distance greater than the bandwidth. In this example we can be reasonably confident that the spectrum has a peak in the region of $f = 0.1$, but a longer record would be needed to establish how sharp this peak might be. In the narrower bandwidth estimate on the right of Figure 3.13, it appears to be more pronounced, but the confidence limits are very wide. In the light of experience we recommended an initial choice of bandwidth of $1/(2\sqrt{n})$, followed by values both twice and one half of this, to observe whether features of the spectrum might be better revealed or smoothed out.

3.9 Tapering and pre-whitening

Before we move on to practical examples of spectral analysis, we describe two further refinements of standard spectral analysis. The first is motivated by the plot, on the left of Figure 3.15, of the cosine harmonic contrasts obtained when we take the series to be a pure cosine cycle $x_t = \cos 2\pi ht$ with frequency $h = 0.201$. We refer to these contrasts as $U_h(f)$. The series is taken over the symmetric time range $t = -50, \ldots, 50$, which helps to make the point more clearly. We would ideally expect the plot of $U_h(f)$ to correspond to the single cycle at frequency h by attaching weight only to this frequency, and to be zero at all other frequencies which we know are not present in the series. But we see that, as the frequency f moves away from h, the plot reduces to zero when the distance of f from h is approximately $(1/n)$, where $n = 101$ is the series length. It then oscillates through a series of lower negative and positive peaks. These are known as *side-lobes*, and they only slowly reduce in magnitude as the distance increases. They do *not* reduce in magnitude as the sample size n increases; only their separation reduces. If x_t contains just two cycles of similar magnitude, with frequencies separated by d, two distinct major peaks at these frequencies are only clearly seen in the contrasts if $d \geq 2/n$. If $d = 1/n$, they merge into a single peak. So at best the frequency *resolution* is $1/n$. However, as the sample length n increases, the peaks arising from any two distinct frequencies present in the data become higher and more narrow, so are better revealed, or *resolved*.

Of possibly more concern than the resolution of two close peaks of similar magnitude is the possibility that evidence for a cycle of relatively small magnitude may be *masked* by a side-lobe of the contrasts arising from a relatively large cycle at some distant frequency. Estimation of a spectrum with a strong peak can be affected by this *leakage of power* to other frequencies. This concern can be alleviated by the use of a *data taper*. This is a scalar weighting sequence h_t that is applied to the series, replacing x_t by $h_t x_t$. We define it initially to have unit value over most of its length, but tapers smoothly to zero towards the start and end of the series. The tapering *proportion* is simply that proportion of the length of the series over which the tapering takes place. The effect of a taper is to improve the orthogonality of more widely separated frequencies, though at the cost of slightly poorer resolution of close frequencies. The sample variance of a series is, however, reduced if tapered in this way, so to compensate for this we will re-scale the taper so that its mean

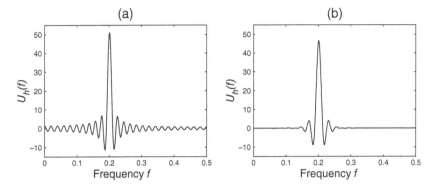

Figure 3.15 *(a) the cosine harmonic contrasts plotted for the series $x_t = \cos 2\pi ht$, with $h = 0.201$, and (b) the contrasts plotted after applying a taper to the same series.*

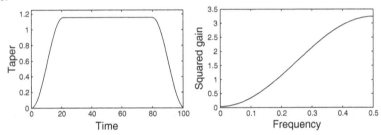

Figure 3.16 *On the left, the split cosine bell taper with tapering proportion 0.4. On the right, the squared gain of the pre-whitening filter (3.39) with $\alpha = 0.8$.*

square value is one. The plot on the left of Figure 3.16 shows the resulting split cosine bell taper with a proportion of 0.4.

The plot on the right of Figure 3.15 shows how the contrasts die away more rapidly from the frequency of the cycle present in the data after applying the taper to the series. Some loss of information follows from applying a taper. For the example used for Figure 3.14, applying a taper of proportion 0.4 reduces the degrees of freedom of the spectrum estimate from 20 to 16.

The argument for tapering is strongest when the spectrum has a huge dynamic range, i.e., varies in magnitude considerably within a relatively narrow frequency band. This is often seen at low frequencies, where the spectrum falls away rapidly from a peak at zero frequency and is commonly described as a red spectrum. It is well to check that the series has been corrected for a fixed linear or quadratic trend, which is a possible reason for this pattern of the spectrum.

The second refinement is known as *pre-whitening* and can be used as an alternative to tapering for series with a red spectrum. Pre-whitening is the application of a simple *filter* to the observed series x_t, of the form

$$\tilde{x}_t = x_t - \alpha x_{t-1}. \tag{3.39}$$

This is a lagged relationship with squared gain $g(f)^2 = 1 + \alpha^2 - 2\alpha \cos(2\pi f)$, illustrated on the right in Figure 3.16 with $\alpha = 0.8$. The spectrum of the filtered series $\tilde{x}_t$ is therefore the product of this squared gain and the spectrum of the original series x_t. It will typically have a very much reduced peak at zero frequency and can be estimated with a lower risk of leakage of power to higher frequencies. On dividing the estimated spectrum of $\tilde{x}_t$ by the squared gain, an improved estimate of the spectrum of the original series x_t is obtained. If the same pre-whitening filter is applied to each of a set of time series, the gain, phase and lagged response of their dependence are unchanged, though the noise spectrum is reduced by the squared gain. The estimates of all these properties can be improved by pre-whitening. The coherency between series is unchanged even when they are pre-whitened by applying different filters, which can help to improve the estimate. It is quite usual to take values of the filter coefficient α close to one, or even equal to one, so that the filter is a simple first order difference. The problem of estimating spectra with a large dynamic range and resolving peaks that are close in frequency has also been addressed using multi-taper techniques based on ideas introduced by Slepian and Pollak (1961), developed by Thomson (1982) and presented in Percival and Walden (1993). These methods combine a set of un-smoothed spectra formed using a corresponding set of tapers. These tapers are designed to concentrate the spectral information in a frequency band of specified width. Although we do not consider that this method is necessary for the examples in this book, multi-taper spectral estimation has been shown to be valuable in many physical applications.

3.10   Practical examples of spectral analysis

We now consider various practical aspects of multivariate spectral analysis, which we illustrate with appropriate examples.

3.10.1   Infant monitoring series

The first of the practical aspects that we consider is the choice of sampling interval, when that is an option. The infant Oxygen level, Pulse rate and Respiration rate series are actually recorded every one tenth of a second. We will illustrate the effects of using a sampling interval of 5 seconds by repeating previous analyses which used an interval of 10 seconds.

Figure 3.17 shows a selection of the results. First, note that the frequency shown as 0.25 (period 4 sampling intervals) in Figure 3.17 corresponds to the frequency of 0.5 (period 2 sampling intervals) shown in Figure 3.10, with both corresponding to a period of 20 seconds. Second, note that the spectrum in Figure 3.17 is distinctly positive over a range of frequencies greater than

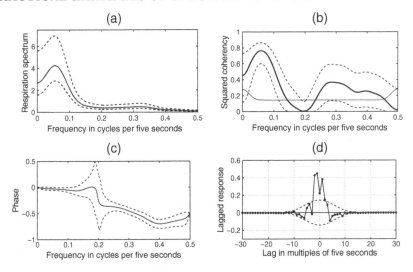

Figure 3.17 *Estimated spectral properties of the dependence of infant Respiration rate on Oxygen level using samples at intervals of 5 seconds, twice the rate used for earlier illustrations, so that the frequency of 0.25 (period 20 seconds) in this figure corresponds to the frequency of 0.5 in Figure 3.10: (a) The smoothed spectrum of the Respiration rate series, with lower and upper 95% confidence limits, (b) the squared coherency between the series, with significance limit and upper and lower confidence limits, (c) the phase of the dependence, with significance limits, (d) the lagged response of the dependence, with significance limits.*

0.25, and moreover the coherency is also strongly significant over this range. This suggests that the shorter sampling interval should be used. By using the longer interval, these frequencies were aliased. The effect of aliasing on the spectrum is described as *folding*. Fold the right hand side of the spectrum in Figure 3.17 over on the left hand half, about a vertical axis through frequency 0.25. The spectrum implied by reducing the sampling rate by a factor of two is obtained by adding the part that has been folded over from the right onto the spectrum on the left. For example, the spectrum at frequency 0.4 is folded onto that at frequency 0.1. The frequency range from 0 to 0.25 is then rescaled from 0 to 0.5, to correspond to the lower sampling rate in cycles per sampling interval. When, as in this example, the spectrum in Figure 3.17(a) has appreciable power at higher frequencies, reducing the sampling rate can result in distortion and loss of information in the dependence between the series. Finally, note in Figure 3.17 that two steps of the time index in the lagged response correspond to one step in Figure 3.10, which is ten seconds. Using the higher sampling rate has revealed further detail in the pattern of response, in particular, the dip at lag zero.

3.10.2   Gas furnace series

We next explain how a simple adjustment to the methods described can improve the estimates of the dependence between two series when there is an appreciable time lag in the relationship between them. The example we use for illustration is the series of input and output measurements on a gas furnace that have been previously modeled by spectral analysis in Jenkins and Watts (1968), and by lagged correlation methods in Box and Jenkins (1970). They are shown in Figure 3.18 but with the Input series negated so that the relationship between the series is positive. It is easy to see a strong similarity in the patterns of the two series, and the slight lag of variations in the Output, behind those in the Input. Figure 3.19(a) shows the lagged response between the series, estimated using a bandwidth of 0.05. The response peaks at a lag of 5 sampling intervals, which is also the lag at which the cross-correlation between the series has its maximum. Figure 3.19(b) shows the corresponding estimated squared coherency, confirming a strongly significant relationship. These estimates can, however, be improved if we first align the series by lagging the Input by 5 sampling intervals. The lagged response estimated after this alignment is shown in Figure 3.19(c) after adjusting the lag to restore the correct alignment.

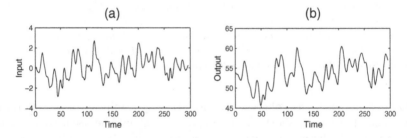

Figure 3.18 *Gas furnace time series: (a) Input and (b) Output.*

Notice that the peak in the response becomes more pronounced and the significance limits are centered on the chosen alignment lag. However, the more dramatic revision is to the estimated squared coherency shown in Figure 3.19(d). This is now much closer to 1 at frequencies below 0.05. The reason for the improvement can be seen from the plots in Figure 3.20 of the estimated gain and phase for this example. The gain reduces by a relatively small factor over this low frequency range, but the phase increases rapidly. The coefficients in the relationship between the harmonic components, as presented in the harmonic regression equation (3.32), are therefore rapidly varying. This regression cannot explain such a high proportion of the variability using a fixed pair of coefficients across the bandwidth, leading to a reduction in the squared coherency.

An appropriate value for the alignment lag is the slope of the estimated phase close to the origin, which in this example, by visual inspection of Figure

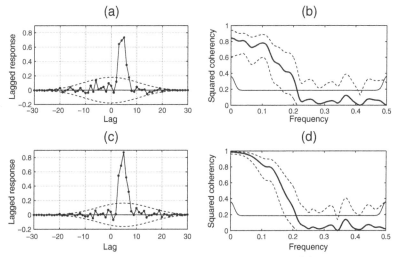

Figure 3.19 *Spectral estimates for the Gas furnace series: (a) the lagged response from Input to Output and (b) the squared coherency between the series, both estimated without alignment. (c) The lagged response from Input to Output and (d) the squared coherency between the series, both estimated with alignment.*

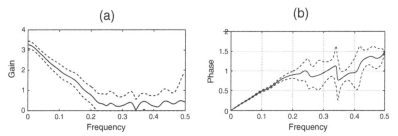

Figure 3.20 *Spectral estimates for the Gas furnace series: (a) the gain of the response from Input to Output and (b) the phase.*

3.20(b), is seen to be 5. Aligning the series by a lag ℓ subtracts the linear term ℓf from the phase, where f is the frequency. The coefficients in the harmonic regression are then more nearly uniform at low frequency, which improves the estimation of squared coherency, gain and phase. The phase estimated after alignment of the series is then corrected by adding on the term ℓf, restoring an improved estimate of the phase for the series prior to alignment. The overall result can also be achieved by centering the lag window w_k on the cross-covariance between the series at lag ℓ instead of lag zero.

3.10.3   Weekly moth trappings

We now consider issues illustrated by the investigation of how the numbers of Lepidoptera (moths) trapped each week are dependent upon two climate variables, temperature and rainfall. Figure 3.21(a) shows the Moth counts for the 345 weeks beginning in the first week of 1993 at Drayton in the UK. These are the daily average numbers of moths trapped. Figure 3.22 shows two climate variables for the same times and place. The Temperature series is the average daily temperature, and the Rainfall series is the average daily rainfall. The main practical point in this example is the importance of correcting for large fixed cycles arising from seasonal patterns in both explanatory and response series. A further consideration, which arises more generally in regression modeling, is the appropriate use of data transformations.

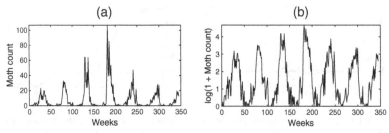

Figure 3.21 *Daily average numbers of moths trapped weekly at Drayton from the first week of 1993: (a) original counts, (b) the counts transformed by adding one then taking logarithms.*

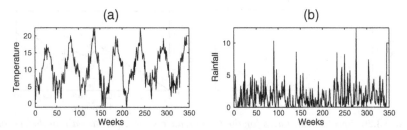

Figure 3.22 *Climate variables recorded weekly at Drayton from the first week of 1993: (a) a daily air temperature in degrees Celsius, (b) average daily rainfall in mm.*

Both the Moth count and the Temperature series show a strong annual cycle with peaks in summer and troughs in winter. Inevitably, this leads to a strong correlation between these two series, but this does *not* necessarily characterize the dependence in which we are interested. A similar strong correlation would be found if we used temperatures from almost any mid-latitude northern hemisphere site. Our interest is in discovering how the local temperature and rainfall affect the variations in moth trappings from week to week, and for this reason we will *correct* both the Moth count and Temperature series for a fixed annual cycle. However, before applying this correction to the Moth count, we note the marked asymmetry of this series, with much higher variability and a sharper peak in the summer than in the winter. The standard statistical approach to this issue is to consider a transformation of the series. Figure 3.21(b) shows the Moth count after applying a logarithmic

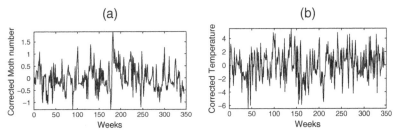

Figure 3.23 *Series corrected for annual cycles: (a) weekly Moth number, (b) weekly Temperature.*

transformation which results in a much more symmetric annual pattern of variation. We will call this transformed series the Moth number.

To correct the Moth number and the Temperature series for the annual cycle, we regress each upon the set of functions $\cos(2\pi kt/p)$ and $\sin(2\pi kt/p)$ and subtract the fitted values. The corrected series are shown in Figure 3.23. The period p used to define the regressors is $365.24/7 = 52.18$ weeks in the year. The regressor for $k = 1$ defines the fundamental annual cycle but the regressor with $k = 2$, for the second harmonic of this cycle, was also included for each series. Figure 3.24 shows how this decision was made for the Moth number; the same applied to the Temperature series. Figure 3.24(a) is the sample spectrum of the series with the prominent peak indicating the fundamental seasonal cycle. Figure 3.24(b) is the sample spectrum after the series has been corrected for this cycle. Again there is a peak that is prominent, though much less so than before, and now at the second seasonal harmonic frequency of 0.0383. One might also imagine peaks at the third and fourth harmonic frequencies (0.0575, 0.0767) of the annual cycle. These are, however, comparable to the general level of the neighboring sample spectra and will not, individually, contribute substantially to the variance of the series or the correlation between them. Correcting for the first fundamental and second harmonic of the seasonal cycle is sufficient to ensure that these features do not dominate in our modeling of the dependence between the series.

These peaks in the spectrum corresponding to the annual cycles are known as discrete spectral components. Failure to correct for these can distort the analysis, as we have just pointed out and will illustrate below for the coherency estimates. Note also that if a series has a non-zero mean that is *not* corrected, this will be evident as a discrete spectral component at frequency zero and can similarly lead to distortion of the analysis.

Figure 3.25 shows the estimated lagged response in the dependence of the corrected Moth number on the corrected Temperature series and the Rainfall series. Note that there is no lag dependence in the first of these. The Temperature has only a simultaneous positive effect on the Moth number. In contrast, the Rainfall has an extended lagged negative effect both in the immediately following weeks and also after a lag of about 7 weeks, associated with the typical summer breeding cycle of the moths.

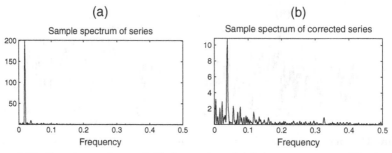

Figure 3.24 *Sample spectrum of Moth number: (a) before correction, (b) after correction by the fundamental annual cycle.*

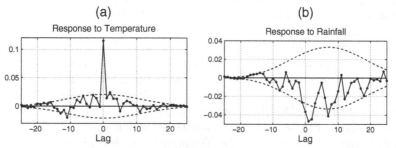

Figure 3.25 *Estimated responses of corrected Moth number to (a) corrected Temperature and (b) Rainfall.*

We do not show the estimated responses using the Moth number and the Temperature series *without* correcting for the seasonal cycle. They do look very similar to those shown in Figure 3.25 for the corrected series. To understand this, Figure 3.26(a) shows a scatter plot of the harmonic contrasts, both cosine and sine, of the Moth number against those of the Temperature, *before* correction. The plot shows only the harmonic frequencies, and we have restricted the series length to 313, corresponding to six annual cycles. The contrasts corresponding to the fundamental frequencies of these cycles are then at exactly the sixth harmonic frequency. The two outliers in this plot correspond to these cycles, all other points clustering close to the origin. Figure 3.26(b) shows the plot when the outliers are removed. In fact, the slope of the regression is quite close for both plots, indicating that the effects of temperature on the larger scale of the annual cycle are very similar to those on the small scale of week to week variations. The wisdom of correcting for annual cycles is that, were the effects *not* similar on these two scales, the estimation of the week to week response would be distorted by including the dominating effect of the large magnitude annual cycle.

To conclude this example, Figure 3.27 shows the partial coherency between the Moth numbers and the Temperature series, both before and after correcting the two series for the annual cycle. The very high coherency between the uncorrected series at low frequencies is typical of that resulting from a strong

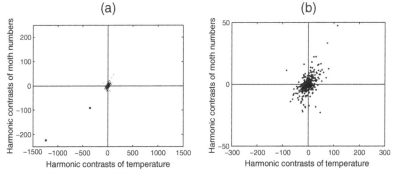

Figure 3.26 *Scatter plot of harmonic contrasts of the Moth numbers against those of the Temperature series: (a) showing all points and (b) scaled to exclude outliers.*

deterministic cycle of the same period in both series. In this example there is also strong coherency at higher frequencies to give assurance of significant dependence between the series. It is, however, always good statistical practice to correct for strong seasonal cycles before investigating the dependence between series.

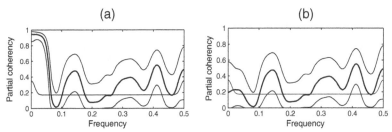

Figure 3.27 *Partial coherency between the Moth number and the Temperature series: (a) before correcting for the annual cycle and (b) after correction.*

3.10.4   Natural gas sendout series

Our next example also involves correction for seasonal effects, but also shows how an apparently small non-causal response at a negative lag could be explained by contamination of the explanatory variable by noise. Figure 3.28(a) shows the natural Gas sendout, as it is known, in a southeast region of England from 1st September 1977 to 31st August 1978. This should be the amount of gas consumed except that some is lost in transmission. Figure 3.28(b) shows a Meteorological indicator which is designed to help explain and possibly predict the Gas sendout. The main component of the indicator is temperature, but wind chill is also considered. The relationship between the series is clearly negative and quite strong, as the plot of one against the other shows in Figure 3.28(e).

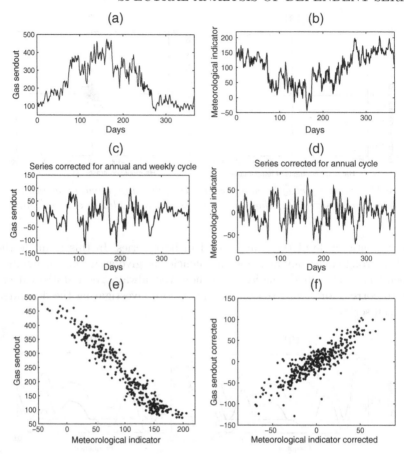

Figure 3.28 *(a) Daily Gas sendout over the period of one year. (b) A Meteorological indicator designed as a predictor of sendout. (c) The Gas sendout corrected for annual and weekly cycles. (d) The Meteorological indicator corrected for an annual cycle and reversed in sign. (e) The scatter plot of Gas sendout against the Meteorological indicator. (f) The scatter plot of the corrected Gas sendout against the corrected Meteorological indicator.*

Both series are corrected for an annual cycle of period 365.24. The Gas sendout series is also corrected for a weekly pattern of period 7 by regression on the complete set of cosine and sine cycles of frequencies $1/7$, $2/7$ and $3/7$. Figures 3.28(c) and (d) show the corrected series with the sign of the Meteorological indicator reversed so that it becomes a direct predictor of Gas sendout, displaying a positive relationship. Figure 3.28(f) shows the graph of the corrected series against each other, with the appearance of a standard bivariate normal scatter. The small number of lower outliers arise from the Christmas period. A further correction could be made for these, but their presence will not greatly affect the analysis. Figures 3.29(a) and (b) show the estimated gain and lagged response function between these corrected series. The dominant feature of the lagged response is the coefficient of 0.915 at lag zero, but note that the gain at frequency zero is 90% greater at 1.75. There is a low level of response at positive lags which, though not apparently significant beyond lag 3, is cumulatively substantial. A unit step in the Meteorological indicator would imply an immediate increase in Gas sendout of 0.915, rising to a level of 1.75 over the next few days as lower temperatures gradually permeate structures. The regression slope of Figure 3.28(f) is 1.3, a compromise between these two values which would over-estimate the immediate impact and under-estimate the longer term.

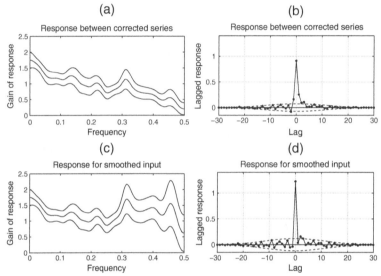

Figure 3.29 *(a) The gain of the frequency response and (b) the lagged response of the corrected Gas sendout to the corrected Meteorological indicator. (c) The gain of the frequency response and (d) the lagged response of the corrected Gas sendout to the smoothed corrected Meteorological indicator.*

However, we also note a small but quite significant response coefficient at a lag of −1. Both non-causal dependence and feedback are implausible explanations of this, but it could be explained if the Meteorological indicator were considered to be affected by noise in the form of measurement error. Figures 3.29(c) and (d) show the gain and response function re-estimated after applying a small amount of smoothing to the Meteorological indicator. The smoothing was a centered symmetric moving average of total span 5 lags, with weights of $(1 - a)$ at lag 0, $4a/6$ at lags ± 1 and $-a/6$ at lags ± 2: the choice of $a = 0.25$ was made following a small amount of experimentation. This approximates a cubic spline smoother and is designed to preserve a cubic through these points whilst reducing the variance of added white noise. Note now that in Figure 3.29(d) there is no longer any significant non-causal response at negative lags and that the response at lag 0 has increased to 1.23.

The gain at zero frequency is unchanged at 1.75, because the smoother leaves the spectrum unchanged at this frequency. There is still therefore a 43% increase beyond the initial response in the eventual response to a step change of the Meteorological indicator.

Consider a response estimated from a stationary input series contaminated by white noise that is uncorrelated with both input and output. The noise-free input can be optimally estimated by a projection operator which reduces to the application of a smoothing filter. The estimated response to the contaminated input is given in the frequency domain as the product of the transfer function of the smoothing filter with the transfer function of the output response from the noise-free input series. In the time domain the lagged responses of the two are convoluted: the response to the contaminated input is given by applying the smoothing weights to the response to the uncontaminated input. This appears to be a plausible explanation of what is seen in this example.

3.10.5 Extrusion process series

Our next example illustrates how understanding the properties of the estimates of transfer functions in the frequency domain can help to improve the estimates of lagged regression. We estimate the response of the times series of Extrusion pressure to the Heater current and Valve setting shown in Figure 1.9 for the plastic Film extrusion process. Figure 3.30(a) shows the partial coherency between Heater current and Extrusion pressure, and Figure 3.30(b) shows the gain of the frequency response from Heater current to Extrusion pressure. Figure 3.30(c) shows the estimated lagged response from Heater current to Extrusion pressure.

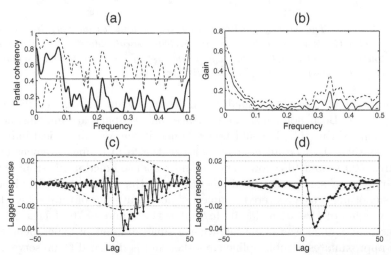

Figure 3.30 (a) The partial coherency and (b) the gain of the frequency response from Heater current to Extrusion pressure. (c) The lagged response of the relationship and (d) the lagged response formed using a cutoff frequency of 0.2.

Note that the partial coherency is only significant below frequency 0.1, and the gain estimates and standard errors fall away rapidly to frequency 0.2. However, above frequency 0.2, and in particular around frequency 0.35, the gain estimate is relatively large but poorly estimated with large standard errors, and not at all significant as assessed by the partial coherency. There is no significant (partial) dependence in this frequency range, but a relatively high variance of the harmonic contrasts in the response (Extrusion pressure) relative to those in the explanatory variable (Heater pressure) leads to these spuriously high estimates of gain. This effect increases the variability of *all* the lagged responses obtained by transformation of the frequency response, but it can be avoided by zeroing the estimated frequency response above a cutoff frequency, which we choose here as 0.2. The cutoff is tapered down to zero smoothly over an interval of twice the chosen bandwidth, centered on the cutoff frequency. The resulting lagged response is shown in Figure 3.30(d). It follows the same general shape as before this modification but is smoother in appearance and has much narrower significance bands. This cutoff procedure may be appropriate for similar process data where the sampling interval for recording the variables is chosen to avoid aliasing. The spectra and cross-spectra in the higher frequency range may then be low in magnitude and contain little information relating to the dependence between the series. The cutoff procedure then provides good estimates of lagged response without recourse to sub-sampling.

3.10.6 Sales series

The time series used in each of the previous examples can be considered to be jointly stationary processes. We now consider an example in which the weekly sales volume of a certain product, over a period of four years, is related to its price and a measure of spending on advertising promotions of the product. These series are shown in Figure 1.5, and none of them appears to be stationary. Both the Sales and the Price series have an obvious trend. The Promotion series varies considerably in magnitude with a twenty week period of zeros partway through. The Sales are also strongly affected by Christmas, and to investigate the impact of this season a further explanatory variable is introduced, an indicator variable taking the value of one in the Christmas week and zero otherwise. Brillinger (1981, Ch. 6) presents a careful treatment of the spectral analysis of series such as these. The methods we have been using can be extended to this context, with explanatory series that may be deterministic rather than stochastic and stationary, under two provisions:

1. The response series is stationary conditional upon the explanatory series; i.e., after removing the effects of the explanatory variable, the part remaining is stationary.

2. The harmonic contrasts of the explanatory variables are non-zero in any given bandwidth within the frequency range. These are required to explain the harmonic contrasts of the response, and a regression fails if the

regressors are zero. A clear exception to this provision is when an explanatory variable consists of one pure cycle (or the sum of a small number of cycles). The contrasts will be zero in any bandwidth that does not contain the frequency of this cycle. In this example the Christmas indicator can be represented as the sum of cycles at frequencies $k/52$ for integer k, but the chosen bandwidth of 0.1 contains several such frequencies so no problem arises.

Expressions for the large sample properties of the estimates of the transfer function are very slightly different for dependence upon deterministic series; see Brillinger (1981, p. 200, 201). The difference relates to the smoothing weights that should be applied to the sample spectrum of the dependent series. We will, however, use the same expressions as for dependence upon stationary series, because this makes very little difference in practical applications.

The Sales, Price and Promotion series are shown in Figure 1.5. It is not wise to assume that the downward trend in the Price explains the upward trend in Sales, and in a simple regression context this assumption would be avoided by adding a linear trend as a further explanatory variable. An equivalent step would be to subtract from each series its trend. We have applied that correction before our cross-spectral analysis, the results of which are presented in Figure 3.31.

This example is typical of market response modeling, in that the effects are not strongly significant, even though the Christmas effect may appear strong. We note that:

1. The response to Price appears to be simultaneous with the price variation, with a lag zero response that is just significant at -0.6. However, the gain in Figure 3.31(d) shows that the magnitude of the response is close to 2 at higher frequencies, suggesting that the reaction to frequent price changes is more substantial. At low frequency the gain is much smaller and barely significant, suggesting a longer term insensitivity to price change, possibly reflecting brand loyalty. A simple regression of Sales on Price, Promotion, the Christmas indicator and a linear trend gives a coefficient of -0.7 for price, whilst a regression between the *differenced series*, which accentuates the effect of changes, gives a coefficient of -1.1 on price. Simple regression, omitting lags and neglecting error correlation, delivers a compromise coefficient. The gain gives the more comprehensive picture of the response.

2. The response to Promotions is again barely significant but remains positive up to lag 10. The corresponding low frequency gain supports a more strongly significant long term effect, which is valuable marketing knowledge.

3. The response to Christmas is clearly significant, with the pattern indicating a build up for a week or two before hand and a rebound in sales a week or two afterwards. This is not surprising, but the gain at low frequency shows considerable uncertainty as to whether Christmas had any net effect upon total sales over the season.

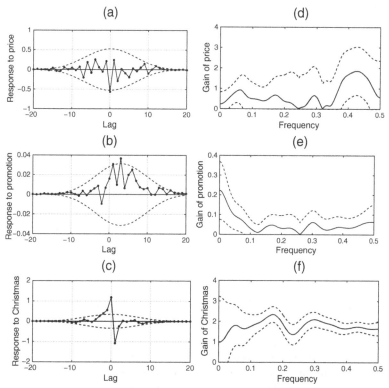

Figure 3.31 *The estimated responses of Sales to (a) Price, (b) Promotion and (c) Christmas, together with the frequency domain gain of (d) Price, (e) Promotion and (f) Christmas.*

There are many other factors, such as weather, that influence sales, resulting in much variability that is not accounted for by the chosen explanatory variables. We can derive the residual contrasts from the frequency domain regression and reconstitute the corresponding residual series which estimates the noise term n_t in the lagged relationship (3.23). This is the *discrepancy* between the Sales series and the part explained by the Price, Promotion and Christmas effect. Its estimated spectrum and autocorrelations can also be constructed. Figure 3.32(a) shows the Sales series corrected for linear trend together with this series of discrepancies, which confirms that there is much variability for which the model fails to account. The estimated autocorrelations of these discrepancies in Figure 3.32(b) decay over a period of two months. The model assumes time invariance for its coefficients, which may in fact vary from year to year. Other possible sources of variability of the discrepancy series may be suggested by its inspection. But even without further investigation, this analysis has provided a useful summary of the effects of the explanatory variables over the period of the data.

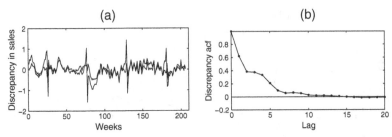

Figure 3.32 *(a) The discrepancy between the Sales series and the model (thick line) together with the mean and trend corrected Sales (thin line), and (b) the estimated autocorrelations of the discrepancy.*

An alternative to trend correction of the series is to use their (mean corrected) first differences in the analysis. The differencing must be applied to all the series to preserve the dynamic relationship between them. In large samples the estimated coherency, gain, phase and lagged responses are insensitive to this transformation; the differencing will, however, carry through to the discrepancy series. In this example, this alternative leads to very similar results. Whether trend correction or differencing is used, low frequency information is reduced, resulting in greater uncertainty in the estimation of the zero frequency gain that is of particular interest. This is unavoidable if we do not wish to risk distorting the estimated dependence of a trending series upon a trending explanatory variable, because of the omission of unknown trending factors which may also be part of the explanation.

3.10.7 Limitations of spectral analysis

Because spectral analysis is a form of non-parametric estimation, it has been thought necessary that the time series to which it is applied should be of a good length. However, as for other applications of non-parametric estimation, much depends on whether the spectrum has pronounced features which it is important to resolve, or if it is simple in form. In the latter case, spectral analysis might be usefully applied to series with length as low as 20, in which the dependence extends over a small number of lags.

Spectral analysis relies strongly on the assumption that the series are stationary. If applied to series in which the structure of lagged dependence is changing, it may fail to determine a significant relationship, or if it does, it will estimate some form of average dependence. This may be the case in the foregoing Sales example, where the impact of Promotions may change throughout the year. Similarly for the Moth count series, in the winter months when moth numbers are low, the effects of the weather may be quite different. Time varying spectral estimation may be carried out by applying the analysis to a moving window of the series. The graphic equalizer on much audio equipment shows the varying sound spectrum in different frequency bands. The topic of evolutionary spectra has been studied for many years (see, for example, Priestley (1965), Priestley (1981, Chap. 11) and Dahlhaus (1996)), but estimation of spectra is typically restricted to long, high-frequency sampled series. Wavelet analysis (Percival and Walden (2006)) is also used to analyze time-varying relationships. A recent practical approach to testing for second order stationarity and estimating local second order properties using wavelets is found in Nason (2013).

The description *non-stationary* is also applied to time series which are integrated stationary processes. Our approach to series which may be of this form has been to correct them for any regular trend but otherwise to apply the same spectral analysis as for stationary series. Evidence of integrated non-stationarity may be apparent in a spectral peak of very high power at frequency zero, though in a finite sample it is not possible to discriminate with certainty between an integrated process and a stationary process with red spectrum. Pre-whitening, possibly even in the form of differencing, may then improve the analysis. In practice, the estimates of lagged responses may not be greatly affected by this, as may be verified for the Sales series example.

In our examples we have also corrected series for fixed periodic annual and weekly cycles. This is not the same as applying a seasonal adjustment filter, which corrects further for slow variations in the seasonal pattern from year to year. Our correction removes harmonic components at the seasonal harmonic frequencies, corresponding to cycles with periods equal to the seasonal period or a simple fraction of that period. Seasonal adjustment also reduces the amplitude of harmonic components in the neighborhood of the seasonal harmonic frequencies. These may contain useful information about the relationship between the series. For example, if there is a trend to warmer summers, the moth trappings may increase correspondingly, but seasonal adjustment may remove this evidence. Many economic time series are seasonally adjusted, and the spectral estimation of their dependence may be adversely affected by this.

Spectral analysis is also directed at estimating linear lagged relationships. We determined a significant lagged response between variables of the Film extrusion process example, but this system almost certainly contains non-linearities. These may not be large provided the variables do not depart far from their set point, and the linear relationships we identified may be sufficient to design successful control schemes.

Overall, spectral analysis is useful for demonstrating and quantifying linear dependence between time series. It is often of value for specifying the form of a linear parametric model for this dependence; this model may then be developed to allow for time variation and non-linearity if these features are found to be important. We refer again to Priestley (1981, Chap. 11) for models that may be considered in this context.

3.11 Harmonic contrasts in large samples

In the preceding sections we have presented the distributional properties of estimates of spectral quantities which are applicable in large samples. This final section is intended to give a simple explanation of the basis on which these are derived. In classical statistics the properties of estimates derived from large, though finite, samples, depend on the distributional assumptions of the whole population. Though not directly analogous, in the time series context there is a similar point, that properties of spectral estimates derived from finite samples depend on the spectral representation of the whole, infinite series. For a formal treatment, see Priestley (1981, p. 246).

Our informal introduction is to examine in what sense the harmonic contrasts of a finite sample converge when the length of this sample increases. We now take a sample of x_t from $t = -m$ to $t = m$ and correspondingly modify the limits in the definitions (3.1) of the harmonic contrasts. The only meaningful sense in which these contrasts converge as m increases is that their cumulative values, or *integrals*, have limits $A(f)$ and $B(f)$, which we will call the *spectral coefficients* of x_t:

$$A(f) = \lim_{n \to \infty} \int_0^f a(h)dh \ \text{ and } \ B(f) = \lim_{n \to \infty} \int_0^f b(h)dh. \qquad (3.40)$$

Figure 3.33 illustrates this convergence for the Respiration rate series. The solid line shows the values of the integrated harmonic contrasts for a series of length 101 (an extra point has been added for the symmetrical time index), and the broken line shows the values for the central 31 points of this sample. As the sample size increases, finer detail is added to these plots, and the functions converge. In the limit, the processes are continuous but not smooth; they do not have well defined derivatives. They are plotted only for positive f, but $A(f)$ is anti-symmetric and $B(f)$ is symmetric, with $A(0) = B(0) = 0$. They can take positive or negative values.

The values of $A(f)$ and $B(f)$ at each frequency f are random variables that are linear functions of the whole realized series x_t, $-\infty < t < \infty$, and any

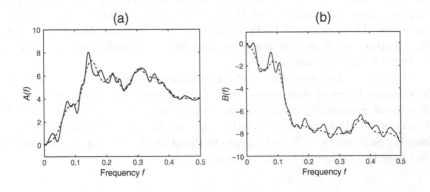

Figure 3.33 *(a) The integrated contrasts a(f), and (b) the integrated contrasts b(f), for the infant Respiration rate series. The solid lines show the plots for a sample of length 101, and the broken lines for a sample of length 31.*

series value x_t can likewise be determined as linear functions of the spectral coefficients:

$$x_t = \int_{-0.5}^{0.5} \{\cos(2\pi ft)dA(f) + \sin(2\pi ft)dB(f)\}. \qquad (3.41)$$

We explain this integral as the limit of approximating sums obtained by breaking up $A(f)$ and $B(f)$ into their jumps or differences over a discrete grid of frequencies $f_k = k/N$, for any choice of (large) N:

$$\delta A_k = A(f_{k+1}) - A(f_k) \quad \text{and} \quad \delta B_k = B(f_{k+1}) - B(f_k). \qquad (3.42)$$

The series is then approximated as a sum of cycles

$$x_t \approx \sum_{k=0}^{N-1} \{\cos(2\pi f_k t)\delta A_k + \sin(2\pi f_k t)\delta B_k\}. \qquad (3.43)$$

This approximation converges to x_t as N increases, in the sense that the mean square error of the approximation tends to zero for each point in time, being represented formally in the limit by (3.41).

Although (3.43) is very similar in appearance to the representation (3.3) of a finite sample, the harmonic contrasts a_k and b_k of a finite series do not simply equate to the differences δA_k and δB_k. In fact, the contrasts can be expressed as

$$a(f) = 2\int_0^{0.5} \{U_h(f)dA(f)\}$$

$$b(f) = 2\int_0^{0.5} \{V_h(f)dB(f)\} \qquad (3.44)$$

where $U_h(f)$ is the weight function plotted in Figure 3.15(a) (or (b) if tapering is used) and $V_h(f)$ is similar. It is from the expression (3.41) that the properties of all the spectral estimates constructed from the contrasts can be derived, using the following properties of the spectral coefficients:

1. The variances of their *increments* over a frequency range from f to g are

$$\text{Var}\,\{A(g) - A(f)\} = \text{Var}\,\{B(g) - B(f)\} = 0.5\int_f^g S(h)dh. \qquad (3.45)$$

Note in Figure 3.33 that the processes are much more variable in the frequency range from 0.10 to 0.15 corresponding to a peak in the spectrum.

2. An increment in $A(f)$ over one frequency range is uncorrelated with an increment in $A(f)$ over any other non-overlapping frequency range, and similarly for $B(f)$. We say they are processes with *uncorrelated increments*. If x_t is Gaussian, then so are $A(f)$ and $B(f)$ and they are known as Brownian processes (see Doob, 1953, p. 97) with variable diffusion rate equal to $\sqrt{0.5S(f)}$ at frequency f. The *increments* in $A(f)$ are *symmetric* and those in $B(f)$ are anti-symmetric.

3. The value of $A(f)$ is uncorrelated with $B(g)$ for any frequencies f and g; the two processes are uncorrelated with each other.

It is usual to express these properties by writing the increments as *spectral differentials* $dA(f) = A(f + df) - A(f)$, and similarly $dB(f)$, and writing the variance matrix of the pair as

$$\text{Var} \begin{pmatrix} dA(f) \\ dB(f) \end{pmatrix} = 0.5 \begin{pmatrix} S(f) & 0 \\ 0 & S(f) \end{pmatrix} df, \qquad (3.46)$$

in the sense that terms of smaller order than df are neglected. For two or more time series, (3.46) is extended in a manner corresponding precisely to the properties of the harmonic contrasts presented in (3.22). Moreover, in the same sense as for (3.46), the increments $dA(f)$ and $dB(f)$ for all the series are uncorrelated with those at *any* other frequency. These properties enable us to numerically determine quantities relating to the estimation of spectral properties, such as the plots shown in Figure 3.14 of degrees of freedom and correlations of smoothed spectral estimates, and the confidence limits of coherency, gain, phase and lagged responses shown in our various examples.

Chapter 4

Estimation of vector autoregressions

4.1 Methods of estimation

We introduced the multivariate autoregressive or VAR(p) model (2.7) in Chapter 2. It is the most widely used practical tool for modeling multivariate time series and in this chapter we explain and illustrate how it can be fitted to observed series by estimating the model parameters. We consider three methods of estimation:

1. solving the Yule–Walker equations presented in (2.5), but replacing the autocovariances $\Gamma_{i,j,k}$ with the sample autocovariances $C_{i,j,k}$ defined in (3.35).

2. by least squares regression of each current observation $x_{i,t}$ on past observations $x_{j,t-k}$ of all the series $j = 1, \ldots, m$ and lags $k = 1, \ldots, p$.

3. by Gaussian maximum likelihood estimation, i.e., by maximizing the likelihood computed on the assumption that the observations have a joint normal distribution.

All three methods give very similar parameter estimates and are asymptotically efficient in very large samples, but there may be important differences between the estimates even in moderate to large sample sizes. The first two methods generally require much less computation than the third. However, we only implement the Yule–Walker method when the model is saturated, i.e., all $m^2 p$ lagged coefficients in the model are estimated, together with the full error covariance matrix. When using lagged regression, if some coefficients are omitted or constrained to be zero, the estimates may not be fully efficient. However, the method can be modified to use generalized least squares, which does give efficient estimates. There are no restrictions on applying constraints when Gaussian maximum likelihood estimation is used.

We will assume in our subsequent treatment that the series have zero mean, so we will not introduce any term for the mean into the expressions we use or methods that we describe. If the series have non-zero mean and require correction by linear regression upon level, trend or seasonality, we will do this by ordinary least squares for the first two methods. This is fully efficient in large samples, but we can also estimate the trend terms along with the other parameters when using maximum likelihood. If the saturated model is fitted, convenient expressions can be established for the large sample covariance matrix of the estimated parameters. The maximum likelihood method of estimation also provides numerical values of this covariance matrix based upon likelihood theory, and these are particularly useful when models are fitted with parameter constraints. For all three methods we select the order p of the model based on an information criterion. We generally use the AIC (Akaike (1973)), but we may also comment on the use of two other criteria, the HQC (Hannan and Quinn (1979)) and SIC (Schwarz (1978)).

We will illustrate various properties of the fitted VAR(p) model. One that is widely used is the spectrum of the series as implied by the model, known as the *autoregressive spectrum estimate* (Akaike (1969)). We will refer to it as the VAR(p) spectrum. Various aspects of this estimate, such as gain and phase properties, may be compared with those given by the smoothed sample spectrum of the series. This can provide a check that the fitted model provides an adequate representation of its structure.

For a univariate series, the estimated AR spectrum and smoothed sample spectrum can be shown to have very similar properties under certain conditions. The first of these is that the series length n is large. The order of the AR must also be large, though small compared with n, and the smooth spectral estimator must have both a narrow bandwidth and small variance. Precise conditions are given by Berk (1974), who shows that the ratio of the variances approaches one for autoregressive spectral estimates from a univariate AR(K) model and smoothed spectral estimates with a rectangular lag window $w_k = 1$ for $k \leq K$. Kromer (1969) derives the same result for a smoothed spectral estimator with general lag window w_k and an autoregressive spectral estimator of order $L = \sum w_k^2$. The conditions require that K and L are large. Shibata (1981) presented an optimal method of autoregressive spectral estimation. Autoregressive spectral estimators have attracted interest (Marple and Nuttall, 1983) as improving the estimation of sharp peaks in the spectrum of series with strong cyclical components. In practice, a smoothed spectral estimator with the same nominal variance as an autoregressive spectral estimator will typically have lower, more rounded peaks when the series shows evidence of cyclical behavior.

One criticism of the autoregressive spectrum estimate in the multivariate context is that the univariate spectrum estimates of one series in the set depend on which other series are included. This is because the fitted VAR model will vary with the selection of series even for a fixed order p of the model. To understand this, we view the VAR model as a means of extrapolating the sample covariances beyond lag p. Up to this lag, the implied covariances of a fitted VAR(p) coincide with the sample covariances of the series, except for very minor differences that depend on the particular method by which the model was fitted. Beyond lag p, the precise values of these implied autocovariances of one series will depend upon the other series, but for large p these values will generally decay quite rapidly. If, therefore, as the series length n increases, the order p of the approximating VAR model is also increased, the sensitivity of the implied model covariances to which other series are included in the model estimation is reduced. The model spectrum is just that determined by these covariances, so it also becomes insensitive. The qualitative difference between the smoothed spectrum estimate and the autoregressive spectrum estimate is that the former windows the covariances and the latter extrapolates them.

4.2   The spectrum of a VAR model

The expression for the VAR(p) spectrum in terms of its coefficients is derived from the frequency domain relationships we have already presented in Chapter 3. It is, however, convenient at this stage to use complex variable notation, including the use of the complex exponential $\exp(i\theta) = \cos\theta + i\sin\theta$ (taking i as the square root of -1). We can then reduce earlier expressions given in terms of cosine and sine functions to much more concise forms.

We require the spectrum of one multivariate time series that has a linear lagged dependence upon another. The building block for this is derived from the relationship of one univariate series y_t to another x_t by the noise-free lagged dependence:

$$y_t = \sum_{k=-\infty}^{\infty} v_{y\,x,k} x_{t-k}. \tag{4.1}$$

We presented this previously, though with a noise term, in (3.6) and then in (3.23). From (3.24), therefore, the relationship between the harmonic contrasts of the series is given in terms of the transfer function $\tau_{y\,x}(f)$ and $\nu_{y\,x}(f)$ as

$$\begin{pmatrix} a_y(f) \\ b_y(f) \end{pmatrix} = \begin{pmatrix} \tau_{y\,x}(f) & -\nu_{y\,x}(f) \\ \nu_{y\,x}(f) & \tau_{y\,x}(f) \end{pmatrix} \begin{pmatrix} a_x(f) \\ b_x(f) \end{pmatrix}. \tag{4.2}$$

We now introduce the complex variable notation by defining the *complex contrasts* of x_t as

$$\alpha_x(f) = a_x(f) - ib_x(f) = \sum_{t=1}^{n} x_t \exp(-2\pi i f t), \tag{4.3}$$

and similarly $\alpha_y(f)$ for y_t.

Also define the *complex transfer function* as

$$T_{y\,x}(f) = \tau_{y\,x}(f) - i\nu_{y\,x}(f) = \sum_{k=-\infty}^{\infty} v_{y\,x,k} \exp(-2\pi i f k). \tag{4.4}$$

Then the relationship (4.2) can be expressed more concisely as

$$\alpha_y(f) = T_{y\,x}(f)\alpha_x(f). \tag{4.5}$$

This last expression would remain unchanged if we replaced $-i$ by $+i$ in the definitions (4.3) and (4.4). The choice of sign conforms with conventions in several texts, such as Brockwell and Davis (1987), and our further definitions of complex spectral quantities will be consistent with this. The definitions (4.3) and (4.4) may be inverted as

$$x_t = \int_{-0.5}^{0.5} \alpha_x(f) \exp(2\pi i f t) df \tag{4.6}$$

and

$$v_{y\,x,k} = \int_{-0.5}^{0.5} T_{y\,x}(f) \exp(2\pi i f k) df. \tag{4.7}$$

We also note here that if we write the lagged relationship (4.1) between y_t and x_t using the backward shift operator (2.27) as

$$y_t = v_{y\,x}(B) x_t \tag{4.8}$$

with the operator power series

$$v_{y\,x}(B) = \sum_{k=-\infty}^{\infty} v_{y\,x,k} B^k, \tag{4.9}$$

then by formally substituting B by $\exp -2\pi i f$ we can also express the complex transfer function as

$$T_{y\,x}(f) = v_{y\,x}(\exp -2\pi i f). \tag{4.10}$$

Now take x_t to be a multivariate time series having components $x_{i,t}$ with multivariate complex contrasts defined by

$$\alpha_x(f) = \sum_{t=1}^{n} x_t \exp(-2\pi i f t). \tag{4.11}$$

Define the *complex cross-spectrum* between two different component series $x_{i,t}$ and $x_{j,t}$ in terms of their co-spectrum and quadrature-spectrum as

$$S_{x\,i,j}(f) = P_{x\,i,j}(f) - i Q_{x\,i,j}(f), \tag{4.12}$$

and define the *complex spectral density matrix* $S_x(f)$ to have diagonal elements given by the univariate or auto-spectra $S_{x\,i}(f)$ and off diagonal elements $S_{x\,i,j}(f)$. Then we can express the auto- and cross-spectra of x_t in terms of the lagged covariance matrices (2.2) in the single expression

$$S_x(f) = \sum_{k=-\infty}^{\infty} \Gamma_{x\,k} \exp(-2\pi i f k). \tag{4.13}$$

This can be inverted to give

$$\Gamma_{x\,k} = \int_{-0.5}^{0.5} S_x(f) \exp(2\pi i f k) df. \tag{4.14}$$

Following (3.36) and (3.37), the *complex sample matrix spectrum* $S_x^*(f)$ of a multivariate series x_t may be similarly defined concisely in terms of the sample covariances or complex contrasts, using the bar to denote complex conjugate, as

$$S_x^*(f) = \sum_{k=-(n-1)}^{(n-1)} C_{x\,k} \exp(-2\pi i f k) = \frac{1}{n} \alpha_x(f) \overline{\alpha_x(f)}'. \qquad (4.15)$$

This may be inverted either as

$$C_{x\,k} = \int_{-0.5}^{0.5} S_x^*(f) \exp(2\pi i f k) df \qquad (4.16)$$

or, for any $N \geq 2n$ and setting $f_j = j/n$, by the finite sum

$$C_{x\,k} = \frac{1}{N} \sum_{j=0}^{N-1} S_x^*(f_j) \exp(2\pi i f_j k). \qquad (4.17)$$

By applying the fast (finite) Fourier transformation algorithm to evaluate the sums in these equations, (4.3), the second equation of (4.15) and (4.17) provide an efficient method for calculating the sample covariances $C_{x\,k}$. After weighting these with a lag window, the first equation of (4.15) provides a smoothed spectrum estimate. The practicality of spectral analysis was enormously advanced by this development in computation.

We approach our objective of deriving the autoregressive spectrum by considering the more general Wold representation (2.12), which we express in operator terms as

$$x_t = \Psi(B)e_t. \qquad (4.18)$$

By extension of (4.5) and (4.10), the relationship between the complex contrasts $\alpha_x(f)$ and $\alpha_e(f)$ of the respective multivariate time series is

$$\alpha_x(f) = \Psi\{\exp(-2\pi i f)\} \alpha_e(f). \qquad (4.19)$$

Using (4.15), we then have the relationship between sample spectra:

$$S_x^*(f) = \Psi\{\exp(-2\pi i f)\} S_e^*(f) \Psi\{\exp(2\pi i f)\}', \qquad (4.20)$$

and correspondingly, on taking expectations for large n, the relationship between the series spectra:

$$S_x(f) = \Psi\{\exp(-2\pi i f)\} S_e(f) \Psi\{\exp(2\pi i f)\}'. \qquad (4.21)$$

But the white noise innovation series e_t has constant spectrum $S_e(f) = V_e$, because all its lagged covariances are zero. Thus,

$$S_x(f) = \Psi\{\exp(-2\pi i f)\} V_e \Psi\{\exp(2\pi i f)\}'. \qquad (4.22)$$

Applying this last expression to the VAR(p) model (2.32), we directly obtain, writing $S_\Phi(f)$ for $S_x(f)$,

$$S_\Phi(f) = [\Phi\{\exp(-2\pi i f)\}]^{-1} V_e [\Phi\{\exp(2\pi i f)\}']^{-1}. \qquad (4.23)$$

There is an alternative route to deriving this expression, based upon noting that the spectral density matrix (4.13) is equal to the covariance generating function (2.54) of x_t with the substitution of $z = \exp(-2\pi i f)$:

$$S_x(f) = \Gamma\{\exp(-2\pi i f k)\}. \qquad (4.24)$$

Applied to the covariance generating function of the VAR(p) model (2.56), this also directly gives (4.23). However, the former derivation provides greater insight into this important result.

A useful extension of the development from (4.18) to (4.23) is that, given two, possibly multivariate, series, $y_t = T(B)x_t$ and $z_t = R(B)x_t$, both derived by applying operators to the same series x_t, their covariance matrix is given in terms of the spectrum $S_x(f)$ of x_t by

$$\text{Cov}(y_t, z_t) = \int_{-0.5}^{0.5} T\{\exp(-2\pi i f)\} S_x(f) R\{\exp(2\pi i f)\}' df. \qquad (4.25)$$

The sample covariance is also given in terms of the sample spectrum by this relationship. These results are of value in later chapters.

4.3 Yule–Walker estimation of the VAR(p) model

As stated in the introduction to this chapter, these estimates are obtained by solving the Yule–Walker equations (2.5), but replacing the autocovariances $\Gamma_{i,j,k}$ with the sample autocovariances $C_{i,j,k}$ defined in (3.35). This method has advantages and drawbacks. An important advantage is that the estimated VAR(p) model necessarily satisfies the stationarity condition. This can be shown to follow from the fact that the matrix of the Yule–Walker equations, formed from the sample covariances, is positive definite for all orders p. This is clearly desirable because it is an estimate of the variance matrix of the set of variables $x_{t-1}, x_{t-2}, \ldots, x_{t-p}$. However, this is not ensured if the matrix is formed from *bias corrected* estimates of the sample covariances obtained by replacing $1/n$ with $1/(n-k)$ in the definition (3.35). We do not therefore use these corrected estimates, but the bias of the sample covariances that we do use may constitute a drawback. Although it becomes negligible as the sample size increases, for some models it may produce an unacceptable bias in the estimates of the model parameters even in quite large samples. We illustrate this in the simple case of a univariate AR(1) model with parameters ϕ_1 and σ_e^2, for ϕ_1 close to 1, which is not uncommon in many applications. The Yule–Walker parameter estimates are

$$\hat{\phi}_1 = C_{x1}/C_{x0} \quad = \quad \left\{\frac{1}{n}\sum_{t=2}^{n}x_t x_{t-1}\right\}\Bigg/\left\{\frac{1}{n}\sum_{t=1}^{n}x_t^2\right\} = r_1 \qquad (4.26)$$

$$\hat{\sigma}_e^2 = C_{x0} - \hat{\phi}_1 C_{x1} \quad = \quad \left\{\frac{1}{n}\sum_{t=1}^{n}x_t^2\right\}(1-\hat{\phi}_1^2).$$

For $\phi_1 = 1 - \delta$ where δ is small and of the order $1/n$, the bias in $\hat{\phi}_1$ is approximately $-(1/n)$. This is comparable with the standard error of $\hat{\phi}_1$, which for large n is $\sqrt{(1-\phi_1^2)/n} \approx \sqrt{2\delta/n}$. It therefore leads to an appreciable, but not excessive, further increase in the contribution of parameter estimation to prediction error. However, the relative bias in $\hat{\sigma}_e^2$ can be close to 100%, because $(1-\hat{\phi}_1^2) \approx 2(1-\hat{\phi}_1)$ will have approximate bias $2/n$, which is directly comparable with $(1-\phi_1^2) \approx 2(1-\phi_1) = 2\delta$. This can be important when models of increasing order are fitted and $\hat{\sigma}_e^2$ is used in a criterion for model order selection. We will illustrate the effect in comparison with estimation by lagged regression using a real bivariate example in the next section. A comparison of Yule–Walker estimation with lagged regression for the univariate AR(1) model is, however, useful here to help understand the underlying reason for the bias.

For this comparison, we will take the sample of x_t to be observed from $t = 0$ to $t = n + 1$. The standard regression of x_t upon x_{t-1} then gives estimates for the AR(1) model:

$$\hat{\phi}_1 \quad = \quad \left\{\sum_{t=1}^{n+1}x_t x_{t-1}\right\}\Bigg/\left\{\sum_{t=1}^{n+1}x_{t-1}^2\right\} \qquad (4.27)$$

$$\hat{\sigma}_e^2 \quad = \quad \left\{\frac{1}{n}\sum_{t=1}^{n+1}x_t^2\right\}(1-\hat{\phi}_1^2).$$

It may be quickly confirmed that these are identical to the Yule–Walker estimates (4.26) *if* we set $x_0 = x_{n+1} = 0$. The Yule–Walker estimates are the regression estimates obtained by extending the sample with a zero at either end. For series lacking strong correlation, this will not have a substantial effect, but for values of ϕ_1 close to 1, these zeros will generally introduce large discontinuity in the appearance of the series and large residual errors at the two ends, contributing to the substantial inflation of the estimate of σ_e^2. This is another instance of the *end-effect* problem in time series. Maximum likelihood estimation of the model provides a rigorous solution to this problem, but a simple and effective treatment applicable to Yule–Walker estimation is to taper the series. This eliminates the discontinuity that would arise from attaching zeros to the ends of the series and substantially reduces bias of the Yule–Walker estimate at a modest cost to statistical efficiency. It is recommended by Zhang (1992). We will illustrate its usage for the real bivariate example in the next section.

4.4   Estimation of the VAR(p) by lagged regression

Least squares regression can be used to estimate the VAR(p) model, taking
the response vectors, separately for each series, as

$$Y_i = \begin{pmatrix} x_{i,n} \\ x_{i,n-1} \\ \vdots \\ x_{i,p+1} \end{pmatrix} \tag{4.28}$$

but each with the same regression matrix of lagged values of all the series:

$$X = \begin{pmatrix} x_{1,n-1} & \cdots & x_{1,n-p} & \cdots & x_{m,n-1} & \cdots & x_{m,n-p} \\ x_{1,n-2} & \cdots & x_{1,n-p-1} & \cdots & x_{m,n-2} & \cdots & x_{m,n-p-1} \\ \vdots & \vdots & \vdots & \vdots & \vdots & \vdots & \vdots \\ x_{1,p} & \cdots & x_{1,1} & \cdots & x_{m,p} & \cdots & x_{m,1} \end{pmatrix}, \tag{4.29}$$

to estimate the vector of coefficients

$$\phi_i' = \begin{pmatrix} \Phi_{i,1,1} & \cdots & \Phi_{i,1,p} & \cdots & \Phi_{i,m,1} & \cdots & \Phi_{i,m,p} \end{pmatrix}'. \tag{4.30}$$

We use ϕ_i to make the distinction between these vectors and the model coefficient matrices Φ_k.

These m separate regressions are not unrelated because their error vectors
will in general be correlated. This would generally require that, for full statistical efficiency, they are estimated simultaneously, taking this correlation into
account. However (see Anderson (2003)), the regressions can be decoupled
and fitted separately because they have a common regression matrix. We will
later demonstrate this property.

Compared with the Yule–Walker equations, these regression estimates have
the advantage of not suffering from the same small-sample bias. The least
squares equations are *unbiased estimating equations* (Cox (1993)), in the sense
that on taking expectations and dividing by $n - p$ they reduce exactly to
the Yule–Walker equations. However, this method does not ensure that the
fitted model is stationary, though this property will only tend to be at risk,
through sampling variability, when the true model parameters are close to the
boundary of the region defined by the stationarity condition. An unavoidable
drawback of the method is that the regression vector is reduced to length $n-p$
in order to accommodate all the regressors up to lag p.

We now illustrate the Yule–Walker and regression methods as applied to
the Gas furnace series shown in Figure 3.18. We will choose the order p of
the model using the AIC (Akaike (1973)). This is a criterion evaluated and
plotted for models of increasing hypothesized order $p = k = 1, 2, \ldots, K$ up to
some specified maximum, which we will set as $K = 20$. The model order is
selected as that for which the AIC is a minimum.

This is an appropriate place to remark that it may be possible, by aligning the components of a multivariate time series, to select a lower order of model using the AIC. We used alignment by a lag of 5 in the spectral analysis of the Gas furnace series in Chapter 3 to improve the estimation of the cross-spectrum and lagged response. The alignment lag can be suggested by the lag at which the series cross-correlation achieves its (absolute) maximum value, but as the spectral analysis reveals, the response of the Output to the Input is first significant at a lag of 3. By using an alignment lag of 5, we would lose the causal relationship between the Input and Output, which we will confirm in the structure of the VAR model selected without alignment. Alignment of series in a VAR model should be applied with caution.

The AIC is defined for a model with hypothesized order $p = k$ as

$$\text{AIC}(k) = -2 \log \text{likelihood} + 2M, \tag{4.31}$$

where $M = m^2\, k$ is the number of autoregressive parameters. We will use the term *deviance* for the term $-2 \log \text{likelihood}$ in (4.31), and this will decrease as (nested) models of increasing order k are fitted. However, the predictive accuracy of the model consequent on this reduction in deviance is eventually outweighed by the statistical variability due to the excessive number of estimated parameters, beyond that needed to represent the structure of the data. This statistical variability is measured by a *penalty term* given by the term $2M$ in (4.31). The AIC measures the overall prediction error allowing for this variability. Selecting a model for which it is a minimum leads to a compromise between a well-fitted and over-fitted model. In a model with M parameters, the penalty term is generally $2M$, which becomes $2\, m^2\, k$ for the VAR(k) model.

If the series is generated by a true VAR(p) model, then using AIC to choose the model order will, in large samples, select the true order p with a probability that is high, but appreciably less than 1 (Shibata (1976)). The orders $p + 1$, $p + 2$, ... may be selected with rapidly decreasing probability. Other criteria used to select model orders are the HQC, presented for univariate autoregressions in Hannan and Quinn (1979) and for multivariate autoregressions in Quinn (1980), and the SIC (Schwarz (1978)). These specify a larger penalty term and will, in large samples, select the true order with probability approaching 1. But for samples that are small or of moderate size, their probability of underestimating the true order is appreciably greater than if the AIC is used. In practice, a VAR model of any finite order may only approximate the covariance structure of the series, and the use of the AIC as a means of selecting a good approximation is supported by Shibata (1980).

When estimating the VAR(p) by lagged regression, the deviance term as used in (4.31) with hypothesized order p is given by that used for multivariate regression, which is

$$-2 \log \text{likelihood} = \text{Dev}_p = (n - p) \log \det \widehat{V}_e, \tag{4.32}$$

where

$$\widehat{V}_e = \frac{1}{n-p} \sum_{t=1+p}^{n} \hat{e}_t \hat{e}_t' \qquad (4.33)$$

is the sample variance matrix of the residuals $\hat{e}_t$ from the regressions. It is *not* corrected for the residual degrees of freedom, which would further reduce the divisor $n - p$ by the number pm of coefficients fitted to each response. We will also justify the deviance definition (4.32) when we consider maximum likelihood estimation.

There is a further concern when using the AIC to select the order of a VAR(p) fitted by lagged regression: the length $n - p$ of the response vector as given by (4.28) will reduce as the hypothesized order $p = k$ increases from 1 to its maximum K. This will introduce further undesirable statistical variability into the AIC. To avoid this, we will apply the strategy of using, for all orders k of fitted model, the response vector defined for the maximum order $p = K$. The vectors constituting the regression matrix in (4.29) are correspondingly shortened, so that p is replaced by K in (4.33) and the last term of (4.32).

If the Yule–Walker method is used to fit the VAR(p) model, the deviance term in the AIC is very similarly defined as

$$\mathrm{Dev}_p = \gamma n \log \det \widehat{V}_e, \qquad (4.34)$$

where now $\widehat{V}_e$ is derived from the solution of the Yule–Walker equations. The factor γ represents the loss of information, or efficiency, due to the application of the normalized taper h_t, to which we referred in Section 3.9. This is equivalent to a reduction in the series length. It is given by

$$\gamma = n / \sum_{t=1}^{n} h_t^4. \qquad (4.35)$$

The four lines in Figure 4.1 are plots of the AIC for the Gas furnace series, which has length $n = 296$. The plots start at order $p = 2$ in order to display their pattern more clearly, because the reduction in AIC from order 1 is so substantial. The uppermost line is the AIC derived using Yule–Walker estimation without tapering, and the next using Yule–Walker estimation with 8% tapering. The difference in level is due to the bias in $\widehat{V}_e$, which is substantially reduced by the use of tapering. The loss of efficiency due to the taper is equivalent, in this example, to a reduction in length of the series by 12. Both lines have their minimum at order $p = 4$. The line just below these, the next to the lowest, is the AIC derived using lagged regression with the maximum order $K = 20$, so that the data length is reduced by this number. The further reduction in bias explains why it is set lower than the line for the tapered Yule–Walker method. The lowest line is the AIC derived using maximum likelihood estimation, which is explained in the next section, but we point out that its use of the full data length explains why it is lower than the line for the regression method, which it otherwise parallels very closely.

The levels of the different lines are not otherwise important. Provided that each is consistently defined, it is the location of the minimum in which we are interested. For the lower two lines this is at $p = 6$, rather than $p = 4$, for the upper two lines. In this particular example, Yule–Walker estimation gives rise to considerable bias. However, if the tapering proportion is increased to 50%, it also gives a minimum AIC at lag 6, but with a loss of efficiency equivalent to 25% of the data length. Even so, the difference in the AIC between these two lags is small for all the methods, and whichever lag is chosen as the order has little effect upon inference regarding the properties of the model.

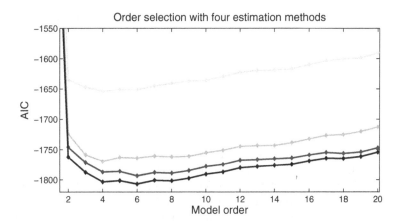

Figure 4.1 *Plots of the AIC derived for the Gas furnace example using four different estimation methods. From the highest line (lightest gray) to the lowest line (black) the methods used are Yule–Walker, Yule–Walker with 8% tapering, regression and maximum likelihood.*

We now turn to consider the properties of estimates of the coefficients Φ_k of a VAR(p) model obtained by lagged regression, under the assumption that it is fitted to a series x_t of length n that follows this true model. The property that is of main practical value is that the procedures of ordinary linear regression can be used reliably to derive standard errors and construct confidence intervals for the parameters. This gives the large sample variance matrix of the estimate of the coefficient vector ϕ_i in (4.30) to be $(X'X)^{-1}\hat{V}_{e\,i,i}$, where X is the regression matrix given in (4.29). However, $(X'X)/(n-p)$ is the sample value of the matrix G_p in the Yule–Walker equations (2.5), to which it converges for n large. Neglecting p in comparison with n, the distribution of the suitably scaled estimate of ϕ_i is

$$\sqrt{n}\left(\hat{\phi}_i - \phi_i\right) \sim N\left(0, G_p^{-1}V_{e\,i,i}\right). \tag{4.36}$$

The formal derivation of this result is given in Lütkepohl (1993) and Reinsel (1993). It requires care because the regressors are stochastic, and it is not possible in the time series context to condition upon them, as is usual in regression analysis, because that would also condition the response vector. The covariance between the scaled estimates of ϕ_i and ϕ_j is given by replacing $V_{e\,i,i}$ by $V_{e\,i,j}$ in the variance term of (4.36). The variance matrix of the whole set of parameters ϕ_i stacked in a single vector ϕ is given by assembling these blocks of variance and covariance matrices in a single $mp \times mp$ matrix given formally by the Kronecker product:

$$\sqrt{n}\left(\hat{\phi} - \phi\right) \sim N(0, G_p^{-1} \otimes V_e). \tag{4.37}$$

Table 4.1 gives the estimates of selected coefficients in the VAR(6) model fitted to the Gas furnace series.

Table 4.1 *Selected estimated coefficients with t values for a VAR(6) model fitted to the Gas furnace series. The first two pairs of columns of estimates and t values are derived from a saturated model and the last pair is derived from a subset model.*

k	$\Phi_{2,1,k}$	t	$\Phi_{1,2,k}$	t	$\Phi_{2,1,k}$	t
1	0.0631	0.8579	−0.0508	−1.1230	0.0630	0.8565
2	−0.1335	−0.8365	0.1000	1.1980	−0.1332	−0.8344
3	−0.4413	−2.4165	−0.0796	−0.9124	−0.4425	−2.4231
4	0.1521	0.8264	0.0269	0.3094	0.1499	0.8142
5	−0.1195	−0.6882	−0.0415	−0.5441	−0.1127	−0.6494
6	0.2482	2.3376	0.0306	0.9423	0.2416	2.2772

The distribution of the coefficients allows us to test whether some might be zero and omitted from the model. The dependence of series 2, the Output, upon series 1, the Input, is significant (at lags 3 and 6) according to the first column of t values shown in Table 4.1. As one would expect, the dependence of the Input upon the Output appears not to be significant on inspecting the second column of t values. In order to test more formally for causal dependence of $x_{i,t}$ upon $x_{j,t}$, i.e., whether all the coefficients $\Phi_{i,j,k} = 0$ for $k = 1, \ldots, p$ for some specified i and j, we can re-fit the model with selected coefficients omitted; see also Granger (1969). However, this is a particular case of subset regression known as *seemingly unrelated regression* (SUR) (Zellner (1962)), for which, in general, we should no longer use ordinary least squares (OLS) regressions to estimate the coefficients separately for each response series. For full efficiency and correct evaluation of likelihood for comparing models with selected parameters omitted, we need to estimate them simultaneously using generalized least squares (GLS).

To do this we modify the parameter vector ϕ_i in (4.30) to retain only those coefficients that are to be estimated in the dependence of $x_{i,t}$ upon lagged values of all the series. We also define, for each such dependent variable, a distinct regression matrix X_i by including in X, as defined in (4.29), only the columns corresponding to these retained coefficients. The response vectors defined in (4.28) remain unchanged and the estimating equations for the whole set of coefficients become

$$\sum_{j=1}^{m} W_{i,j} X_i' Y_j = \sum_{j=1}^{m} W_{i,j} X_i' X_j \phi_j. \tag{4.38}$$

Note that if all the X_i are the same, this is satisfied by the separate solutions to $X_i'Y_j = X_i'X_i\phi_j$, justifying the earlier use of OLS. Again, we refer to the subsequent section on maximum likelihood estimation for justification of GLS estimation.

The weights $W_{i,j}$ in (4.38) are the elements of $W = V_e^{-1}$, which also need to be estimated. The initial step is to use separate OLS estimates of ϕ_i to generate the estimated innovation series $\hat{e}_{t,i}$ and thereby the sample estimate of V_e using 4.33 with $k = p$. This is equivalent to solving (4.38) using $W = I$. The second step is to determine W from this initial estimate of V_e and solve the GLS equations. The process may be iterated to convergence, which is very rapid. The deviance of the fitted model is then given by $(n-p)\log\det V_e$. The variance matrix V_Φ of the estimated parameters is estimated by the inverse of the matrix of the estimating equations for the full vector of parameters on the RHS of (4.38).

We illustrate the procedure by removing from the model all the coefficients $\Phi_{1,2,k}$ for $k = 1, \ldots, p$. The last two columns in Table 4.1 show the new estimates of $\Phi_{2,1,k}$ for $k = 1, \ldots, p$ with their t values. These are very little changed because V_e and hence W is very nearly diagonal; the correlation between the two innovation series is only -0.0528. The increase in the deviance from omitting the six parameters is 5.95. This is very close to the mean of the chi-squared distribution on 6 degrees of freedom, to which it should be referred to test the hypothesis that the reduced model is acceptable. Clearly, this hypothesis is accepted.

We complete this section by deriving various properties of the fitted VAR(6) model for the Gas furnace series. Figure 4.2 shows the squared coherency, gain and phase which are derived from the model spectrum given

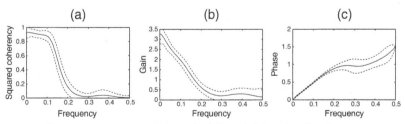

Figure 4.2 *Spectral properties of the VAR(6) model for the Gas furnace series: (a) the squared coherency between the series, (b) the gain from Input to Output and (c) the phase between Output and Input, with approximate two standard error limits.*

by (4.23). These may be compared with those in Figures 3.19 and 3.20 estimated by spectral analysis. The estimated properties are similar, confirming that the VAR model captures the spectral properties of the series. However, the VAR model estimates are smoother and have narrower error limits. The smoothed spectral estimates are constructed using a bandwidth of 0.05, which corresponds to autoregressive spectral estimates formed from a model of order $p = 10$, according to the equivalence discussed in the introduction to this chapter. Spectral estimates constructed from a VAR(6) model can therefore be expected to have lower variance.

Figure 4.3 shows the open loop impulse response function (2.34) from the Input series $x_{1,t}$ to the Output $x_{2,t}$. Again, the pattern of the impulse response is very similar to the lagged response coefficients derived by spectral analysis in Figure 3.19(a) and (c). The step response rises to a level very close to the gain at frequency zero shown in Figure 3.20(a). These results are very little changed whether we use the saturated model or that with the coefficients $\Phi_{1,2,k}$ constrained to zero.

All the estimated properties of the VAR model are shown with error limits of plus or minus two standard deviations. These are large sample approximations derived using an implementation of the delta method based upon simulation. The delta method assumes that a property of the model, i.e., a function of interest F of the model parameters, may be adequately represented by a local linear approximation about the true values of the parameters. This approximation may be used to derive directly the mean and variance of the estimated function. A second order approximation may be used to modify the approximation to the mean, but we do not implement that. In practice, the linear approximation is centered upon the estimated parameters. Analytic expressions for the linearization may be avoided by a numerical simulation procedure described here for our example:

1. Choose a small scale factor α, say $1/100$.

2. Obtain a sample δ_r of a multivariate normal random variable with mean zero and variance matrix V_Φ of the estimated coefficients.

3. Evaluate the function of interest $F_r = F(\hat{\Phi} + \alpha\delta_r)$ for the perturbed parameter value.

4. Repeat for a large number of independent samples δ_r.

5. Use the rescaled sample standard deviation $\mathrm{SD}(F_r)/\alpha$ as the standard deviation of $F(\hat{\Phi})$.

The scaling ensures that the linearization is local. Without it, the perturbations may take the parameter outside the region defined by the stationarity condition, leading to unbounded variation in the property of interest. The accuracy of the derived limits will not be very reliable if that property is a highly non-linear function of the parameters and the sample size is small. However, they will be valid in large samples under wide conditions, and in practice they provide a useful indication of precision in typical applications.

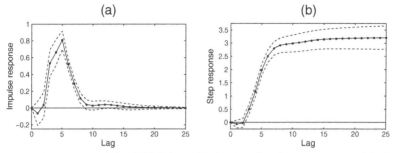

Figure 4.3 *Properties of the VAR(6) model for the Gas furnace series: (a) the impulse response from Input to Output and (b) the step response, with approximate two standard error limits.*

4.5 Maximum likelihood estimation, MLE

Before describing the specific application to the VAR(p) model, it is useful to present some generalities of maximum likelihood (ML) estimation.

The likelihood of a set of model parameters, given the observed data, is the probability density function evaluated at these data and using the parameters. For a data set x from the zero-mean multivariate normal distribution with $\text{Var}(x) = V$, this density is given, up to a multiplicative constant, by

$$f(x) = \frac{1}{\sqrt{\det(V)}} \exp\left\{ -\frac{1}{2} x' V^{-1} x \right\}. \tag{4.39}$$

So for observations $x_1, \ldots, x_n$ from a VAR(p) model, in (4.39) we set x and V, respectively, to the stacked vector X_n and variance matrix $\tilde{G}_n$ in (2.38). Maximum likelihood estimation is simply to choose the parameter values which maximize the likelihood. However, it is usual to work with the logarithm of the likelihood, or the quantity minus twice the log likelihood, which we will call the deviance. The maximum likelihood estimates (MLEs) are then found by minimizing the deviance. For multivariate normal observations, this is, from (4.39),

$$\text{Dev}(\theta) = \log \det(V) + x' V^{-1} x. \tag{4.40}$$

We here emphasize that the deviance is a function of the parameters θ which determine the variance matrix V, with the observations x treated as given quantities.

The general forms of the likelihood and deviance expressions (4.39) and (4.40) will prove useful, but for the VAR(p) model we can derive these quantities more directly by considering the conditional distribution of $x_{p+1}, \ldots, x_n$ given $x_1, \ldots, x_p$ and factorizing the density as

$$f(X_n) = f(x_{p+1}, \ldots, x_n \mid x_1, \ldots, x_p) \, f(x_1, \ldots, x_p). \tag{4.41}$$

In terms of the innovations

$$e_t = x_t - \Phi_1 x_{t-1} - \cdots - \Phi_p x_{t-p} \quad \text{for } t = p+1, \ldots, n, \tag{4.42}$$

which are uncorrelated with variance V_e, we can write the first factor on the RHS of (4.41) as

$$f(x_{p+1}, \ldots, x_n \mid x_1, \ldots, x_p) = \prod_{t=p+1}^{n} \frac{1}{\sqrt{\det(V_e)}} \exp\left\{-\frac{1}{2} e_t' V_e^{-1} e_t\right\}. \quad (4.43)$$

The second factor is

$$f(x_1, \ldots, x_p) = \frac{1}{\sqrt{\det(\tilde{G}_p)}} \exp\left\{-\frac{1}{2} X_p' \tilde{G}_p^{-1} X_p\right\}. \quad (4.44)$$

This can be further simplified in a similar manner as (4.43), using the innovations $e_{k,t}$ and their variance matrices $V_{k,e}$ defined in (2.39) and (2.40) as

$$\prod_{t=1}^{p} \frac{1}{\sqrt{\det(V_{k,e})}} \exp\left\{-\frac{1}{2} e_{k,t}' V_{k,e}^{-1} e_{k,t}\right\}. \quad (4.45)$$

Here, k is taken to be $t-1$ within the product to indicate that $e_{k,t}$ is the innovation in x_t, with variance $V_{k,e}$, based upon the finite set of past values $x_1, \ldots, x_{t-1}$. Together, these expressions give the deviance as a function of the whole set of coefficients Φ and the innovation variance V_e:

$$\text{Dev}(\Phi, V_e) = (n-p) \log \det(V_e) + \sum_{t=p+1}^{n} e_t' V_e^{-1} e_t \quad (4.46)$$

$$+ \sum_{t=1}^{p} \left\{\log \det(V_{k,e}) + e_{k,t}' V_{k,e}^{-1} e_{k,t}\right\}.$$

We remark that:

1. We call the quantity on the right of the first line of (4.46) the GLS deviance of the model. Minimizing this alone with respect to the coefficients Φ, in which it is quadratic, gives the generalized least squares equations (4.38) described in the previous section.

2. The quantities on the second line of (4.46) are all functions of the same parameters Φ and V_e from which they may be derived using methods described in Section 2.8. However, their sum is not quadratic in Φ. In fact, it increases without limit towards the boundary of the stationarity region of the model. This ensures that the deviance has a minimum within this region.

3. Although the likelihood is based upon the distributional assumption of normality, it may be used under much wider conditions, and we will refer to it generally as the Gaussian likelihood. In particular, assuming only that the series x_t has the second moment properties of a true VAR(p) model, for large n the deviance will have its minimum at the parameter values of this model. Under more, but not too restrictive, conditions, the MLEs will have the large sample distribution presented in (4.36).

A useful procedure which may be used to maximize a likelihood is to find the maximum with respect to a subset of the parameters, the resulting values of which will generally depend upon the remaining parameters. The likelihood so evaluated will be a function only of the remaining parameters. We call it the *profile*, or *concentrated*, likelihood of these parameters, and say that we have concentrated out the original subset. This procedure is particularly useful if it can be carried out analytically. The profile likelihood might then be maximized numerically with respect to the remaining parameters and the overall set of MLEs thereby obtained. The same procedure can be applied to the deviance, but using minimization.

Consider the GLS deviance term on the right of the first line of (4.46). This may also be expressed in terms of the sample value $\widehat{V}_e$ of V_e, given in (4.33), as

$$\text{Dev}_{GLS}(\Phi, V_e) = (n - p) \log \det(V_e) + (n - p)\text{trace}(\widehat{V}_e V_e^{-1}). \qquad (4.47)$$

This deviance is minimized with respect to V_e for any given Φ by setting V_e equal to $\widehat{V}_e$. Omitting a term $m(n-p)$ as being constant, we obtain the profile deviance of Φ as

$$\text{Dev}(\Phi) = (n - p) \log \det \widehat{V}_e. \qquad (4.48)$$

Consequently, the MLE of Φ may be found by minimizing $\det(\widehat{V}_e)$ as a function of Φ. The practical procedure we proposed in the previous section was, however, based on the fact that the GLS deviance can also be analytically minimized with respect to Φ for fixed V_e by solving the GLS equations (4.38). Alternately minimizing with respect to V_e and Φ rapidly locates the overall MLEs. The final value of $\det(\widehat{V}_e)$, evaluated at $\widehat{\Phi}$, is a commonly used *measure of fit* of the model and is a natural extension of the sum of squares function used to assess models in ordinary least squares estimation. It is used in the deviance (4.32) to define the AIC (4.31) for a model of order k.

There is no analytic means of readily minimizing the total deviance (4.46) with respect to either V_e or Φ, and their MLEs are found by numerical minimization of the deviance. The deviance can, however, be further simplified for this purpose as

$$\text{Dev}(\Phi, V_e) = \sum_{r=1}^{N} \left\{ 2 \log \sigma_r + \frac{a_r^2}{\sigma_r^2} \right\} \qquad (4.49)$$

in which $N = mn$ and a_r is the scalar sequence of orthogonalized terms of the innovation vectors $e_{0,1}, \ldots, e_{p-1,p}, e_{p+1}, \ldots, e_n$, with variance σ_r^2. For a typical vector e with variance V, orthogonalization is achieved by factorizing $V = LL'$ with lower triangular L and setting $D = \text{diagonal}(L)$. Then $a = DL^{-1}e$ is the orthogonalized vector with variance D^2. Each element of a is the error from the projection of the corresponding element of e onto its previous elements.

The values a_r and σ_r^2 can also be derived by direct orthogonalization of the stacked series vector X_n with variance matrix $\tilde{G}_n$ as given in (2.38). This is not as efficient numerically, though it can be quite practical for a short series. Application of the Kalman filter to the state space form (2.72) of the VAR(p) model can also be used to generate these values. One advantage of both of these methods is that they work just as well when the series contains missing values, whereas the previous method fails. The elements corresponding to the missing values are simply omitted from X_n and $\tilde{G}_n$ when using the direct orthogonalization and the sequence length N reduced accordingly. The Kalman filter can be used to construct the projection of each element of the series x_t upon previous values in the sequence, with unobserved values simply omitted from the sequence. The orthogonalized values are the projection, or prediction, errors. This form of the likelihood or deviance as expressed in (4.49) is known as the *prediction error decomposition* form.

Although the total deviance (4.46) cannot be readily concentrated with respect to the full matrix V_e, it can be concentrated with respect to a scalar factor of V_e. This is most usefully described using the form (4.49). If we express $V_e = \sigma^2 T_e$, then $\sigma_r = \sigma\,\tau_r$, and (4.49) is minimized by $\hat{\sigma}^2 = \sum(a_r^2/\tau_r^2)/N$. The profile deviance value $\mathrm{Dev_{pr}}$ so obtained can be expressed in terms of the original a_r and σ_r^2, and omitting the constant $N - N\log N$, as

$$\mathrm{Dev_{pr}}(\Phi, V_e) = N\log\left\{\sum_{r=1}^{N}\left(\frac{G\,a_r}{\sigma_r}\right)^2\right\} \tag{4.50}$$

where the geometric mean of the standard deviations is

$$G = \left(\prod_{r=1}^{N}\sigma_r\right)^{1/N}. \tag{4.51}$$

The argument of the logarithmic function in (4.50) is a sum of squares function of the parameters which again can serve as a useful natural measure of fit of the model. We will call it the pseudo-deviance $\mathrm{Dev_{ps}}$. It reduces to the classic sum of squares of residuals if all the standard deviations are identical. It is also amenable to the application of standard non-linear least squares methods for efficient numerical minimization, though it should be noted that it is invariant to simple scaling of V_e. It is usual therefore to constrain one element of V_e to any fixed value. Then the value of V_e at the minimum is multiplied by $\hat{\sigma}^2 = \sum(a_r^2/\sigma_r^2)/N$ to obtain its MLE. In practice, we will use the expression in (4.49) in efficient iterative numerical minimization procedures based on scoring techniques, which also give good estimates of the variance matrix of the estimated parameters using large sample maximum likelihood theory. These procedures usually converge rapidly, especially if initial parameter values are supplied using the Yule–Walker estimates. It can be useful to monitor the progress of the iterations using both the deviance and pseudo-deviance. Upon convergence these are related by

$$\mathrm{Dev}(\hat{\Phi}, \hat{V}_e) = N\log\mathrm{Dev_{ps}}(\hat{\Phi}, \hat{V}_e) + N - N\log N. \tag{4.52}$$

This is useful because $N \log \mathrm{Dev_{ps}}(\widehat{\Phi}, \widehat{V}_e)$ may be used instead of the deviance in the AIC of (4.31) to select the model order and is a natural generalization of using $N \log$(residual sum of squares) in applications of the AIC to the selection of multiple regressions. Omitting the term $N - N \log N$ in (4.52) does not affect the order selection.

We illustrate maximum likelihood estimation using the Pig market series. This is an appropriate example because the series are short, with only 48 points in each, and the number of series is relatively large. The number of coefficients in the prediction of each series by a VAR(p) model is therefore $5p$. So for a model of order $p = 6$ there are only $48 - 30 = 18$ residual degrees of freedom for each series. Using lagged regression would reduce this by a further 6, which is avoided by using maximum likelihood. The upper line in Figure 4.4(a) shows the AIC plotted for the VAR(p) models fitted up to a maximum order of 6. However, the AIC represented by this line has been modified by a factor developed by Hurvich and Tsai (1989). This replaces the penalty term $2M$ used in (4.31) to define the AIC by

$$N \frac{1 + M/N}{1 - (M + 2)/N} \qquad (4.53)$$

where N is the total number of data, which is nm for our VAR model. For large N, this penalty is close to $N + 2M$, which is equivalent to the AIC in (4.31), because N is constant and does not influence the minimum order. The lower line in Figure 4.4(a) shows the AIC without this modification, demonstrating its inadequacy as an order selection criterion. The modification to the criterion is closely related to the use of residual degrees of freedom in place of the sample size in estimating the residual variance from a multiple linear regression. However, the AIC, even when modified, has a further limitation. Suppose, as the number of model parameters increases, a long succession of

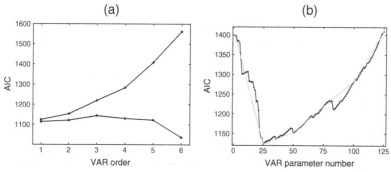

Figure 4.4 *The AIC for VAR(p) models fitted to the five Pig market series: (a) the modified AIC (upper line) and unmodified AIC (lower line) for orders p from 1 to 6 and (b) the modified AIC for models increased successively by one coefficient in turn up to a VAR(5), with the values for complete coefficient matrices linked by the gray line.*

insignificant parameters is fitted, typically parameters that are truly zero, though this would not be known. Then the AIC will increase steadily through this succession. Then, if a highly significant parameter is reached, one that has an important role in the model, the AIC will reduce substantially. This reduction may not, however, be sufficient to offset the previous increase. The AIC minimum may then occur before that parameter is included, so that it is excluded by this strategy of model selection. This pattern can occur particularly in the context of time series. It is illustrated in the present example by Figure 4.4(b), which shows the modified AIC for models increased successively by one coefficient in turn up to a VAR(5). The coefficients are introduced row by row for each successive coefficient, so the plot would change with the choice of sequence. However, it is strongly suggestive of significant parameters within the lag 2 and lag 4 coefficients. Fitting a VAR(4) model results in two coefficients at lag 2, one at lag 3 and one at lag 4, with t values greater than 1.9, the largest of these being the t value of -2.45 for the estimate of $\Phi_{2,3,4}$. These are saturated models fitted to highly collinear variables. The sparse structural model fitted to these series in Chapter 5 contains several more significant coefficients at these lags.

An alternative to the AIC is to plot the reduction in deviance, ΔDev, associated with the introduction of each successive coefficient. In large samples, this is distributed as the square of a standard normal variable, under the hypothesis that the coefficient has been added to an already complete model, and so has true value zero. Again, in small samples this distribution needs modification similar to that of the AIC. We transform the deviance reduction to the half-normal variable $\sqrt{\Delta\text{Dev}(N-M)/N}$, where N and M are as in (4.53). Figure 4.5 shows the plot of these values with the limit of 2.0 to indicate approximate significance. This procedure is very similar to plotting the sample partial autocorrelations of a univariate time series, as advocated by Box and Jenkins (1970). Although most points above the limit in Figure 4.5 correspond to lag 1 coefficients, this plot suggests that a model for this series should include further selected coefficients up to lag 4.

In the next chapter we describe an approach to the selection of the coefficients in a sparse structural VAR(p) model for these series, using the estimated

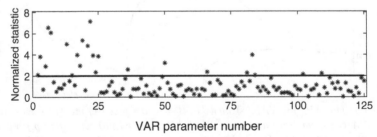

Figure 4.5 *The reduction in deviance for models increased successively by one coefficient in turn up to a VAR(5), transformed to half normal statistics.*

partial correlation graph between the variables. However, even this depends upon the specification of the total set of variables to be considered, i.e., the overall model order p. From the evidence of this investigation, we will specify variables up to a maximum lag of $p = 4$ for the Pig market analysis.

We conclude this section by showing in Figure 4.6 forecasts of the Pig market series that are obtained from fitting a VAR(2) model to the first 8 years of the series using maximum likelihood. The forecasts are reasonably accurate for several quarters into the future, with realistic error limits. This model has a total of 65 parameters, including the innovation variance matrix. In the next chapter we develop a model with a total of just 30 coefficients, including some at lags 3 and 4. The advantage of this is not especially in improved prediction, but in a more efficient and meaningful representation of the structure of the relationships between the series.

4.6   VAR models with exogenous variables, VARX

Among our examples is that of the weekly series of average daily moth trappings shown in Figure 1.4, in which the main variable of interest, the moth trapping counts, are believed to be influenced by three environmental variables: solar radiation, temperature and rainfall. These three variables are described as *exogenous* to the life cycle of the moths: they are generated outside that system. The environmental variables may be interdependent, but they are not themselves influenced by the moth trapping numbers, which are described as *endogenous*: generated within the system of the life cycle of moths. These descriptive terms, exogenous and endogenous, are widely used in econometric modeling and elsewhere, though more formal definitions may be presented. For example, Hamilton (1994, p. 225) defines an endogenous variable to be one that is correlated with the regression error term. This is illustrated by a cautionary note regarding estimation bias that can arise in regression with autocorrelated errors. The series of infant monitoring records shown in Figure 1.3 are all endogenous to the circulatory system. Their representation by a VAR(p) model avoids this difficulty yet completely characterizes the system.

Another example with exogenous variables is that of the Sales series shown in Figure 1.5. The Price and Promotion series can be considered as exogenous to explain the endogenous Sales series. There is some possibility that Price and Promotion could be influenced by recent innovations in Sales, which would cause them to become endogenous. However, this would give rise to apparent non-causal dependence of Sales upon future values of Price and Promotions, and the spectral analysis shown in Figure 3.31 gives no obvious indication of this. In the analysis of the Gas furnace example in the previous section, with estimated responses shown in Figure 4.3, the input and output series can be considered to be, respectively, exogenous and endogenous because the data were generated in an open loop experiment. For our purposes, a series is exogenous if there is no feedback to it from the endogenous series.

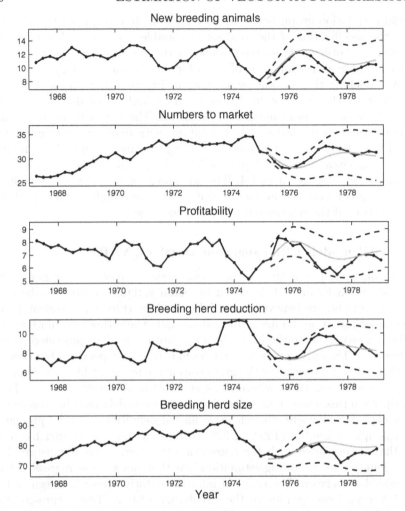

Figure 4.6 *Forecasts of the Pig market series using a VAR(2) model fitted to the first 8 years. The series are shown by the solid black lines, forecasts of the last 4 years by the solid gray lines and limits by the broken lines.*

We therefore consider now how we formally modify our model building when the series are partitioned into a set of exogenous and a set of endogenous series, which we will call, respectively, x_t and y_t, each of which may have one or more components. If the combined set of series is stationary, the main implication of the assumption that x_t is exogenous is that it has no causal dependence upon y_t, i.e., the present and lagged values of y_t are uncorrelated with the innovations in x_t when predicted from its own past alone. A VAR model representing them is then of the form

$$\begin{pmatrix} \Phi_{yy}(B) & \Phi_{yx}(B) \\ 0 & \Phi_{xx}(B) \end{pmatrix} \begin{pmatrix} y_t \\ x_t \end{pmatrix} = \begin{pmatrix} e_t \\ f_t \end{pmatrix}, \tag{4.54}$$

where e_t and f_t are mutually uncorrelated white noise series. From the stationarity condition, we may deduce that $\Phi_{y,y}(B)$ and $\Phi_{x,x}(B)$ are, individually, stationary VAR operators. However, it may not be reasonable to assume that x_t has the structure of stationary time series. For the Sales series example, the Price and Promotion series may be determined by a range of marketing strategies. It might, even so, be reasonable to assume that Price is trend stationary, i.e., stationary after correction by a linear trend. The Promotion series has a generally similar appearance throughout, but varies considerably in magnitude, with a 20 week period of zeros partway through, so cannot be claimed to have the truly homogeneous behavior typical of a stationary series. The indicator series introduced to account for the Christmas effect in the spectral analysis is deterministic.

In general, the main interest is in the dependence of the endogenous series on the exogenous series and its own past, so that we restrict attention to estimation of the VARX model, as it is known, extracted from the first line of (4.54) as

$$\Phi_{yy}(B)y_t = -\Phi_{yx}(B)x_t + e_t. \tag{4.55}$$

This may be readily estimated by the lagged regression method for VAR models, as described in Section 4.4. Provided the same explanatory terms are present for each component of y_t, OLS regression may be used; otherwise the GLS equations (4.38) should be used, but with the weighting matrix and parameter sets restricted to the response and regressors present in (4.55).

The first step of model building is modified by applying the AIC to the likelihood based upon the estimated innovations e_t derived from fitting the VARX model (4.55). This is in contrast to the likelihood from fitting the VAR model (4.54). We will use this AIC to select the same order p of $\Phi_{yy}(B)$ and $\Phi_{yx}(B)$, although it is possible to allow and select different orders, say p and q, for these operators. The same selection of orders *can* be made using the AIC from fitting the full VAR model (4.54) provided the operator $\Phi_{xx}(B)$ is fixed and not estimated and the correlations between e_t and f_t constrained to zero. The log-likelihood or deviance will then differ from that obtained by fitting (4.55) only by a constant due to the separate contribution from the model for x_t. The selection of order by minimum AIC will then be the same. This approach can be useful if y_t and x_t are jointly stationary but have some missing values. The prediction error decomposition form of the likelihood (4.49) can then still be constructed by application of the Kalman filter jointly to y_t and x_t.

We first illustrate the construction of VARX models using the series shown in Figure 1.9 for the plastic Film extrusion process. The data were collected from an open loop experiment in which the two input series were manually set to follow independent white noise sequences. All the series were sampled automatically at a somewhat higher rate than the input sequence was varied, so that the recorded values of the input were correlated to low lag. The order $p = 8$ was selected using AIC for the VARX model relating the two output series to the two input series. In this context the model properties of greatest interest are the open loop impulse responses from the inputs to the outputs. Calculation of the open loop response from a multivariate input x_t to a multivariate output x_t is a simple generalization of that for a single input and output given in (2.36). The relationship (4.55) can be re-written as

$$
\begin{aligned}
y_t &= -\Phi_{yy}(B)^{-1}\Phi_{yx}(B)x_t + \Phi_{yy}(B)^{-1}e_t \\
&= \nu(B)x_t + n_t
\end{aligned}
\tag{4.56}
$$

where

$$
\nu(B)x_t = \nu_0 x_t + \nu_1 x_{t-1} + \nu_2 x_{t-2} + \cdots
\tag{4.57}
$$

and n_t is the noise in the output series that is present even in the absence of any variation in the input series.

In our example ν_k is a 2×2 matrix and Figure 4.7(a) shows the estimated sequences $\nu_{1,1,k}$ for the response from the Heater current to Extrusion pressure, with the estimated two standard error limits plotted around the zero axis. This response is similar to the lagged regression coefficients obtained

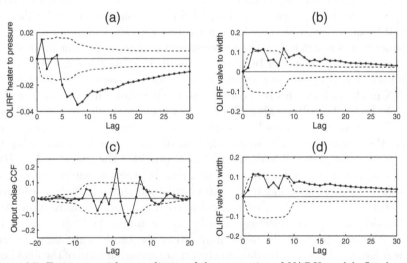

Figure 4.7 *Estimates with error limits of the properties of VARX models fitted to the Extrusion process series: (a) the OLIRF from Heater to Pressure, (b) the OLIRF from Valve to Width, (c) the cross-correlations (CCF) of the output noise in Pressure and Width, (d) the OLIRF from Valve to Width from the model excluding Pressure.*

by spectral analysis and shown in Figure 3.30. Figure 4.7(b) shows the estimated sequences $\nu_{2,2,k}$ for the response from the Valve to extrusion Width. This same sequence can also be estimated by fitting a VARX model to the *single* output of Width, with the same two inputs, and this is shown in Figure 4.7(d). The two estimates are very similar, although the computational steps in their derivation are rather different. In the VARX model that produced Figure 4.7(b), Width is regressed upon lagged values of the two inputs *and* the output series Pressure, besides itself. In the VARX model that produced Figure 4.7(d), lagged values of Pressure were *not* included, and the order of model selected by AIC was $p = 9$.

Fitting the model with both inputs and outputs has the advantage of characterizing also the bivariate structure of the output noise series n_t in (4.56) as a VAR model with operator $\Phi_{yy}(B)$. If the components of the noise series are correlated, this can be important in the design of an efficient control scheme for the system. In fact, there are significant cross terms in the estimated coefficients of $\Phi_{yy}(B)$, relating the noise in the output Width to lagged values of the noise in the output Pressure. Figure 4.7(c) shows the lagged correlations between these series, inferred from this model, after each series has had its low frequency trend removed. This adjustment reveals significant low lag relationships between the output noise series which would otherwise be masked by the low frequency movements.

This Extrusion process had proved difficult to control, the objective being to maintain a steady state target value for the output Width by feedback control adjustment of the inputs. The responses in Figures 4.7(a,b) are slow to decay, indicating that control actions have a persistent long run effect. Control actions also have a quite variable short term effect. Because of the lagged cross-correlation between the two output noise series, control action to compensate for short term disturbance in Pressure needs to anticipate a consequent disturbance in Width, but avoid a shift away from the target value that may arise from the long run effect of the action. The process model may be used to design optimal control schemes that minimize mean square deviation of the outputs from target values, or to provide computer simulation experiments for developing standard robust control schemes.

Our second example to illustrate the construction of VARX models is very different: to model the response of the Sales series shown in Figure 1.5 to the Price and Promotion series, taking into account also the effect of Christmas. There is one practical difference in this example compared with the previous example, that we will allow for the possible simultaneous response of Sales to Price and Promotion, i.e., coefficients at lag zero in $\Phi_{yx}(B)$ and $\nu(B)$ in (4.56). This could be avoided by re-aligning the series so that the Price and Promotion at time t are instead associated with time $t - 1$, but it is simple and natural to allow for simultaneous as well as lagged dependence in the regression on exogenous variables. This also anticipates the structural models of the next chapter which allow for more general simultaneous dependence between multivariate time series. We will also allow simultaneous dependence upon an indicator variable for Christmas and introduce indicator variables for the significant effects shown by the spectral analysis in Figure 3.31, for each of the three weeks previous to Christmas. Finally, we will introduce a single indicator variable for the downturn in sales in the weeks following Christmas, with decay in this effect being modeled by the dynamics of lagged dependence.

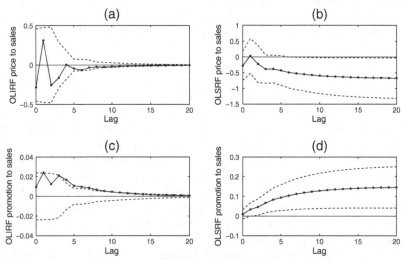

Figure 4.8 *Estimates with error limits of the properties of the VARX model fitted to the Sales series: (a) the OLIRF from Price to Sales, (b) the OLSRF from Price to Sales, (c) the OLIRF from Promotions to Sales and (d) the OLSRF from Promotions to Sales.*

Figure 4.8 shows the estimated response functions from Price and Promotion to Sales, derived from VARX modeling of these series and using the order $p = 3$ selected by the AIC. There was also a pronounced dip in the AIC at lag 7, associated with a significant coefficient ($t = 2.6$) in the dependence of Sales on the post-Christmas effect, but the estimated responses from Price and Promotion using this order are very similar to those using the order 3 model, which are shown in the figure. This shows a marginally significant negative eventual response to a step increase in Price and a more clearly positive eventual response to a step increase in Promotion. The model can be used to assess the practical and statistical significance of the sensitivity of Sales to Price and Promotion by experimental variation of these inputs. As discussed when we modeled these series by spectral analysis in Chapter 3, it is important to include lagged effects to get a reliable indication of the eventual step response or gain.

In both these examples, the open loop impulse and step responses are important for the objectives of control in the first example and assessing the impact of marketing strategy in the second. The transfer function models espoused by Box and Jenkins (1970) have been found to be very effective for parametric modeling of the pattern of such responses, together with the

structure of the output noise. The patterns of the response sequences estimated by spectral analysis and VARX modeling are valuable for indicating
the parametric forms of these transfer function models. The parametric form
of the VARX model may sometimes provide an efficient representation of the
dependence of the output on the inputs, but it is limited in its capacity to
represent a range of different response patterns from several inputs. This is
because of the common factor $\Phi_{yy}(B)$ that divides through the responses of
all the input series and the noise on the RHS of (4.56). Thus, in the simple case
that $\Phi_{yy}(B) = 1 - \alpha B$, the implied responses to each input will eventually
decay away geometrically with decay factor α, as will the autocorrelations of
the noise.

In practice, it is quite possible to have very different patterns of response to
the inputs, for example, as shown in Figure 3.25, the response of Moth count
to Temperature is simultaneous with no lag, whereas the response to Rainfall
is significant over a period of many weeks. The correlations of the remaining
model error, or noise, are not shown, but decay geometrically over two or three
weeks. Our working principle, that a sufficiently high order VAR model will
provide a sufficiently precise approximation to the dependence structure of
the series, is valid, but for some examples, such as the moth trapping series,
a sufficiently high order model may not be selected by the AIC unless the
series are very long. The selected model may then give a poor approximation
to the structure. A particularly difficult scenario arises when the output noise
is close to white and accounts for a substantial proportion of the output series
variance. An effective approach to building transfer function models, of the
same form as Box and Jenkins (1970), is then to use the instrumental variable
methodology presented in Young (2011).

Nevertheless, there are many examples, such as the Pig market series,
in which the VAR and VARX models provide a natural and quite adequate
representation of the dependence between time series using models of relatively
low order. The further methods introduced in the next two chapters extend
the scope and efficiency of the VAR models so far discussed.

We end this section with a comment on how we model integrated processes,
also known as unit root processes. We introduce these as borderline cases of
stationary VAR models in which one or more of the zeros of $\det \Phi(z)$, or roots
of $\det \Phi(z) = 0$, approaches unity. The classic and most simple such case is
the univariate random walk model $x_t = \phi x_{t-1} + e_t$, in which ϕ approaches
1. Such processes have no fixed mean, and their variance increases through
time. The presence of a unit root implies that the first difference of the series
should be modeled as stationary. Very considerable effort has been devoted to
the formulation and evaluation of tests of the hypothesis that a process can
be represented by an integrated, rather than a stationary model; see Said and
Dickey (1984) and Phillips and Perron (1988). Furthermore, in the multivariate context, there is the possibility of co-integration, where each of several
series may be integrated yet there are one or more linear combinations of
them that are stationary; see Engle and Granger (1987). These concepts are

particularly important in econometrics, where they describe long term relationships between time series. If series are integrated or co-integrated, then some advantage can be gained in applying these properties in the context of long term prediction.

A cascade of research and publications has followed upon the appearance of the papers just cited, so it may be questioned why we give little attention to these issues. Our strategy is to recommend that, where there is concern that a process may possibly be integrated, the model is fitted by maximum likelihood rather than simple regression. More particularly, we recommend that a restricted form of (maximum) likelihood is used, referred to as REML or sometimes as marginal likelihood. Recent research, see Chen and Deo (2009, 2012) shows that this approach results in reliable inference on and close to the integration boundary of the stationarity region where the model becomes integrating. Earlier demonstrations of the value of using this form of likelihood in the time series context are Cooper and Thompson (1987), Corduas (1987) and Tunnicliffe Wilson (1989). In the first example of this section, the Extrusion process, the output series appear to be drifting away from their original set point, in the manner of an integrated process, but we model them as stationary. In the second example, the Sales and Price series are trending, but we model them as stationary about their trends. An alternative would be to model their first differences as series with non-zero means to account for the trend. In a series of finite length, no certain inference can be made as to which of these alternatives is correct. However, using restricted likelihood, the first of these models, known as trend-stationary, encompasses the second model and can be used for reliable inference even as the boundary of integration is approached.

4.7 The Whittle likelihood of a time series model

Whittle (1963, Thm. 6) expresses the likelihood for the parameters in a model for a stationary multivariate time series of length n in terms of the sample spectrum of the observed series. Using our notation, this theorem presents a close approximation to the deviance (4.46) we defined for the VAR model, in terms of the sample spectrum $S_x^*(f)$ and the model spectrum $S_\Phi(f)$ given, respectively, by (4.15) and (4.23). We will express it using a generic parameter set β as

$$\text{Dev}_W = n \int_{-0.5}^{0.5} \log \det S_\beta(f) + \text{trace}\left\{S_x^*(f)S_\beta(f)^{-1}\right\} df. \qquad (4.58)$$

This expression, known as the *Whittle likelihood*, but which we will refer to as the *Whittle deviance*, is valid for a wide range of models with spectrum $S_\beta(f)$. It has both theoretical and practical value, with the integral evaluated as a finite sum over a discrete grid of $N \geq n$ equally spaced frequencies. As in Chapter 3, we recommend a choice of N close to $4n$, but a useful intuitive explanation of (4.58) derives from taking $N = n$. It is then the deviance of the discrete set of harmonic contrasts of the series, defined in (3.2), under the Gaussian assumption and assuming independence at the harmonic frequencies.

For the VAR(p) model with spectrum given by (4.23), if the innovation variance matrix V_e is a free parameter, it may be concentrated out of (4.58) giving (after removal of a constant) the deviance in terms of the remaining parameters Φ_k as

$$\text{Dev}_W = n \log \det \int_{-0.5}^{0.5} \Phi(\exp -2\pi i f) S_x^*(f) \Phi(\exp 2\pi i f)' df. \qquad (4.59)$$

Minimizing this expression with respect to the coefficients Φ_k and utilizing (4.16) gives the Yule–Walker equations (2.5) in the sample covariances $C_{x\,i,j,k}$ for estimation of the model.

The Whittle likelihood has been used recently, for example, by Robinson (2008), to study estimation of models using information in the spectrum at low frequencies. We will interpret an estimation procedure we use in Chapters 6 and 7 in terms of minimizing the Whittle deviance (4.58) but with greater weight being given to lower frequencies in the spectrum.

If we replace the sample spectrum $S_x^*(f)$ in (4.58) by the true spectrum $S_x(f)$ of x_t and remove the series length n, we obtain a quantity that measures the discrepancy between the true spectrum and the model spectrum $S_\beta(f)$. This quantity takes its minimum only when the true and the model spectrum coincide, and after correcting for this minimum we obtain

$$\int_{-0.5}^{0.5} \text{trace} \left\{ S_x(f) S_\beta(f)^{-1} \right\} - \log \det \left\{ S_x(f) S_\beta(f)^{-1} \right\} df - m. \qquad (4.60)$$

We call (4.60) the *information divergence per observation*, or simply, divergence, of the model spectrum $S_\beta(f)$ from the true spectrum $S_x(f)$. It is positive except that it takes the value zero when the true and the model spectrum coincide. If the model is a VAR(p), minimizing (4.60) with respect to the coefficients Φ_k gives the Yule–Walker equations (2.5) in the true covariances $\Gamma_{x\,i,j,k}$ for the autoregressive approximation to the series. If the series x_t has Wold representation (2.12) with innovation variance V_e and an approximating VAR(p) model has error variance $V_{p,e}$, then the divergence can also be expressed as

$$\text{trace} \left\{ V_e V_{p,e}^{-1} \right\} - \log \det \left\{ V_e V_{p,e}^{-1} \right\} - m. \qquad (4.61)$$

Chapter 5

Graphical modeling of structural VARs

5.1 The Structural VAR, SVAR

In this chapter we describe structural VAR models and how they can be constructed using the methods of *graphical modeling*. The form of the structural VAR (SVAR) that we will use differs primarily from that of the canonical VAR model by including regression terms for the dependence between current variables in addition to dependence of current on past variables. These terms explain any correlation between the innovations of the canonical VAR so that the residual series, or structural model innovations, from the SVAR model will be uncorrelated with each other. Our aim is also that structural models be sparse, in the sense that they represent the dependence using a relatively small number of model terms or coefficients. It is reasonable to believe that in a model that reflects the true structure of a system of inter-related series, the current value of each series should depend on a relatively small number of other current and lagged values.

Identifiability is a major concern with structural models. For the canonical VAR model, it was necessary only to identify the order to identify a unique model. Structural modeling in econometrics has traditionally relied strongly on economic theory to specify the terms in the model, with constraints to ensure that it can be uniquely identified. This is because there are many possible structural models which could equally well represent the observable statistical properties of the time series. However, our approach to the identification of structural VAR models is largely empirical; we would use contextual information where available and relevant to help guide us to an appropriate model, but our procedure is mainly to use appropriate statistics to identify the sparse structure of a well-fitting model. It is based on the theory of graphical modeling, and the statistics it uses are the partial correlations between current and lagged values of the series.

5.2 The directed acyclic graph, DAG

It is natural to represent causal relationships among variables in a diagrammatic manner using *directed graphs* in which an arrow *links* each variable to the others which it affects causally. The books by Whittaker (1990), Lauritzen

(1996) and Edwards (2000) provide accessible accounts of general statistical modeling procedures that are based on these representations of dependence. This topic has been an extremely active subject of research in recent years, but we limit ourselves to considering the specific application to structural modeling of multivariate time series. An application in this area which makes reference to the wider literature is found in Awokuse and Bessler (2003).

The canonical VAR, presented in the previous chapter, models the current observations of the multivariate series at time t through the dependence on previous (lagged) observations of the time series themselves. A visually effective way to present this model for a set of three series is by the graph in Figure 5.1, where each of the six variables is a *node*.

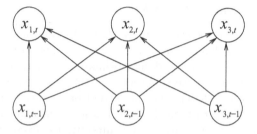

Figure 5.1 *Graphical representation of a saturated canonical VAR(1).*

In a VAR model, all current variables at time t, e.g., $x_{1,t}$, $x_{2,t}$ and $x_{3,t}$, depend on all past variables: in the case of a VAR(1), the variables at time $t - 1$. Such dependence in the graph is represented by the arrows or *directed edges* linking the nodes representing current and past variables, where the direction is suggested by the flow of time. We can simply indicate this complete dependence on the past by saying that the VAR is *saturated* on past variables. As discussed in the previous chapter and as is apparent from the graph, in the VAR there are no terms describing dependence between current variables and, because of that, typically the innovations are correlated. The dependence between the innovations in the VAR reflects the conditioning of the current variables upon all the past variables. Including explicit terms for the dependence between current variables provides an alternative way to capture the dependence between innovations. We use as an initial example the monthly Flour price series shown in Figure 1.2, for which the criteria AIC, HQC and SIC all selected an order 2 autoregression. The series were of simultaneous prices in the three cities of Buffalo, Minneapolis and Kansas City, and it is unsurprising that the innovations from the canonical VAR(2) were highly correlated, as shown by their correlation matrix:

$$\begin{pmatrix} 1.0000 & 0.9664 & 0.8700 \\ 0.9664 & 1.0000 & 0.8976 \\ 0.8700 & 0.8976 & 1.0000 \end{pmatrix}$$

By a structural VAR (SVAR) we will therefore mean a multivariate time series model of the form

$$\Phi_0 x_t = \Phi_1 x_{t-1} + \Phi_2 x_{t-2} + \cdots + \Phi_p x_{t-p} + a_t, \tag{5.1}$$

where in contrast to the (canonical) VAR, the non-singular matrix Φ_0 has rows by which current values of the elements of x_t are related, and the elements of the residuals a_t are assumed to be uncorrelated. It is immediate that every model of the form (5.1) can be transformed to a canonical VAR by dividing through by Φ_0, i.e.,

$$
\begin{aligned}
x_t &= \Phi_0^{-1}\Phi_1 x_{t-1} + \Phi_0^{-1}\Phi_2 x_{t-2} + \cdots + \Phi_0^{-1}\Phi_p x_{t-p} + \Phi_0^{-1}a_t \quad (5.2)\\
&= \Phi_1^* x_{t-1} + \Phi_2^* x_{t-2} + \cdots + \Phi_p^* x_{t-p} + e_t. \quad (5.3)
\end{aligned}
$$

The result is known as the *reduced form* VAR (Lütkepohl, 1993, pp. 54, 325). The residuals a_t of the SVAR are therefore related to the innovations e_t of the VAR, and the general covariance matrix V of e_t to the diagonal covariance matrix D of a_t, by

$$\Phi_0 e_t = a_t \tag{5.4}$$

$$\Phi_0 V \Phi_0' = D. \tag{5.5}$$

Although any given SVAR determines a unique corresponding VAR reduced form, there are many SVAR models that correspond to a given VAR, resulting from the possible choices of Φ_0 that diagonalize V in (5.5). It is therefore our hope that the sparse parameterization of the SVAR will help us to identify just one, or possibly a small number of these, which can adequately represent the structure of the series. Note that, because the transformation (5.2) from the SVAR to the VAR depends on the inverse of Φ_0, a sparse SVAR is likely to have a reduced form VAR with much less sparse coefficient matrices. However, the inverse of the innovation covariance matrix is $V^{-1} = \Phi_0' D^{-1} \Phi_0$, which will reflect in part any sparse aspects of Φ_0. This is actually a special case of a property arising in the general theory of graphical modeling which we will shortly introduce and apply to the identification of SVAR models.

The word *structural*, as we use it for the SVAR (5.1), has a meaning somewhat distinct from its original use for a structural model in econometrics (Hurwitz, 1962). In that context a structural model should be able to predict the effects of interventions which are of the nature of changes in a variable or parameter. In fact, our SVAR has the form of a *simultaneous equation model*; see Zellner and Theil (1962). We will refer to this as a SimEM, because the abbreviation SEM is generally used for the wider class of structural equation models.

In the general SimEM, the elements of a_t are not explicitly required to be uncorrelated, but the model is only generally identifiable if a large number of constraints are imposed on the coefficients, typically zero values. Without these constraints, many different sets of parameters would equally well represent the observable statistical properties of the data. The dependencies are suggested by theoretical considerations underpinning the analyzed system of equations and are not empirically identified. For example, Zellner and Theil (1962) model the current relationships between six economic variables by using economic theory.

For our SVAR model we have assumed that the elements of a_t are uncorrelated, i.e., the covariance matrix D of a_t is diagonal, otherwise the model would certainly not be identifiable. However, we will also assume that

- each row of Φ_0 describes how one element of x_t depends on some (if any) of the other current elements;

- this dependence is recursive, not cyclical. This means that we could re-order the elements of x_t so that Φ_0 is upper triangular with unit diagonals: each element in turn depends on none, one or more elements that are *lower* in the ordering, taking the first as the highest. The ordering need not be unique; indeed, if there were no dependence at all, the ordering could be arbitrary.

The dependence between variables in an SVAR is conveniently illustrated by the sparse SVAR shown in Figure 5.2. We contrast this with the saturated VAR in the graph in Figure 5.1, noting that it includes directed edges linking current variables. The number of coefficients in this SVAR is the number, 8, of links, whereas the VAR in Figure 5.1 has 12 coefficients, including the 3 correlation coefficients between current variables, which are not explicit in the figure. These correlations are modeled in the SVAR by the links between current variables.

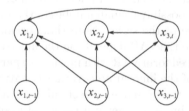

Figure 5.2 *An example of a graph representing a sparse SVAR(1).*

The resulting graph should be acyclic, i.e., it should *not* have any cyclic dependence between current variables. The graph in Figure 5.3 gives an example of the cyclic dependence that we exclude: it is possible by following the arrows from one variable eventually to return to the same variable; thus $x_{1,t}$ affects $x_{3,t}$, which affects $x_{2,t}$, which affects $x_{1,t}$.

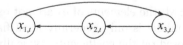

Figure 5.3 *A graph representing cyclic dependence.*

The diagram representing the SVAR, such as that in Figure 5.2, consisting of *nodes* with *directed edges*, is an example of a *directed acyclic graph* (DAG). Nodes (or their associated variables) that have a directed edge leading to a given node are known as the *parents* of that node, and the node itself as their *child*. Thus, in Figure 5.2, $x_{2,t-1}$, $x_{3,t-1}$ and $x_{3,t}$ are the parents of $x_{2,t}$. In general, a DAG completely determines the distribution of a set of variables, given, for each node, its distribution conditional upon its parents. Because we are assuming the normal linear model, this means that the linear regression of each current variable upon its parents fully determines the properties of the multivariate time series x_t. As we will see in Section 5.6, it also greatly simplifies the estimation of the model, allowing the use of ordinary least squares separately for the regression equation of each current variable. It is essential for this purpose that the graph be acyclic. However, we do reconsider in Section 5.9 the possibility of describing the dependence between current variables using a simultaneous equation model, for which an acyclic directed graph is not an appropriate description.

To confirm that the SVAR represented by a DAG does completely determine the series properties, consider starting at some initial time t_0 with a given set of values (say zeros) for the past variables $x_{t_0-1}, \ldots, x_{t_0-p}$. Then x_t may be simulated using the regression equation for each component current variable on other current and lagged values, provided this is carried out in the order determined by the DAG. For the SVAR illustrated in Figure 5.2, $x_{3,t}$ is first generated from $x_{2,t-1}$ and $x_{3,t-1}$, then $x_{2,t}$ from $x_{2,t-1}$, $x_{3,t-1}$ and $x_{3,t}$ and finally $x_{1,t}$ from $x_{1,t-1}$, $x_{2,t-1}$, $x_{3,t-1}$ and $x_{3,t}$. In this case it does not matter in which order $x_{1,t}$ and $x_{2,t}$ are generated, only that an acyclic ordering exists. Successive values x_{t_0}, x_{t_0+1}, x_{t_0+2}, $\ldots$ may be determined in this way, and, provided the reduced form of the model is stationary, the process so generated will reach the stationary equilibrium. This thought experiment also draws attention to the fact that the DAG as exemplified in Figure 5.2 is partial. To be complete, the same edges between the current components of x_t should also be shown between the components of x_{t-1}, and also the component variables of x_{t-2} should be shown as predictive for x_{t-1}. This extension could be continued indefinitely into the past and the future, but, of course, it is not necessary to do so. The DAG showing only the links between current variables and from lagged to current variables is sufficient for a process model that is stationary or invariant to a time shift.

The key to constructing an SVAR with a sparse structure that adequately represents the observed series is the determination of the (or an) ordering of the dependence of current variables using appropriate statistical summaries of the data. The theory of graphical models assists us once more; in the next section we describe how to estimate a conditional independence graph which can enable us to achieve this aim. To conclude this section, we remark that a saturated SVAR model is one which, for the chosen ordering of current variables, has no sparse structure, i.e., each current variable is regressed upon all those current variables that are lower in the ordering, and on all the lagged values up to the specified order. Such a model is exactly equivalent to the saturated VAR model, so it confers no advantage over the VAR, and any one such model is equivalent to any other based upon a different ordering of the current variables.

5.3 The conditional independence graph, CIG

We introduce this section by developing a DAG model for the dependence between the innovations of a multivariate time series, using the Flour prices as an example. Given that we have identified the order of a VAR model for the series and estimated its parameters, the innovations are effectively observable (subject to some sampling variability of the parameter estimates). We will go on to show how this can be extended to a DAG representing an SVAR model for the Flour prices themselves. This strategy is possible because, given a DAG representation of a multivariate times series x_t, such as illustrated in Figure 5.2, the innovations e_t from a canonical VAR fitted to the series can be described by a DAG having the subgraph restricted to the current variables. We are simply conditioning on all the lagged values; it is not necessary to know which lagged values are actually linked to current values. Figure 5.4 shows the subgraph of Figure 5.2 to illustrate this point, which we will shortly apply to the Flour price series.

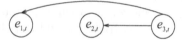

Figure 5.4 *The DAG representing the innovations of the SVAR(1) described by the DAG in Figure 5.2.*

In Section 5.1 we displayed the correlation matrix between the innovations of the Flour price series. We noted that the innovations were all highly correlated, suggesting that any one series of innovations can be predicted from the others. However, it is often the case with highly correlated sets of variables that some variables do not make a significant contribution to a prediction in the presence of other predictors, although they are correlated with the predicted variable. Taking each variable in turn as the predicted or response, and taking all the other variables together as predictors, we can determine which are relevant in the presence of all the others and, just as important, which are not. A graph may then be constructed in which the nodes (as for a DAG) correspond to the set of variables, and any two nodes (variables) are connected by an edge only if the first variable is dependent on the second when all the other variables are also included as predictors. In the general context of graphical modeling, we say that the edge between two nodes is *absent* if and only if the corresponding pair of variables are independent *conditional upon all the remaining variables in the set*. The resulting graph is then called a *conditional independence graph* (CIG). We note that the edges are not directed because by this definition the relationship between two nodes is symmetric, even though we introduced the idea by considering one variable as predicted and the other as one of a set of predictors. The CIG clearly represents structure of the variables that differs from that specified in a DAG, but we will see how it can be used to draw conclusions about possible DAG representations by exploiting the conditional independence property. Most important, it can be directly estimated from the data, with no requirement to specify an ordering of the variables.

In our context of the normal linear model, the conditional independence graph is fully determined by the covariance matrix V of the variables (or by its correlation matrix). In order to construct the CIG from this matrix, it is not necessary to determine the linear predictor of each variable in turn, from all the others, though in principle this could be done. The conditional independence between two normal variables, given the remainder, is determined by their partial correlation being zero.

We introduced partial correlations in Section 2.9. In a system of several variables, the partial correlation between two of them is just their correlation once the linear dependence of each of them on the remaining variables has been subtracted. This definition extends to non-normal variables, but it is convenient to retain the notation of multivariate normality where subtracting the linear dependence on other variables is equivalent to conditioning upon them. More formally, given a set of random variables $X = X_1, \ldots, X_n$, the partial correlation $\pi(X_i, X_j)$ between two random variables X_i and X_j included in X is given by

$$\pi(X_i, X_j) = \rho(X_i, X_j \mid X \setminus \{X_i, X_j\}), \tag{5.6}$$

where $X \setminus \{X_i, X_j\}$ is the set of all variables except X_i and X_j. We have the following connection with prediction, that in the minimum mean square error linear predictor of X_i from $X \setminus X_i$, the coefficient of X_j is zero if and only if $\pi(X_i, X_j) = 0$ and symmetrically the coefficient of X_i is zero in predicting X_j.

As previously presented in Section 2.9, for a set of random variables $X = X_1, \ldots, X_n$ with covariance matrix $V = \mathrm{Var}(X)$, we can directly construct the whole matrix of partial correlations as

$$\pi(X_i, X_j) = \frac{-W_{i,j}}{\sqrt{W_{i,i} W_{j,j}}}, \tag{5.7}$$

where $W_{i,j}$ are the elements of the matrix $W = V^{-1}$.

We can represent a summary of all the non-zero partial correlations by a diagram in which the nodes representing the random variables are linked if their partial correlation is different from zero. Such a diagram is called a *partial correlation graph*. If the variables are normally distributed, a zero partial correlation corresponds to conditional independence, i.e.,

$$\pi(X_i, X_j) = 0 \Leftrightarrow X_i \perp X_j \mid X \setminus \{X_i, X_j\}, \tag{5.8}$$

and in such a case the diagram is also the conditional independence graph (CIG). Figure 5.5 shows the CIG that we identify below for the Flour price series innovations. The implications of the CIG for determining a DAG representation of the dependence between the variables will be considered in the next section. The important point is that we can estimate the CIG using the sample partial correlation matrix between the variables and identifying those entries which, subject to sampling variability, may be judged to be zero.

We need then a test procedure to decide whether a non-zero partial correlation can be considered to be zero. A large value for a sample partial correlation suggests a non-zero partial correlation, so we need to determine critical values, i.e., thresholds above which a value is to be considered significantly different from zero, for a prescribed level of significance. The matrix of sample partial correlations is readily constructed from the sample covariance matrix of the data and the critical values for these are determined as follows:

1. Given a random sample of a set of normal random variables $X = \{X_1, \ldots, X_m\}$ with covariance matrix V, let the *data matrix* $\mathbf{X}$ consist of columns of length n of the mean corrected samples of X. The sample covariance matrix is then $\widehat{V} = \frac{1}{n}\mathbf{X}'\mathbf{X}$. The sample inverse covariance matrix is computed as $\widehat{W} = \widehat{V}^{-1}$ and by its entries, $\widehat{W}_{i,j}$, the sample partial correlations can be calculated as

$$\widehat{\pi}(X_i, X_j) = \frac{-\widehat{W}_{i,j}}{\sqrt{\widehat{W}_{i,i}\widehat{W}_{j,j}}}. \tag{5.9}$$

2. Under the hypothesis that $\pi(X_i, X_j) = 0$, the ratio

$$\frac{\widehat{\pi}(X_i, X_j)\sqrt{n - m + 1}}{\sqrt{1 - \widehat{\pi}(X_i, X_j)^2}} \tag{5.10}$$

is distributed as a t_{n-m+1} variable where $n-m+1$ is the number of degrees of freedom, because this ratio is the t-value of the estimate of the coefficient β_j in the regression

$$X_i = \beta_1 X_1 + \cdots + \beta_{i-1}X_{i-1} + \beta_{i+1}X_{i+1} + \cdots + \beta_m X_m + a_i \tag{5.11}$$

of X_i on $X \setminus \{X_i\}$.

3. We reject the null hypothesis that $\pi(X_i, X_j) = 0$ at level α if

$$|\widehat{\pi}(X_i, X_j)| > \frac{t_{\alpha/2, n-m+1}}{\sqrt{t^2_{\alpha/2, n-m+1} + (n - m + 1)}}, \tag{5.12}$$

where $t_{\alpha/2, n-m+1}$ is the corresponding critical value of the t_{n-m+1} distribution.

The estimated CIG is then constructed with links between the nodes X_i

and X_j only if this test is rejected. In large samples, say when $n - m > 30$, (5.12) can be well approximated by

$$|\widehat{\pi}(X_i, X_j)| > \frac{z_{\alpha/2}}{\sqrt{z_{\alpha/2}^2 + (n - m + 1)}}, \tag{5.13}$$

where z is the standard normal distribution.

In the graphical modeling literature, other tests for significance of partial correlations are used. For example, Whittaker (1990, p. 189) considers a test procedure based on the large sample properties of maximum likelihood estimates which leads to the asymptotically equivalent large sample distribution

$$-n \, \log[1 - \widehat{\pi}(X_i, X_j)^2] \sim \chi_1^2.$$

Tests like the ones described above strictly apply only to testing a single partial correlation where the probability of making a type I error, i.e., of *wrongly* concluding that a value is non-zero, is α. Because we test several partial correlations simultaneously, we are in a multiple testing situation where the probability of making at least one type I error is greater than α. Similarly, the overall probability of making at least one type II error, of *wrongly failing* to conclude that a value is non-zero, is greater than it would be for any single test. In practice, we will use these tests to screen all the partial correlations to identify a preliminary CIG structure. This is similar to the use of the partial autocorrelation function for identifying the AR model order of a univariate time series. The graphs so identified will be used to formulate models which will then be fitted and tested rigorously. We have found that a good strategy to cope with the problem of multiple testing in identifying a CIG is to consider the presence of edges with different significance levels, e.g., 0.01, 0.05 and 0.1, and to indicate these by the lines used in presenting the graph. The use of the 0.01 level reduces the overall type I error and that of the 0.1 level reduces the overall type II error. A recent approach to controlling the error levels is given by Drton and Perlman (2008).

The difference between a CIG and a partial correlation graph is due to the assumption of normality of the set of variables X in the CIG. This gives the presence of edges in a CIG the stronger meaning of conditional dependence rather than partial correlation, as in the latter type of graph. However, from the modeling point of view, they both give an indication of the explanatory variables to include in a linear model: when the variables we are dealing with are not Gaussian, the null hypothesis of the test is not independence but lack of linear predictability.

To illustrate the construction of a CIG, let us consider the example of the Flour price series innovations. The sample partial correlation matrix is

$$\widehat{\pi} = \begin{pmatrix} 1.0000 & & \\ 0.8532 & 1.0000 & \\ 0.0231 & 0.4483 & 1.0000 \end{pmatrix}$$

where the diagonal elements, corresponding to the partial correlations of each variable with itself, are equal to 1 and where only the lower triangular part is needed as the matrix is symmetric.

We can observe that $\hat{\pi}_{1,2}$, corresponding to the partial correlation of the innovations of $x_{1,t}$ and $x_{2,t}$, given the innovations of $x_{3,t}$, is large (0.8532) while $\hat{\pi}_{1,3}$ is small (0.0231). We can formally test which partial correlations are significantly different from zero by using (5.13). There are three series of innovations, indicated, respectively, as $e_{1,t}$, $e_{2,t}$ and $e_{3,t}$, and each one of them has 98 observations, so $n = 3$ and $m = 98$. If we set the significance at a level of $\alpha = 0.05$, then $z_{0.05/2} = 1.96$, and we can set the threshold for the significance of partial autocorrelations as

$$|\hat{\pi}(X_i, X_j)| > \frac{1.96}{\sqrt{1.96^2 + (98 - 3 + 1)}} = 0.196.$$

This confirms our initial impression that $\hat{\pi}_{1,2}$ is significantly different from zero as $0.8532 > 0.196$ while $\hat{\pi}_{1,3}$ is not, because $0.0231 < 0.196$. The computed threshold also indicates that $\hat{\pi}_{2,3}$ is significantly different from zero.

We could consider other levels of significance, e.g., $\alpha = 0.01$ and $\alpha = 0.1$ with corresponding threshold values $|\hat{\pi}(X_i, X_j)| > 0.254$ and $|\hat{\pi}(X_i, X_j)| > 0.165$. Hence both $\hat{\pi}_{1,2}$ and $\hat{\pi}_{2,3}$ are significant even at a 0.01 level of significance, while $\hat{\pi}_{1,3}$ is not significant even at a 0.1 level of significance.

All this information is efficiently conveyed in the CIG, where we link variables with significant partial correlations, i.e., between $e_{t,1}$ and $e_{t,2}$ and between $e_{t,2}$ and $e_{t,3}$. In general, we will represent the strength of the significance of the partial correlations by the thickness of the lines representing the edges, so in this case we could indicate a strong partial correlation (significant at $\alpha = 0.01$) with a thick continuous line. We will indicate a mid-strength partial correlation (significant at $\alpha = 0.05$) with a continuous line and a weak partial correlation (significant at $\alpha = 0.1$) with a broken line. No link will appear for a partial correlation not significant at any of the levels α considered.

In this example, all the significant partial correlations are strong, and consequently all the edges are represented by thick continuous lines. Figure 5.5 shows the resulting graph for the innovations of the Flour price series. Because the three variables are all normal, the presence of edges indicates conditional dependence among the variables and the graph is a CIG.

A note of caution should be attached to estimation of the CIG. The significance of an estimated partial correlation is equivalent to that of a t-value

Figure 5.5 *Estimated conditional independence graph of the VAR(2) innovations of the Flour price series.*

in a regression and can only be interpreted as indicating whether or not a variable should be included, given the other variables in the model. If a variable (equivalently a link in the CIG) is removed because of a low absolute t-value and the regression re-fitted, the t-values of the coefficients of the remaining variables will in general change, possibly substantially if there is high collinearity in the data. So a second variable with a relatively low absolute t-value in the original regression may acquire a high absolute t-value when the first is removed. This will be an important consideration when we come to fitting the DAG structures that we identify on the basis of a CIG. Nevertheless, we will take the edges in the estimated CIG as a first indication of the dependency within the set of variables considered. We will go on to model these dependencies using regression equations described by a directed acyclic graph. But having fitted a directed acyclic graph model, we will carry out diagnostic correlation checks on the residuals, designed to detect inadequacy in the model. So if we note significant correlation remaining, we will use this to indicate possible extensions of the model with new edges. This approach to modeling dependence is in the spirit of estimation and diagnostic checking advocated by Box and Jenkins (1970).

In the next section, we illustrate how a CIG, such as shown in Figure 5.5, enables us to draw conclusions about possible DAG representations of the variables.

5.4   Interpretation of CIGs

In this section, we will restrict ourselves to the non-time series context of a set of random variables such as observed time series innovations, for which we have a complete CIG. We will show first how the information in a CIG can in some examples help us to directly determine a DAG representation of the variables. We then explain how, more generally, we can determine which DAG representations are consistent with a given CIG, and so by exploring these possibilities, find a suitable model.

We start with the example of the CIG in Figure 5.5 for which the predictive interpretations of the graph may be set out explicitly. Consider in turn the interpretation for $e_{1,t}$, $e_{2,t}$ and $e_{3,t}$. The general prediction equation for the first of these is

$$e_{1,t} = \beta_{1,2}e_{2,t} + \beta_{1,3}e_{3,t} + \alpha_{1,t}, \tag{5.14}$$

where $\alpha_{1,t}$ is independent of the predictors $e_{2,t}$ and $e_{3,t}$. But because there is no link between $e_{1,t}$ and $e_{3,t}$, i.e., their partial autocorrelation is zero, we may *omit* $e_{3,t}$ from this equation and write

$$e_{1,t} = \beta_{1,2}e_{2,t} + \alpha_{1,t}. \tag{5.15}$$

For the second variable,

$$e_{2,t} = \beta_{2,1}e_{1,t} + \beta_{2,3}e_{3,t} + \alpha_{2,t}, \tag{5.16}$$

for which there is no simplification because $e_{2,t}$ is linked to both $e_{1,t}$ and $e_{3,t}$. However, the third variable is similar to the first, giving

$$e_{3,t} = \beta_{3,2}e_{2,t} + \alpha_{3,t}. \tag{5.17}$$

The key to constructing a DAG representation for the variables is that they must be ordered in such a way that each variable is only predicted by others which are lower in the ordering. This is necessary so that their joint distribution may be expressed as the product of conditional distributions: for example, for random variables X, Y and Z, the joint distribution may be expressed as

$$f(x, y, z) = f(x)f(y|x)f(z|x, y). \tag{5.18}$$

In the normal linear context (with mean corrected variables) this means specifying a marginal variance for X, the regression equation for Y on X with a residual variance and the regression equation for Z on X and Y with a residual variance.

Applying this to $e_{1,t}$, $e_{2,t}$ and $e_{3,t}$, we require the marginal variance of $e_{1,t}$, a regression of $e_{2,t}$ on $e_{1,t}$ and a regression of $e_{3,t}$ on $e_{1,t}$ and $e_{2,t}$. But we know from (5.17) that $e_{1,t}$ can be omitted from this last equation, giving

$$\begin{aligned}
e_{1,t} &= a_{1,t} \\
e_{2,t} &= \theta_{2,1}e_{1,t} + a_{2,t} \\
e_{3,t} &= \theta_{3,2}e_{2,t} + a_{3,t}.
\end{aligned} \tag{5.19}$$

Note that we have used different notation for the coefficients and error terms from those in (5.14), (5.16) and (5.17) above, because $a_{1,t}$, $a_{2,t}$ and $a_{3,t}$ are now *orthogonal* residuals, a property not shared by the regression errors in (5.14), (5.16) and (5.17), which are in general correlated. The DAG representation for these regressions is shown in Figure 5.6.

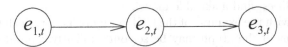

Figure 5.6 *A possible DAG representation of the VAR(2) innovations of the Flour price series.*

However, we may instead take $e_{2,t}$ lowest in our ordering of the variables, followed by $e_{1,t}$ then $e_{3,t}$. This simply interchanges the roles of these variables in the first two equations of (5.19) and is represented in Figure 5.7.

A third possibility is to reverse the original ordering, which would reverse the direction of both arrows in Figure 5.6. However, these are the only three DAGs consistent with the CIG in Figure 5.5. For example, the DAG shown in Figure 5.8, obtained by reversing both arrows in Figure 5.7, *cannot* be deduced from the CIG.

Figure 5.7 *A further DAG representation of the VAR(2) innovations of the Flour price series.*

Figure 5.8 *An incompatible DAG representation of the VAR(2) innovations of the Flour price series.*

The dependence of $e_{2,t}$ on $e_{1,t}$ and $e_{3,t}$, as shown, could only be obtained by taking $e_{2,t}$ as highest in the ordering, but then a dependence of either $e_{3,t}$ on $e_{1,t}$, as shown in Figure 5.9, or the reverse, could not be avoided. The DAG in Figure 5.9 could be used to represent the joint distribution of the variables, but fails to take advantage of the simplicity of structure revealed in the CIG.

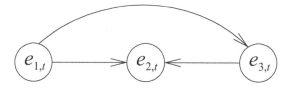

Figure 5.9 *A DAG representation of the VAR(2) innovations of the Flour price series that fails to take advantage of the simplicity of the CIG.*

5.5   Properties of CIGs

The previous illustration with just three variables was particularly simple. For a general CIG, there are several properties that are useful for interpreting its structure and determining the possible DAG models by which this structure may be explained. A useful notion is the set of *neighbors* of a node (or variable). These are simply the variables to which that node is linked. The *pairwise Markov property* of a CIG is that by which it is constructed—any two variables that are not linked are conditionally independent given all the remaining variables. There are two further properties that may be deduced from this:

- The *local Markov property* is that each variable, given its neighbors, is conditionally independent of all the remaining variables.

- The *global Markov property*. This generalizes the pairwise property to the case when we consider two groups (or blocks) of variables. If there are no links between any pair of variables, one taken from each block, then the two blocks are conditionally independent of each other given all the remaining variables.

The local property is a special case of the global property, which may be deduced from the pairwise property. In the context of the global property, the two blocks are said to be *separated* by the remaining variables, and the result is known as the separation theorem. Its general proof is not simple, but it is very straightforward if all the variables are assumed to be jointly normal. Whittaker (1990) provides a general proof with reference to original sources and acknowledgment of the developers of these ideas. We illustrate the properties using the CIG shown in Figure 5.10. The neighbors of X_2 are X_1, X_3 and X_4, so the local Markov property states that conditional upon these, X_2 is independent of X_5 and X_6. This independence is not just pairwise, between the pair X_2, X_5 and the pair X_2, X_6, but between X_2 and the variables $\{X_5, X_6\}$ taken jointly, which is in general a stronger statement (although not for multivariate normal variables). The global Markov property states that $\{X_1, X_2\}$ are independent of $\{X_5, X_6\}$ conditional upon $\{X_3, X_4\}$. This has the useful consequence that X_2 alone is independent of $\{X_5, X_6\}$ conditional upon $\{X_3, X_4\}$. These properties then enable us to build up one

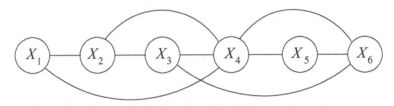

Figure 5.10 *A CIG illustrating local and block independence.*

possible model for these variables, represented by the DAG in Figure 5.11, based upon a choice of ordering from X_1 as the highest through X_6 as the lowest. There is no simplification of the relationship between X_4, X_5 and X_6, but X_5 can be omitted as an explanatory variable for X_3, since it is separated from $\{X_1, X_2, X_3\}$ by $\{X_4, X_6\}$. By the earlier argument, both X_5 and X_6 can be omitted as explanatory variables for X_2. The only explanatory variables for X_1 are X_2 and X_4, because these are its neighbors.

Given the CIG between a set of variables, the task of formulating a DAG representation of the relationship between them is considerably enhanced by use of a theorem, known as the *moralization theorem* or *moralization rule*, by which we may derive the CIG that is implied by any proposed DAG. It is then possible to postulate DAG representations and check whether they are consistent with the given CIG. The theorem gives the following simple steps for deriving the CIG:

1. For each node of the DAG, insert an undirected edge between all pairs of its parent nodes, unless they are already linked by an edge.

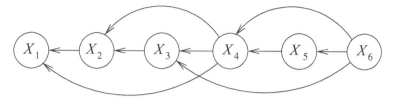

Figure 5.11 *A DAG representation of a model that may be deduced from the CIG in Figure 5.10.*

2. Replace each directed edge in the DAG by an undirected edge.

The first step is described as *marrying the parents,* and hence the construction of the CIG is called *moralization* of the graph (Lauritzen and Spiegelhalter, 1988).

As an example, consider the DAG in Figure 5.11. In this example, moralization introduces no new edges and the resulting CIG is just that in Figure 5.10. However, consider next a trimmed version of this DAG, as shown in Figure 5.12. Moralization of the parents of X_1 will lead to a link between X_2 and X_4. Also, moralization of the parents of X_4 leads to a link between X_5 and X_6. The resulting CIG is therefore the same as that in Figure 5.10. In practice, it is not always straightforward to formulate a DAG, or set of DAGs, that are consistent with a given CIG, and there are simple examples of CIGs with which no DAG is consistent. This can occur due to omission of important variables from the set considered. In the multivariate normal case, it is always possible to fit a regression of each variable on those lower in any chosen ordering, to obtain a model represented by a DAG and which will reproduce the partial correlation graph from which the CIG is constructed. However, there is no reason why such a DAG should reflect any sparse structure in the CIG.

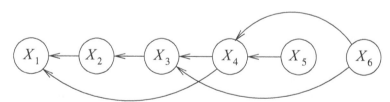

Figure 5.12 *A DAG representation of a reduced model for which the corresponding CIG is also that shown in Figure 5.10.*

In general, however, the moralization theorem can be very helpful in selecting one or more DAG representations of the variables and excluding certain possibilities. For example, from the DAG in Figure 5.11, it is possible to remove either the edge between X_4 and X_6 or that between X_5 and X_6 and retain consistency with the CIG in Figure 5.10, but not to remove both together. For the Flour price innovations, the moralization theorem tells us immediately that the DAG in Figure 5.8 is not consistent with the CIG in Figure 5.5, because it implies a moralization link between $e_{1,t}$ and $e_{3,t}$. In the case of multivariate normal variables, it is possible to determine some simple quantitative rules for the magnitudes of links in the partial correlation graph that arise due to moralization. These would, for example, help to suggest whether the edge between X_4 and X_6 or that between X_5 and X_6 might be a moralization link in the CIG in Figure 5.10. We set out these rules in Section 5.8, but reproduce the most simple of them here. The others extend this rule to situations where the three variables have parents and children.

The DAG on the left of Figure 5.13 represents a variable z dependent on the independent variables x and y. The CIG on the right represents the moralized CIG. The partial correlation of the moralized link in this graph is given in terms of the other two by the product rule, which is developed further in Section 5.8:

$$\pi(x,y) = -\pi(x,z) \times \pi(y,z). \qquad (5.20)$$

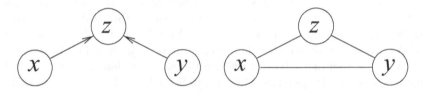

Figure 5.13 *A simple directed graph and its corresponding conditional independence graph.*

In practice, the selection between different possible DAG explanations of an estimated CIG is assisted by comparing the goodness of fit of the estimated models and testing for significance of the coefficients that correspond to links in the competing models. We illustrate this procedure in the next section.

5.6 Estimation and selection of DAGs

We will continue to work in the context of a random sample from a multivariate normal distribution, the innovations from the Flour price series providing an appropriate example which we use for illustration. The main ideas of this section are readily extended to the time series context, including lagged values, in the next section. As in Section 5.3, let the data matrix $\mathbf{X}$ with elements $x_{i,j}$ consist of columns $x_1, x_2, \dots, x_m$ of the mean corrected samples of the variables $X = \{X_1, X_2, \dots, X_m\}$. Any proposed DAG representing these variables specifies a regression equation for each variable in terms of a subset of the remaining variables, with the assumption that the errors from any one regression are uncorrelated with those from another. To fit the DAG, we therefore carry out ordinary least squares (OLS) regression of each data column x_i on the explanatory columns specified by the DAG. This provides estimates of the coefficients corresponding to each link in the DAG, and their t value, and columns a_i of residuals constituting a matrix A. Figure 5.14 shows one of the DAG models considered for the Flour price series innovations, with the estimated coefficients and t values (in parentheses) adjacent to the links.

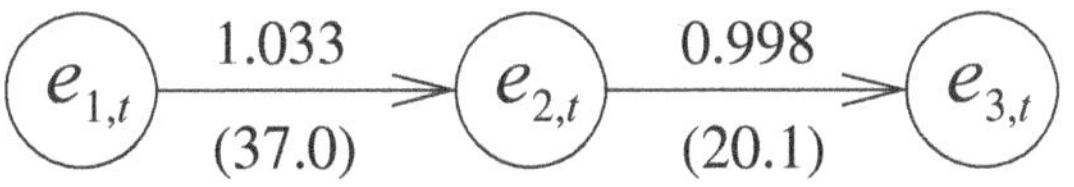

Figure 5.14 *A fitted DAG model of the VAR(2) innovations of the Flour price series, with the estimated coefficients and t values (in parentheses) adjacent to the links.*

The estimated coefficients are clearly highly significant. The next step is to assess the adequacy of this model as a representation of the relationships between the variables. For this purpose, we first use the likelihood of the model and its associated information criteria. In practice, we will use the deviance of the model, i.e., minus twice the log likelihood, which is given by

$$n \left(\log s_1 + \log s_2 + \log s_3 \right) \tag{5.21}$$

where s_1, s_2 and s_3 are the sample variances of the residuals (*not* corrected for degrees of freedom)

$$s_i = \frac{1}{n} \sum_{t=1}^{n} a_{i,t}^2. \tag{5.22}$$

The difference between this and the likelihood of the saturated model, which also includes $e_{1,t}$ in the regression for $e_{3,t}$, we will call the deviance reduction D. We will compare this with the difference in degrees of freedom, k, between the proposed model and the saturated model, which is just 1 in this example. For a formal test of the null hypothesis that coefficients of the omitted links in the DAG are all zero, we can refer D to the chi-squared distribution on k degrees of freedom. The information criteria that we will use are defined by $D+2k$ for the AIC, $D+2(\log \log n)k$ for the HQC and $D+(\log n)k$ for the SIC. The values of these for this example are displayed in Table 5.1. The deviance difference is less than the degrees of freedom (df), so there is no reason not to prefer the proposed model to the saturated model. The information criteria are all negative, indicating preference for the proposed model. The further step

Table 5.1 *Likelihood assessment of the DAG in Figure 5.14.*

df k	deviance D	AIC	HQC	SIC
1	0.05	-1.95	-2.99	-4.53

to assess the adequacy of the fitted model is to check the cross-correlations between the residuals, as shown in Table 5.2.

The correlation between $a_{1,t}$ and $a_{2,t}$ is necessarily zero as a result of the regression, but the other two are not. The approximate standard error of

Table 5.2 *Correlations between residuals of the DAG fitted to the Flour price inno-vation series.*

	$a_{1,t}$	$a_{2,t}$	$a_{3,t}$
$a_{1,t}$	1.000		
$a_{2,t}$	0.000	1.000	
$a_{3,t}$	0.006	−0.022	1.000

these two correlations is bounded by $1/\sqrt{n} = 0.101$, and we deem them to be acceptable, i.e., not suggestive of any model inadequacy.

In Section 5.4 we showed that the DAG in Figure 5.7 was also consistent with the estimated CIG of these variables, and so was the DAG in Figure 5.6 with the edge directions reversed. If we fit these, we find that they give exactly the same deviance and information criteria as in Table 5.1. We say that these models are likelihood equivalent—they imply exactly the same distributional properties of the variables, although the residual series are different because of the different orderings of the variables in the DAG. All saturated models are also likelihood equivalent to each other, so any one of them may be specified to obtain the saturated model deviance. However, on fitting the model shown in Figure 5.8, which is *not* consistent with the CIG, a deviance difference of 139.6 shows how extremely inadequate is that representation of the dependence between the variables. This very simple example has been a platform for introducing the main ideas, which we will now demonstrate in the time series context.

5.7　Building a structural VAR, SVAR

There are six possible saturated DAGs representing the relationships between three variables, corresponding to the number of distinct orderings. Consideration of the CIG of the Flour prices series innovations reduced the number under consideration to three. Each of those three models extends naturally to an SVAR(2) model for the original series, in which the dependency between the current variables is exactly the same as for that between the innovations, but with the additional dependence on all values up to lag 2. Thus the innovations model in Figure 5.14 extends to the SVAR in Figure 5.15. When fitted, the estimated coefficients between current variables in the extended SVAR(2) are exactly the same as for the innovations model, and so are the deviance and information criteria.

To continue the analysis, we now estimate the CIG between all the current values and the variables lagged up to the model order of 2. Recall that this order was determined by application of the AIC to the saturated VAR. The implication is that, given the variables up to lag 2, no variables at a greater lag have predictive value. However, we must restrict the lags used to construct the CIG to those up to the model order. If we were to include values up to greater lags, the partial correlations would in general be reduced in magnitude (see

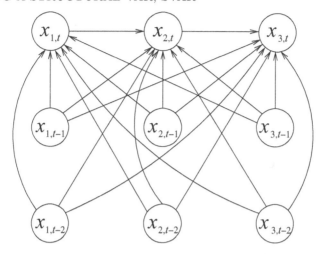

Figure 5.15 *The SVAR(2) series model equivalent to the innovations model in Figure 5.14.*

Chapter 9). In fact, we need only the partial correlations between the current variables and between the current and lagged variables. These correspond to the dependencies which are considered for inclusion in the SVAR. We do not need the partial correlations between any pair of lagged values, although these will be produced as part of the calculations.

To estimate the CIG, we assemble a data matrix X in which the columns in general consist of $\{x_{i,t-u}; t = p+1, \ldots, n\}$ for $i = 1, \ldots, m$ and $u = 0, \ldots, p$, where p is the model order. The estimated CIG is then constructed exactly as before, and from this we extract the required values. For the Flour price series, these are shown in Table 5.3.

The critical values for significance at the levels 0.10, 0.05 and 0.01 are, respectively, 0.171, 0.202 and 0.262. These are based on transforming the critical

Table 5.3 *Partial autocorrelations, up to lag 2, between the Flour price series.*

	$x_{1,t}$	$x_{2,t}$	$x_{3,t}$
$x_{1,t}$			
$x_{2,t}$	**0.853**		
$x_{3,t}$	0.023	**0.448**	
$x_{1,t-1}$	**0.452**	**−0.497**	0.130
$x_{2,t-1}$	**−0.288**	**0.522**	**−0.402**
$x_{3,t-1}$	−0.012	**−0.299**	**0.658**
$x_{1,t-2}$	**0.478**	**−0.301**	−0.132
$x_{2,t-2}$	**−0.412**	**0.264**	0.054
$x_{3,t-2}$	−0.036	0.036	0.058

values of the normal distribution, as described in Section 5.3, rather than the t distribution, not only because the degrees of freedom are sufficiently high, but because the small sample theory underlying the t distribution does not fully extend to the context of multivariate time series regression. However, the asymptotic use of the normal distribution is justified; see Reale and Tunnicliffe Wilson (2002) and Tunnicliffe Wilson and Reale (2008). It will be noted that there are 13 significant values according to these critical values, shown in bold type in the table, and they are all significant at the 0.01 level. Note that the partial correlations between current variables are necessarily identical to those between the innovations.

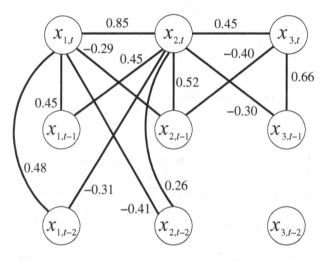

Figure 5.16 *The estimated CIG constructed for the Flour price series with partial correlations shown adjacent to the links.*

Figure 5.16 shows the resulting CIG. The links between current and lagged values provide further information beyond that in the CIG for the innovations alone for identifying the directions of dependence between current variables in the DAG representation of an SVAR model for the series. They also inform us about the links between current and lagged values in the DAG:

1. A link is certainly required between $x_{2,t}$ and $x_{3,t}$ in this DAG. There is strong evidence that this is *not* in the direction $x_{2,t} \leftarrow x_{3,t}$, because this would lead us to expect moralization links in the CIG, between $x_{3,t}$ and all the other parents of $x_{2,t}$ in a possible DAG representation. Three strong candidates for such parents, which have no links with $x_{3,t}$, are the lagged values $x_{1,t-1}$, $x_{1,t-2}$ and $x_{2,t-2}$, and we would have to suppose that their links with $x_{2,t}$ all arose from moralization if our argument were to fail. Our working hypothesis is therefore to assume the direction of this link to be $x_{2,t} \rightarrow x_{3,t}$.

2. The link between $x_{2,t}$ and $x_{3,t-1}$, with a partial correlation of -0.299, is

now a strong candidate for explanation as a moralization link. Applying the product rule, this partial correlation is very close to -0.295, which is the negative product of the partial correlations of 0.448 and 0.658 between, respectively, the hypothesised parents $x_{2,t}$ and $x_{3,t-1}$ and their common child $x_{3,t}$.

The result of these arguments is that $x_{3,t}$ depends in a simple manner on only $x_{2,t}$, $x_{2,t-1}$ and its own lagged value $x_{3,t-1}$. The series $x_{1,t}$ and $x_{2,t}$ are *not* dependent on $x_{3,t}$ at all, and their interdependence appears to be qualitatively symmetrical in the CIG of Figure 5.16. There are two likelihood equivalent DAG interpretations, in both of which all the links between these two series and their lagged values are included. The only difference lies in the choice of direction of the dependence between $x_{1,t}$ and $x_{2,t}$. However, in the model where this direction is chosen to be $x_{1,t} \to x_{2,t}$, the coefficient of the link $x_{1,t-1} \to x_{1,t}$ in the estimated DAG has a t value of 0.95. We therefore choose the direction $x_{1,t} \to x_{2,t}$ and exclude the link $x_{1,t-1} \to x_{1,t}$, which therefore suggests that the corresponding link in the CIG is due to moralization. This gives the more parsimonious SVAR as that represented in Figure 5.17.

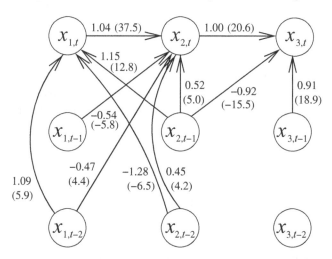

Figure 5.17 *The DAG representing a parsimonious SVAR for the Flour price series, with estimated model coefficients, and t values in parentheses, adjacent to the links.*

This model was again estimated by OLS regression of each current variable in turn, upon the explanatory variables indicated by the DAG, with the regression vectors taken from the same data matrix of current and lagged values used in the construction of the CIG. The estimated coefficients and their t values are shown adjacent to the corresponding links in Figure 5.17. Re-estimation of the model using full maximum likelihood only has a small gain in efficiency for a series of this length. This comes from making use of the relatively small amount of extra information in the first two months of data that were necessarily trimmed from the vectors of current variables in the data matrix. The maximum likelihood estimates, only slightly different from the OLS estimates, are actually shown in (5.23), which displays the coefficient matrices Φ_0, Φ_1 and Φ_2 of the SVAR as used in the model equation (5.1).

$$
\begin{matrix}
\Phi_0 & \Phi_1 & \Phi_2
\end{matrix}
$$

$$
\begin{pmatrix} 1 & 0 & 0 \\ -1.04 & 1 & 0 \\ 0 & -1.01 & 1 \end{pmatrix}
\begin{pmatrix} 0 & 1.16 & 0 \\ -0.54 & 0.52 & 0 \\ 0 & -0.91 & 0.89 \end{pmatrix}
\begin{pmatrix} 1.14 & -1.13 & 0 \\ -0.45 & 0.43 & 0 \\ 0 & 0 & 0 \end{pmatrix}. \tag{5.23}
$$

Table 5.4 gives the values of the deviance difference and information criteria relative to the saturated model as estimated by OLS. This gives 0.11 for the p-value of testing the null hypothesis that the omitted coefficients are zero. The information criteria also favor the chosen SVAR.

Table 5.4 *Likelihood assessment of the DAG in Figure 5.17.*

df k	deviance D	AIC	HQC	SIC
10.00	15.69	-4.31	-14.77	-30.16

The SVAR model, for these series, lends itself better to interpretation than the canonical VAR. First, the representation for $x_{3,t}$ can be re-arranged with only a minor approximation as a simple regression on $x_{2,t}$ with a univariate AR(1) disturbance:

$$
\begin{aligned}
x_{3,t} &= x_{2,t} + n_t \tag{5.24} \\
n_t &= 0.91 n_{t-1} + a_t.
\end{aligned}
$$

Second, again with only minor approximation, we can represent

$$
x_{1,t} = x_{1,t-2} + (x_{2,t-1} - x_{2,t-2}) + a_{1,t}, \tag{5.25}
$$

in which the current value of $x_{1,t}$ is the value two months previously, but with a correction equal to the change in price of $x_{2,t}$ over the previous two months. Third, we can approximate

$$
x_{2,t} = x_{1,t} + 0.5 (x_{2,t-1} + x_{2,t-2}) - 0.5 (x_{1,t-1} + x_{1,t-2}) + a_{2,t}, \tag{5.26}
$$

in which the current value of $x_{2,t}$ is the current value of $x_{1,t}$ corrected by the difference between the average price of the two series over the previous two months.

A structural model such as we have constructed does not take into account the fact that the Flour price series also reflect cycles of general economic activity which appear to be evident in their plots, and the model we have fitted does not capture any such cyclical behavior. However, the model has a simple interpretable structure. With just 11 coefficients, 10 fewer than a saturated model (and three variance parameters), it competes well with other published models for these series. A structural VARMA(1,1) model has also been developed for these series in Oxley et al. (2009). This also has a simple interpretation, but the construction of a VARMA model is in general much more difficult, and we do not pursue this line of model development in this book.

The construction of the SVAR model is not completed until we have carried out sensible checks. As before, we check the cross-correlations between current residuals, which are shown in Table 5.5. We also check the lagged cross-correlations shown in Figure 5.18, with error limits of $\pm 2/\sqrt{n}$. These limits give a good indication of the adequacy of the cross-correlations, although they do not allow for the effects of parameter estimation.

Table 5.5 *Correlations between residuals of the final SVAR(2) for the Flour price series.*

	$a_{1,t}$	$a_{2,t}$	$a_{3,t}$
$a_{1,t}$	1.000		
$a_{2,t}$	0.000	1.000	
$a_{3,t}$	−0.019	−0.023	1.000

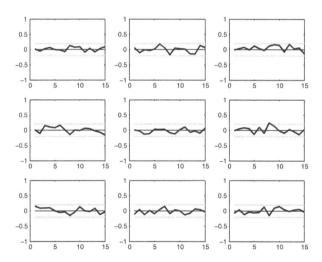

Figure 5.18 *Residual auto- and cross-correlations of the final SVAR(2) model for the Flour price series. Nominal two standard error limits about zero are shown by the gray lines.*

There are no indications of model inadequacy to be gleaned from these checks. Our final comment on this model is that it is stationary, though the reciprocal roots of the operator in the reduced VAR model are 0.9879, 0.9228, 0.9083, -0.4643, 0.2680 and 0.0. The first of these is close to the boundary of stationarity. We pointed out in Chapter 2 that a sparse VAR(p) model fitted by solving Yule–Walker equations does not necessarily reproduce any lagged correlations and it is not necessarily stationary. The same is true of the sparse SVAR models that we have explored by graphical modeling and fitted by OLS. However, what we have aimed to do is to reproduce all the partial correlations up to lag p, by selecting terms in the sparse model that capture the pattern of zero and non-zero partial correlations. As a final check, in Table 5.6 we compare the partial correlations that may be computed for the fitted model with the sample values used to identify the model. The two entries that arise by moralization are between $x_{1,t}$ and $x_{1,t-1}$, and between $x_{2,t}$ and $x_{3,t-1}$. The general impression is that the model captures the sample partial correlations quite well. There is strong correlation both between and within these series. However, the graphical modeling approach we have followed has overcome the potential problems of multicollinearity and generated a model that has simple and interpretable structure.

Table 5.6 *Each pair of columns shows, on the left, the sample partial autocorrelations up to lag 2 between the Flour price series, and on the right, the non-zero partial correlations of the selected model shown alongside. The bold numbers indicate a non-zero model value, with the corresponding sample value to its left.*

	$x_{1,t}$		$x_{2,t}$		$x_{3,t}$	
$x_{1,t}$						
$x_{2,t}$	**0.853**	**0.861**				
$x_{3,t}$	0.023		**0.448**	**0.458**		
$x_{1,t-1}$	**0.452**	**0.482**	**−0.497**	**−0.442**	0.130	
$x_{2,t-1}$	**−0.288**	**−0.329**	**0.522**	**0.524**	**−0.402**	**−0.368**
$x_{3,t-1}$	−0.012		**−0.299**	**−0.308**	**0.658**	**0.672**
$x_{1,t-2}$	**0.478**	**0.486**	**−0.301**	**−0.385**	−0.132	
$x_{2,t-2}$	**−0.412**	**−0.439**	**0.264**	**0.337**	0.054	
$x_{3,t-2}$	−0.036		0.036		0.058	

Another check that is advisable, where appropriate, is to compare the spectral properties of the fitted model with smoothed spectral estimates of the series, and a final check is to hold back some data at the end of the series to compare with out-of-sample forecasts of the same points made using the model. We will illustrate these checks as they are applied to the further example in Section 5.10.

We end this section by summarizing the steps by which we build our SVAR:

- Use the AIC to determine the order of a saturated VAR model which well represents the series.

- Construct the sample CIG of the estimated innovation series derived from this VAR model.

- Use this to explore possible DAG representations of the innovation series.

- Estimate these DAG models and check their likelihood criteria and residual cross-correlations.

- Construct the sample CIG of the series and its values lagged up to the determined order.

- Use this to explore possible DAG representations of an SVAR model for the series, basing the selected dependencies between current variables on those explored for the innovation series.

- Estimate these DAG models and check their likelihood criteria and lagged residual auto- and cross-correlations. If necessary, include or exclude terms from the model to achieve an adequate sparse model.

- Check that the CIG implied by the model reasonably well matches the sample CIG; apply other comparisons of the model and sample spectral properties and examine an out-of-sample set of forecasts from the model to assess its consistency in this respect.

5.8   Properties of partial correlation graphs

The main aim of this chapter is to show how the estimated partial correlation matrix may inform us in selecting the dependencies between the current and lagged terms of a structural vector autoregressive model. An important tool has been the rule for moralization of an hypothesized DAG representation of these variables, to derive the implied CIG of the model, for comparison with the estimated CIG. One *converse* application of the rule is that a pattern in the CIG, of the form shown on the left of Figure 5.19, *excludes* the pattern in the DAG shown on the right of the same figure as a possible interpretation. The three DAGs in which one or both of the directions of the links are reversed

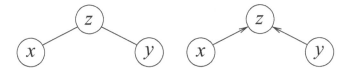

Figure 5.19 *A simple conditional independence graph and the directed graph interpretation that it* excludes.

are consistent with the CIG that is shown. Because the direction of a link from the past is determined by the arrow of time, this simple result can often give a useful indication of the direction of a link between current variables, as shown in the example of the Flour price series.

It will be useful before our next example in Section 5.10, to look further at the relationship between the model structure and the partial correlation matrix that underlies the moralization rule. In this section, we list some quantitative features which may assist in its intelligent application, particularly in the context of a CIG that is subject to sampling fluctuations, which may mask the significance of some of the coefficients. Relatively straightforward derivations are provided. We continue in Section 5.9 to consider the simultaneous equation model (SimEM) for representing the relationship between variables and point out that the moralization rule can also be applied to this further model structure.

In our context of normal linear models, there is a well-known intimate relationship between the matrix of the regression equations for the model and the matrix of partial correlations. First express the regression equations in the form $\boldsymbol{Mx} = \boldsymbol{a}$, where $\boldsymbol{M}$ is a matrix of model coefficients, $\boldsymbol{x}$ is the vector of model variables and $\boldsymbol{a}$ is the vector of uncorrelated model errors. We assume for now that the errors have been normalized to unit variance by dividing each row of the model equations by the standard deviation of its error term. This enables us to express the inverse covariance matrix of x simply as

$$G = M'M. \tag{5.27}$$

The partial correlation between x_i and x_j is then $-G_{ij}/\sqrt{G_{ii}G_{jj}}$, where $G_{i,j}$ is the scalar product of the columns of $\boldsymbol{M}$ associated with the variables x_i and x_j, and $G_{i,i}$, $G_{j,j}$ are the respective squared norms of these columns. This result is therefore zero if the scalar product is zero.

We now set out some of the quantitative properties of partial correlations relating to the application of the moralization rule.

1. Consider first the most simple graph to which this rule may be applied, shown on the left of Figure 5.20, which represents a dependent variable z with two independent variables x and y.

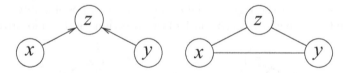

Figure 5.20 *A simple directed graph and its corresponding conditional independence graph.*

This structure may be represented by the equations

$$
\begin{aligned}
z &= ax + by + e \\
x &= f \\
y &= g
\end{aligned}
\tag{5.28}
$$

where e, f and g are uncorrelated mean zero random variables, with respective variances σ_e^2, σ_f^2 and σ_g^2. Then x and y are random with no dependence on any other variable, and z is dependent on both x and y. For this graph, the partial correlations between the three variables are related by the product rule

$$\pi(x, y) = -\pi(x, z) \times \pi(y, z). \tag{5.29}$$

Provided a and b are both non-zero, the CIG has all three variables linked, as shown on the right of Figure 5.20. Furthermore, if partial correlations of this CIG satisfy (5.29), it admits an interpretation, or representation, in the form of the DAG in Figure 5.20.

The rule is directly derived from (5.27). Divide all three equations by the standard deviations of their error terms, and express them in matrix form as

$$
\begin{pmatrix} 1/\sigma_e & -a/\sigma_e & -b/\sigma_e \\ 0 & 1/\sigma_f & 0 \\ 0 & 0 & 1/\sigma_g \end{pmatrix} \begin{pmatrix} z \\ x \\ y \end{pmatrix} \tag{5.30}
$$

$$
= \begin{pmatrix} r & \alpha & \beta \\ 0 & s & 0 \\ 0 & 0 & t \end{pmatrix} \begin{pmatrix} z \\ x \\ y \end{pmatrix} = \begin{pmatrix} \epsilon \\ \nu \\ \gamma \end{pmatrix},
$$

where ϵ, ν and γ are now uncorrelated with mean zero and unit variance. Then

$$
\begin{aligned}
\pi(z, x) &= \frac{-(r \times \alpha + 0 \times s + 0 \times 0)}{(k \times m)} = -\frac{r\,\alpha}{k\,m} \\
\pi(z, y) &= \frac{-(r \times \beta + 0 \times 0 + 0 \times t)}{(k \times n)} = -\frac{r\,\beta}{k\,n} \\
\pi(x, y) &= \frac{-(\alpha \times \beta + s \times 0 + 0 \times t)}{(m \times n)} = -\frac{\alpha\,\beta}{m\,n}
\end{aligned} \tag{5.31}
$$

where

$$
\begin{aligned}
k^2 &= r^2 + 0^2 + 0^2 = r^2, \\
m^2 &= \alpha^2 + s^2 + 0^2 = \alpha^2 + s^2, \\
n^2 &= \beta^2 + 0^2 + t^2 = \beta^2 + t^2,
\end{aligned} \tag{5.32}
$$

are the respective squared norms of the columns associated with z, x and y. Then

$$
\pi(x, y) = -\pi(z, x)\,\pi(z, y)\frac{k^2}{r^2}, \tag{5.33}
$$

but in this example $k^2 = r^2$, so we obtain the stated relationship. It is possible to reverse this argument to show that the product rule admits a representation, in the form of the DAG in Figure 5.20, but only because we exclude cyclical (recursive) dependence between the variables.

Note that, if we reverse either or both of the directions attached to the links in Figure 5.20, then the partial correlation between x and y will be zero, so the absence of a link in the CIG between x and y does admit such representations. The result follows because either α or β or both are replaced by zero in the matrix of equations. The column of M corresponding to z will acquire non-zero coefficients representing its influence on x and/or y, but these will *not* induce a non-zero partial correlation between x and y.

2. Now extend the simple example to the relationships represented in Figure 5.21, where the broken line arrows indicate parents of x, y and z. Exactly

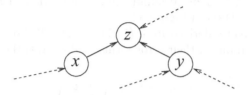

Figure 5.21 *A simple directed graph with further parents of all three variables added.*

the same rule (5.29) applies in this case. The central idea of a partial correlation is that of conditioning on all other variables, so the extra parents are effectively treated as constants in the relationships between x, y and z, so do not affect their (conditional) correlation. The algebraic argument comes from extending the matrix relating the variables to

$$
\begin{pmatrix} r & \alpha & \beta & \cdots \\ 0 & s & 0 & \cdots \\ 0 & 0 & t & \cdots \\ 0 & 0 & 0 & \cdots \end{pmatrix} \begin{pmatrix} z \\ x \\ y \\ \vdots \end{pmatrix} = \begin{pmatrix} \epsilon \\ \nu \\ \gamma \\ \vdots \end{pmatrix}, \tag{5.34}
$$

where the dots at the end of the first three rows represent coefficients of other variables which are parents of x, y and z. Clearly, the calculations of the partial correlations from the first three columns do not depend on these. The zeros at the foot of the first three columns represent the fact that x, y and z are not parents of any other variables. It is immediate that the calculated partial correlations are exactly as in the first example.

3. Extend the first example to include further children of x and y, but with z remaining as the only common child as shown in Figure 5.22. The product rule (5.29) continues to apply.

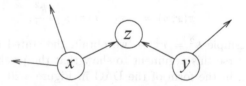

Figure 5.22 *A simple directed graph with further children of x and y added.*

The matrix required for the algebraic derivation is now

$$
\begin{pmatrix} r & \alpha & \beta & \cdots \\ 0 & s & 0 & \cdots \\ 0 & 0 & t & \cdots \\ 0 & u & 0 & \cdots \\ 0 & 0 & v & \cdots \end{pmatrix} \begin{pmatrix} z \\ x \\ y \\ \vdots \end{pmatrix} = \begin{pmatrix} \epsilon \\ \nu \\ \gamma \\ \kappa \\ \lambda \end{pmatrix}, \tag{5.35}
$$

where u and v represent the fact that x and y each have another child, but not in common. In fact, the further children of x and y result in a diminution of the magnitude of the partial correlations $\pi(x, z)$ and $\pi(y, z)$, because of the increase in the norms of the columns for x and y. However, the result holds true because the norm of the column for z remains unchanged.

4. Consider now the addition of a further *common* child w of x and y, as shown in Figure 5.23, to the simple case of the first example.

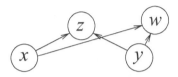

Figure 5.23 *A simple directed graph with a further common child of x and y added.*

The product rule is now extended to include both children:

$$\pi(x, y) = -[\pi(x, z)\,\pi(y, z) + \pi(x, w)\,\pi(y, w)].   \qquad (5.36)$$

It is possible for the strength of the moralization link to be increased or reduced. It could, coincidentally, be reduced to zero. As in the previous example, the magnitude of the partial correlations $\pi(x, z)$ and $\pi(y, z)$ will be diminished by the presence of the further child. The inclusion of the extra product term is evident from the addition of a further row, representing the new common child, in the matrix equations:

$$\begin{pmatrix} r & \alpha & \beta & 0 \\ 0 & s & 0 & 0 \\ 0 & 0 & t & 0 \\ 0 & \tau & \rho & p \end{pmatrix} \begin{pmatrix} z \\ x \\ y \\ w \end{pmatrix} = \begin{pmatrix} \epsilon \\ \nu \\ \gamma \\ \kappa \end{pmatrix}.   \qquad (5.37)$$

As before, $\pi(x, z)\,\pi(y, z) = -(\alpha\,\beta)/(m\,n)$ and now $\pi(x, w)\,\pi(y, w) = -(\tau\,\rho)/(m\,n)$. We see also that $\pi(x, y) = -(\alpha\,\beta + \tau\,\rho)/(m\,n)$, which provides the required result.

5. For our next example, we simply include a child of z, as shown in Figure 5.24. The effect is to modify the simple product rule to

$$\pi(x, y) > -\pi(x, z) \times \pi(y, z),   \qquad (5.38)$$

where the $>$ sign indicates greater in magnitude, irrespective of sign. In fact, the further children of z result in a diminution of the magnitude of the partial correlations $\pi(x, z)$ and $\pi(y, z)$, whereas $\pi(x, y)$ remains the same. This is seen from the matrix equations

$$\begin{pmatrix} r & \alpha & \beta & 0 \\ 0 & s & 0 & 0 \\ 0 & 0 & t & 0 \\ q & 0 & 0 & p \end{pmatrix} \begin{pmatrix} z \\ x \\ y \\ w \end{pmatrix} = \begin{pmatrix} \epsilon \\ \nu \\ \gamma \\ \kappa \end{pmatrix},   \qquad (5.39)$$

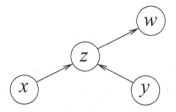

Figure 5.24 *A simple directed graph with a child of z added.*

where the last row represents the dependence of w upon z. The relationship (5.33) still holds but now, instead of $k^2 = r^2$, we have $k^2 = r^2 + q^2 > r^2$, from which the rule (5.38) follows.

6. Finally, we consider a situation, represented in Figure 5.25, where a precise rule cannot be set out. The point is that adding the links to w can com-

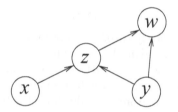

Figure 5.25 *A simple directed graph with a further child of y and z added.*

pletely modify $\pi(y, z)$, in magnitude and sign, whereas the value of $\pi(x, z)$ arising from moralization is little affected. This is seen from the matrix equations

$$\begin{pmatrix} r & \alpha & \beta & 0 \\ 0 & s & 0 & 0 \\ 0 & 0 & t & 0 \\ q & 0 & h & p \end{pmatrix} \begin{pmatrix} z \\ x \\ y \\ w \end{pmatrix} = \begin{pmatrix} \epsilon \\ \nu \\ \gamma \\ \kappa \end{pmatrix}, \qquad (5.40)$$

Compare $\pi(y, z) = -(r\,\beta + q\,h)/(kn)$ with the ratio $-\pi(x, y)/\pi(x, z) = (k^2/r^2)\,(r\,\beta)/(kn)$ to see how the presence of the coefficient h, of the influence of y on w, can cause these to be very dissimilar.

All these rules can be applied directly to the investigation of dependence between the innovations from a VAR model. They can also be applied without modification to the investigation of a SVAR model using a sample CIG constructed from current and lagged values of the series. For a given set of SVAR model coefficients, it is very straightforward, using the relationship (5.27), to calculate the partial correlations for the links between current variables. For the links between current and past variables, the computation is not so direct.

To understand this, we have to write the matrix M for the SVAR model in partitioned form:

$$M = \left(\begin{array}{c|ccc} \Phi_0 & -\Phi_1 & -\cdots & -\Phi_p \\ \hline 0 & & A & \end{array} \right), \tag{5.41}$$

where for simplicity we have omitted the scaling by the error term standard deviations. The partition corresponds to current and past variables. The zero in the lower left partition represents the fact that the past variables are not dependent on the current variables. We see then that the partial correlations between current variables can be determined from Φ_0. The lower right partition A can be any matrix that orthogonalizes the set of past variables. The scalar product term used in the calculation of the partial correlation between a current variable and a past variable is not dependent upon the elements of A, because the relevant products contain a zero from the lower left partition. The elements of A only count in the calculation of the norm of the column vector associated with a past variable, and this does not depend on the particular choice of orthogonalizing A. We have in fact presented, in Section 2.10, the inverse covariance matrix of a VAR model, which is seen to have exactly this structure, and that of an SVAR is readily derived from it.

5.9   Simultaneous equation modeling

We now consider how simultaneous equation models (SimEMs) might be usefully applied to describe the relationship between current variables in an SVAR. For simplicity of exposition, we introduce the idea of these models when no lagged terms are present. We retain the model formulation introduced at the start of the previous section, taking the current variables as x. Thus $Mx = a$ where a is a vector of uncorrelated variables with means of zero and unit variances, but we now permit the possibility that M cannot be transformed to upper triangular form by any re-ordering of the variables (and equations). Such models are of long standing in econometric theory, where each row represents an economic constraint and the solution x of the set of equations represents the equilibrium economic conditions. Simulation of a realization of x cannot now be generated recursively; by its very nature, given M and a realization of a, the simultaneous equations have to be solved for x. Of interest in our context is the fact that the inverse covariance matrix of x is still given by $G = M'M$, so for the example of

$$M = \begin{pmatrix} a & b & 0 & 0 \\ 0 & c & d & 0 \\ 0 & 0 & e & f \\ g & 0 & 0 & h \end{pmatrix},$$

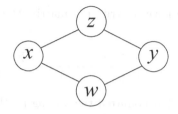

Figure 5.26 *A CIG arising from a simultaneous equation model.*

the inverse covariance matrix has the structure

$$
\mathbf{G} = \begin{pmatrix} A & B & 0 & G \\ B & C & D & 0 \\ 0 & D & E & F \\ G & 0 & F & H \end{pmatrix},
$$

which is represented by the CIG structure in Figure 5.26. This CIG cannot be explained by a DAG with the same degree of sparse structure, although, given any numerical values for the partial correlations, triangular factorization will result in a DAG representation with 5 links. The SimEM may therefore have some value for explaining an estimated CIG of this form, which has no simple DAG explanation. A topic of central interest in simultaneous equation modeling is that of model identification when using samples of x to estimate the model coefficients. A necessary condition, satisfied in the above example, is that the number of non-zero coefficients is less than $m(m+1)/2$, where m is the matrix size, because this is the number of free parameters in a covariance matrix of that size.

We now briefly consider simultaneous equation modeling of current variables in an SVAR. There seems to be no necessity to introduce this into the Flour price example presented earlier in this chapter, but we will illustrate it on this series. The important point is that OLS estimation can no longer be used to fit the SimEM model. We will now assume the convention that the error variances are free parameters and that in the matrix M a single coefficient of a different variable in each equation is constrained to have the value 1; this can always be arranged if M is non-singular. By the very nature of a SimEM, it cannot be said that any equation represents the dependence of one variable on the remainder. Nevertheless, this convention allows us to use a graphical representation of the model in which, for each equation, the variable with a constrained unit coefficient is represented, as in the DAG model, as the child of the remaining variables that are present in that equation. However, the graph is *not* a DAG, as it may contain cycles, and a pair of variables may have links in both directions. Neither is this graph necessarily unique. Nevertheless, it conveys the relationships contained in the SimEM, and we may still apply the moralization rule to deduce the links in the CIG that is implied by this graphical representation of the model. This follows because a link in the

CIG depends only on the presence or absence of elements in the columns of M.

The most important point about estimation of a SimEM is that, in general, incorrect results are obtained using OLS for each equation to regress the unit coefficient variable upon its parents. Methods for their estimation were presented by Zellner and Theil (1962). For a DAG model, re-ordering of the variables and equations could always transform the matrix M to upper triangular with unit diagonals, which has unit determinant. This is not possible for the SimEM. The expression for minus twice the log likelihood of the model still contains the residual sample variances as in (5.21), but from that expression must now be subtracted the quantity $2 \log \det M$, which is zero for any DAG model.

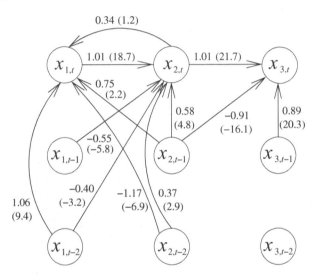

Figure 5.27 *The representation of an SVAR for the Flour price series, with a simultaneous equations model for current variables. Estimated coefficients, and t values in parentheses, are shown adjacent to the links.*

We have used numerical maximization of this likelihood to estimate the model represented in Figure 5.27 to the Flour price series. This model has a SimEM structure between the current variables, which is given by the addition to the selected model of Figure 5.17, of a directed link from $x_{2,t}$ to $x_{1,t}$. This model can be estimated because of the slight asymmetry of the dependence of these two terms on the past variables. If this symmetry is restored by re-introducing the dependence of $x_{1,t}$ on $x_{1,t-1}$, which was previously removed as not significant, there is no unique maximum likelihood estimate. There is a range of parameters which are likelihood equivalent because they give the same value of the likelihood. The new model term in Figure 5.27 is not significant and the reduction in deviance is only 0.64, in comparison with the previously selected model, which appears to give a very satisfactory representation of these series without the introduction of a SimEM model relating current variables. However, this extension to allow SimEM relationships between current variables, rather than restricting these to acyclic dependencies, can be important in some contexts. In Section 9.4 we show how it can be useful for determining the direction of current dependency in a model for the Term rate series illustrated in Figure 1.7.

5.10   An SVAR model for the Pig market: the innovations

We now develop a structural VAR model for the Pig market series. These are quarterly indices for five variables which we take in the following order to constitute the multivariate time series:

1. New breeding animals introduced to increase the breeding herd, and taken from animals otherwise reared for the market.

2. The Number of animals sent to market, reared from the breeding herd.

3. The Profitability of the market, mainly reflecting the market price, but taking into account costs of feed and other items.

4. The Breeding herd reduction from the slaughter of older sows.

5. The Breeding herd size.

A structural model is particularly appropriate for these series. We expect, from prior knowledge of the market context, that a realistic model should contain some dependencies between current variables and dependencies of each current variable on a relatively small number of lagged variables. As for the Flour series example, we begin by identifying a model for the dependencies between current variables based on the estimated CIG of the innovations of a VAR(p) model for the series. We will extend this to an SVAR model of the series themselves by examining the enlarged CIG formed from the series and their values up to lag p. The model initially identified by this means will be provisional. It will be subject to diagnostic auto- and cross-correlation checks of the assumption that the residuals from the model are independent white noise series. These checks suggest some inadequacy of the provisional model and lead us to include a number of further lagged variables in the model. Also, the coefficients of a number of lagged variables are found not to be significant and are removed. Some of these will correspond to links in the CIG that prove to be present through moralization. Once these residual checks are satisfactory, two further confirmatory checks are carried out that the model does adequately capture the important statistical properties of the series themselves. The first of these is to compare the (smoothed) sample spectra of the series with the computed spectral properties of the fitted model. The second is to compare the sample partial correlations, used to construct the series CIG, with the computed partial correlations of the fitted SVAR model. Finally, we refit the model using the first 8 years of data and use it to construct out-of-sample forecasts of the final 4 years. Comparison of the forecasts with the actual values of the series is an informal style of checking, but serves to increase confidence in the model.

We remarked in Chapter 1 on the strong correlations both within and between some of the series. This enables us to fit a model with strongly predictive model terms, but care must be taken in the selection of these terms because of the collinearity of predicting variables. Because the breeding and marketing of pigs is so well understood, there is good prior knowledge of the directions and lags of relationships between the series. For example, the current Number to market depends very much on the Breeding herd size. We will use this knowledge to inform and support, but not to replace, our empirical identification of these relationships by statistical methods. One of the particular facts about the market is the length of the cycle from introducing a new animal into the breeding herd and its offspring being sent to market. This is three to four quarters, consistent with the maximum lag of 4 that we noted in our comments following Figure 4.4 in Section 4.5. We will therefore take the overall model order to be $p = 4$ in the following analysis.

The dependence between current values of the series in an SVAR model is the same as that between the innovations from the canonical VAR model. We identify this dependence by examining the sample correlation matrix of the innovations, shown in Table 5.7, and more usefully, the sample partial correlations derived from these and shown in Table 5.8.

Table 5.7 *Estimated cross-correlations between the Pig market series innovations.*

	$e_{1,t}$	$e_{2,t}$	$e_{3,t}$	$e_{4,t}$	$e_{5,t}$
$e_{1,t}$	1.000				
$e_{2,t}$	0.196	1.000			
$e_{3,t}$	0.255	−0.614	1.000		
$e_{4,t}$	0.099	0.261	−0.383	1.000	
$e_{5,t}$	0.561	0.435	−0.051	0.292	1.000

Table 5.8 *Estimated partial correlations between the Pig market series innovations.*

	$e_{1,t}$	$e_{2,t}$	$e_{3,t}$	$e_{4,t}$	$e_{5,t}$
$e_{2,t}$	0.239				
$e_{3,t}$	0.404	−0.679			
$e_{4,t}$	0.094	−0.159	−0.385		
$e_{5,t}$	0.430	0.358	0.147	0.250	1.000

The correlations are not as large as those in our introductory example of the Flour price series, and because the series are short, these statistics are highly variable, with two standard errors of approximately 0.30 for the correlations in Table 5.7. The partial correlations in Table 5.8 are exactly the same as those we will find between the current values when forming partial correlations from x_t and its values up to lag 4. It is important, because the series are short, that we assess the significance of these partial correlations using the limits given in (5.13), setting the number of estimated coefficients $m = 24$, appropriate to this larger set of variables. This gives limits of 0.345, 0.401 and 0.499 corresponding to significance levels of 10%, 5% and 1%. It is therefore quite possible that sampling variability could mask the significance of some reasonably large coefficients. However, as shown in Figure 5.28, there are several clear links evident in the CIG derived from Table 5.8. The strengths of the links are indicated by significance at the 10% level (broken line), 5% level (thin unbroken line) and 1% level (thick line).

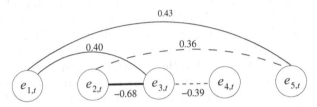

Figure 5.28 *Sample CIG constructed from canonical VAR(4) model residuals of the Pig market series. The strengths of the links are indicated by the adjacent values of the partial correlations and by the significance at the 10% level (broken line), 5% level (thin unbroken line) and 1% level (thick line).*

The immediate value of the CIG is that it selects just 5 out of the 10 possible links between the variables for our consideration. We therefore now consider various possible DAG relationships that might explain this CIG. Prior information is, however, available from the context of the original series, as supplied by those responsible for their compilation. First, we can expect that series $x_{5,t}$, the Breeding herd size, is positively affected by $x_{1,t}$, the New breeding animals introduced into the herd during the previous quarter, implying the link direction $e_{1,t} \rightarrow e_{5,t}$. Then we can expect that $x_{2,t}$, the Numbers to market, will negatively affect the price and therefore the Profitability series $x_{3,t}$, implying the link direction $e_{2,t} \rightarrow e_{3,t}$. We do not expect the reverse direction; the Numbers to market are determined in large part by the Breeding herd size, implying the link direction $e_{5,t} \rightarrow e_{2,t}$. We can, however, expect that Profitability will negatively affect the Breeding herd size reduction, $x_{3,t}$. Lower prices will encourage Breeding herd reduction, implying the link direction $e_{3,t} \rightarrow e_{4,t}$. The remaining partial correlation between $e_{1,t}$ and $e_{3,t}$ has a less clear interpretation. The direction $e_{1,t} \leftarrow e_{3,t}$ appears plausible except that $x_{1,t}$ is the number of New breeding animals introduced during the previous quarter and is more sensibly explained by Profitability at the *previous* quarter. It also results in a cyclic interpretation of the links, $e_{1,t} \rightarrow e_{5,t} \rightarrow e_{2,t} \rightarrow e_{3,t} \rightarrow e_{1,t}$. The direction $e_{1,t} \rightarrow e_{3,t}$ might be explained by the fact that increased demand for New breeding animals could increase the price. Figure 5.29 shows the DAG corresponding to these speculative interpretations of the CIG with the estimated coefficients of the links. These were in fact fitted by both regression and maximum likelihood to the series x_t, including all variables at lags 1 to 4 besides the links shown between current variables. Again, this was to ensure that for such a short series the t values were correctly evaluated, although the estimates were exactly the same from regression between the residuals. The estimation results of regression and maximum likelihood were in good agreement.

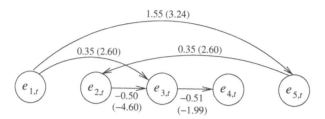

Figure 5.29 *Estimated DAG fitted between the canonical VAR(4) model residuals of the Pig market series with the estimated coefficients (t values) shown adjacent to the directed links.*

We first comment that this DAG implies a moral link, not shown in the CIG of Figure 5.28, between the parents $e_{1,t}$ and $e_{2,t}$ of $e_{3,t}$. However, the product rule indicates that the partial autocorrelation of such a link should be approximately $-(0.404)(-0.609) = 0.246$. This is very close to the observed partial correlation of 0.239 between $e_{1,t}$ and $e_{2,t}$, which is not, however, sufficiently large to be significant. This does, however, add confidence to the chosen interpretation. Our next point is that there are other DAG representations equivalent to that in Figure 5.29. This can be tested by evaluating their deviances. Reversing the links one at a time, $e_{2,t} \leftarrow e_{3,t}$ and $e_{1,t} \leftarrow e_{5,t}$ both give the same deviance and are equivalent to Figure 5.29. However, the reverse links $e_{3,t} \leftarrow e_{4,t}$, $e_{2,t} \rightarrow e_{5,t}$ and $e_{1,t} \leftarrow e_{3,t}$ are not equivalent and all have slightly increased deviance, which indicates that they are not preferred models. The deviance of the last of these, giving a cyclical dependence graph, is evaluated by maximum likelihood because regression estimation is not valid. We have not considered all possible permutations of directions of the links in Figure 5.29; more evidence regarding these may be available from interpretation of the full CIG of x_t and its values up to lag 4.

To reassure ourselves that this is an acceptable fit, we can compare the deviance and AIC of this fit with those of the saturated model. Table 5.9 shows the difference in both the deviance and AIC for these two models. They have both been modified to allow for the short series length, so that, if the DAG model is correct, the deviance difference is approximately chi-squared in distribution with the degrees of freedom being the difference, 5, in the number of parameters. A positive value for the AIC difference supports the DAG model, as is the case here. The deviance difference is less than the degrees of freedom, so both these quantities favor the DAG model.

Table 5.9 *A summary of the fit of the DAG for the innovation series, compared to the saturated model including the reduction in the number of parameters, the increase in the deviance and the reduction in the information criterion.*

Differences in degrees of freedom	Deviance	AIC
5.00	2.32	31.97

To complete this initial investigation, we check the correlation matrix, shown in Table 5.10, of the structural residuals a_t derived from fitting the DAG of Figure 5.29. The standard error of these sample correlations would be $1/\sqrt{(44)} = 0.15$ if the residuals were truly independent random samples of length 44. However, we will call this the nominal standard error; the actual standard errors will be less, as is evident from some of the small or zero entries, because they relate to regression residuals. There is, nevertheless, no reason to doubt the fitted model from the values shown in this table; all are well within the nominal two standard error limits.

Table 5.10 *Correlations between the structural residuals from fitting the DAG of Figure 5.29 to the canonical residuals of the Pig market series.*

	$a_{1,t}$	$a_{2,t}$	$a_{3,t}$	$a_{4,t}$
$a_{2,t}$	−0.053			
$a_{3,t}$	0.000	−0.021		
$a_{4,t}$	0.213	−0.111	−0.093	
$a_{5,t}$	0.000	0.036	0.053	0.212

5.11   The full SVAR model of the Pig market series

To develop the full SVAR model, we now consider the estimated CIG for the original set of series and its values up to lag 4. This is derived from the partial correlations between these variables using the same thresholds of significance for the links as for the CIG in Figure 5.28. It is shown in Figure 5.30 divided into two parts for clarity, as explained in the caption. Note that the sub-diagram restricted to the current variables is the same as that for the innovations in Figure 5.28, and this is included in both parts of the figure to help assess which links might arise from moralization.

Our aim is to construct, between the same variables, a DAG that represents an SVAR model. A starting point is to take the directions of the links between current variables in this DAG to be the same as those in the DAG of the innovations series in Figure 5.29. These should, however, be taken as tentative; consideration of Figure 5.30 may be of further help in either confirming these or suggesting an alternative. A link between a current and past value in Figure 5.30 may correspond to true dependence in the SVAR model. It may also arise from moralization of such a dependent link with one between current

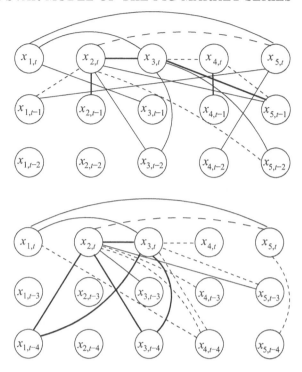

Figure 5.30 *The sample CIG constructed from the Pig market series and values up to lag 4. For clarity, the graph is divided into one for the links from current values to lags 1 and 2 and one for the links from current values to lags 3 and 4.*

variables or be partly explained by both true dependence and moralization. For example, the direction of the link $x_{2,t} \to x_{3,t}$, inherited from Figure 5.29, is supported by the CIG. If it were in the other direction, we would expect to see a moralization link between $x_{3,t}$ and $x_{2,t-1}$, which would then appear to be common parents of $x_{2,t}$. A consequence of taking the direction $x_{2,t} \to x_{3,t}$ is that the links $x_{1,t-4} \to x_{2,t}$ and $x_{3,t-4} \to x_{2,t}$ are open to interpretation as moralization links.

However, because of the complexity of Figure 5.30, it is not always easy to determine, using the rules in Section 5.8, to what extent a particular specification of the *directed* links between current variables affects our interpretation of their links with the past. In this connection, we remark on the fact that for three of the series no link appears in Figure 5.30 between the current variable and its immediate past value. In our experience this is unusual; the immediate past value of a series is usually one of the strongest predictors of the current value. Having raised this point, it is of interest to report that two of these added links are *not* present in our final model: those for $x_{3,t}$, the Profitability series, and $x_{5,t}$, the Breeding herd size. They were introduced, on prior grounds, as a check, but found not to be significant. That for $x_{1,t}$ was, however, highly significant, with a t value of 11.8. The corresponding link in the CIG was absent because it was masked by moralization; both $x_{1,t-1}$ and $x_{1,t}$ are parents of $x_{5,t}$. The partial correlation derived for the link $x_{1,t-1} \to x_{1,t}$ in our final model is $\pi = 0.25$. Removing $x_{1,t-1}$ as a parent of $x_{5,t}$, it would be 0.33. The sample value is 0.18, well within sampling variability of $\pi = 0.25$, but not exceeding our lowest threshold of significance.

We now describe a method of identifying those links which represent true dependence of current on past variables and which are consistent with a particular specification of the directed links between current variables. This is simply to fit an SVAR including all lagged values, but with just the specified directed links between current variables. For any given current variable, the t value of its estimated coefficient on a particular lagged variable is equivalent to the partial correlation between them, given all the other lagged values and *just* those other current values included as predictors. Significant values indicate links representing only true dependence on the past. The link $x_{1,t-1} \to x_{1,t}$ was found to be significant by this means, whereas several links in the CIG were found not to be significant, suggesting that they arose from moralization, for example, $x_{3,t-2} \to x_{2,t}$ and $x_{1,t-4} \to x_{2,t}$.

However, to be conservative, for our initial DAG interpretation we include all those links between current and past values appearing in the CIG in Figure 5.30, together with five further links identified from the foregoing method. The directions of the links between current variables were taken from the DAG for the innovations series in Figure 5.29.

On fitting the corresponding SVAR model, out of 35 estimated coefficients, there were 9 which were not significant, and of these, 6 corresponded to links that were in the CIG but not selected by the foregoing method. Further model development consisted of introducing five other coefficients that were indicated by model checking and removing one other coefficient that was not then significant. Model checking consisted of inspecting the residual acf for any unduly large values, comparing the model and sample spectra, and assessing the out-of-sample forecasting performance of the model. A comparison was also made between the sample partial correlations used in the model selection process and the corresponding partial correlations implied by the fitted model. The residual lagged cross-correlations generally give the best indication as to which new terms are needed in the model to rectify inadequacies detected by these checks and comparisons.

Of these checks, the out-of-sample forecasting assessment was particularly important in this example because poor performance may be a symptom of over-fitting. This may easily occur because the number of potential coefficients is a large proportion of the data size. The out-of-sample forecasts of the VAR(2) model shown in Figure 4.6 are quite acceptable and are, in fact, not very different from the in-sample forecasts (which are not shown) over the same time range derived from the model fitted to the full length of series. This is not true for a (saturated) VAR(4) model. The in-sample forecasts, though not very different from those using the VAR(2) model, are noticeably better, but the out-of-sample forecasts from fitting the model to the first 8 years of the series are very poor. This is because each response vector is reduced to a length of $32 - 4 = 28$ and has $4 \times 5 = 20$ regressors, most of which have no predictive capacity but contribute considerable variability to the estimated coefficients and forecasts.

We do not present the details of the model development but show the DAG representing the final SVAR model, which has 30 coefficients (35, including the structural innovation variances), in Figure 5.31. This is followed by the various model checks.

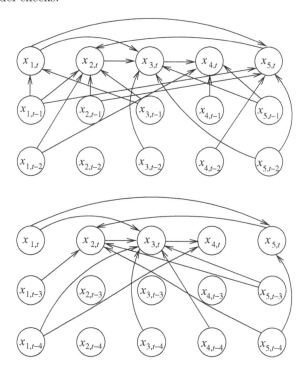

Figure 5.31 *The DAG representing the SVAR(4) model of the Pig market series. For clarity, the graph is divided into one for the links between current values and lags 1 and 2 and one for the links between current values and lags 3 and 4.*

As shown in Table 5.11, the fit of the SVAR(4) model compared well with that of the saturated canonical VAR(4) model when judged in terms of the increase in deviance and reduction in AIC.

The cross-correlations of the residuals at lag 0, as shown in Table 5.12, were acceptable, and Figure 5.32 shows the cross-correlations from lags 1 to 10. A portmanteau statistic (see Hosking (1980)), formed as the sum of squares of all 250 lagged cross-correlations up to lag 10, multiplied by the series length, has the value 197.50. This is somewhat less than the expected value of 220 (250

Table 5.11 *A summary of the fit of the SVAR(4) model of the Pig market series compared to the saturated VAR(4) model, giving the increase in deviance and reduction
in AIC.*

Differences in degrees of freedom	Deviance	AIC
80	81.34	263.4

minus the number of model coefficients) assuming a correct model, confirming
that the overall magnitude of the sample cross-correlations is acceptable.

Table 5.12 *Correlations between the structural residuals from fitting the SVAR(4)
model represented by the DAG of Figure 5.31 to the Pig market series.*

	$a_{1,t}$	$a_{2,t}$	$a_{3,t}$	$a_{4,t}$
$a_{2,t}$	0.012			
$a_{3,t}$	0.032	−0.065		
$a_{4,t}$	0.142	−0.085	−0.137	
$a_{5,t}$	0.053	0.069	−0.066	0.181

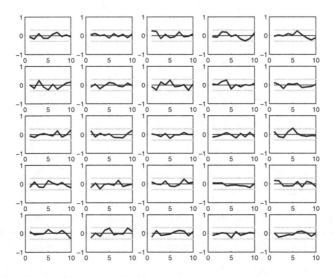

Figure 5.32 *Residual auto- and cross-correlations of the SVAR(4) model of the Pig
market series.*

Figure 5.33 presents a comparison of selected spectral properties of the final model with those of the smoothed sample spectrum. A bandwidth of 0.05
is used to resolve comparable features, though this leads to high variability
in the sample spectra. Series 2 is the Numbers to market and series 3 is the
Profitability. The model and sample properties appear to be reasonably consistent with each other, though the model and sample phase differ at frequency

zero. Here, the sample phase indicates a negative relationship, whereas the model phase indicates a positive relationship. These series interact through a feedback cycle, with positive lagged dependence of Number to market on Profitability through the process of increasing the Breeding herd size, and with negative dependence of Profitability on Numbers to market. Consequently, the sign of the low frequency, or long term, dependence is, understandably, challenging to identify.

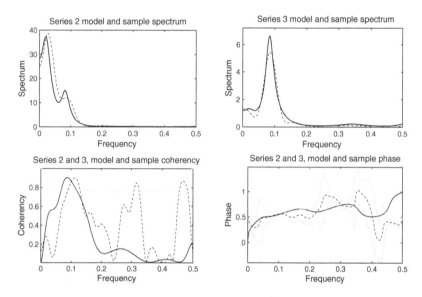

Figure 5.33 *Selected spectral properties of the SVAR(4) model fitted to the Pig market series. The model properties are shown by the solid line and the sample values by the broken line. The significance limit for the squared coherency is shown as a gray line, and the limits for phase by gray lines.*

The comparison between the model and sample partial correlations, for the lags used in identifying the SVAR(4), are shown in Figure 5.34. We tabulated these for the Flour series example, but with 110 values in this example it is easier to visualize these by plotting them against each other. The sample and model values for the saturated VAR(4), which are not shown, lie very close to the line of equality. This is because the VAR(p) captures the acf and therefore the partial correlations up to the model order. Figure 5.34 reveals the variability of the sample partial correlations when a model with far fewer parameters is fitted, but within this limitation there is a reasonably good correspondence between the model and data. The horizontal line in the center corresponds to the zero value partial correlations implied by the model.

There are only 20 of these, of which just 2 of the corresponding sample partial correlations lie outside the 10% signifcance threshold of 0.345. So, although the model has only 30 coefficients, there are in fact 90 model partial

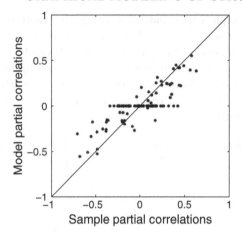

Figure 5.34 *The partial correlations of the SVAR(4) model plotted against the sample partial correlations of the Pig market series.*

correlations that are strictly non-zero, though most are very close to zero. It is reassuring that the graphical modeling methodology, through identifying the important sample partial correlations, leads to a satisfactory sparse model with sensible interpretation, for a set of very short series with complex interdependence.

Our final comment is on the forecasting assessment of the model. The out-of-sample forecasts of the last 4 years of the series based on the SVAR model we have identified, but fitted to the first 8 years, are in fact again very similar to those shown for the VAR(2) model in Figure 4.6. Instead of these, we therefore show the corresponding in-sample forecasts in Figure 5.35. These are noticeably closer to observed values. This is to be expected because it is the later part of the series that most clearly follows a cycle with defined period. The corresponding in-sample forecasts of the saturated VAR(4) model are very similar to these. The SVAR with only 30 coefficients does, however, produce much better out-of-sample forecasts than the VAR(4) model with 100.

The use of an SVAR model of order 4, rather than a saturated VAR of order 2, is strongly justified by significant coefficients beyond lag 2, including one at lag 3 and two at lag 4 with absolute t values greater than 4.0. But the series are highly correlated, and it would appear that lower lag variables can in large part compensate as predictors for the the higher lag variables when the model is restricted to a VAR(2). In fact, we have explored a simplified SVAR(2) model which does not represent some of the spectral properties quite as well as the SVAR(4) but still represents the cyclical structure. That is the model used to generate the multivariate acf for illustration of multivariate autoregressive modeling in Chapter 2.

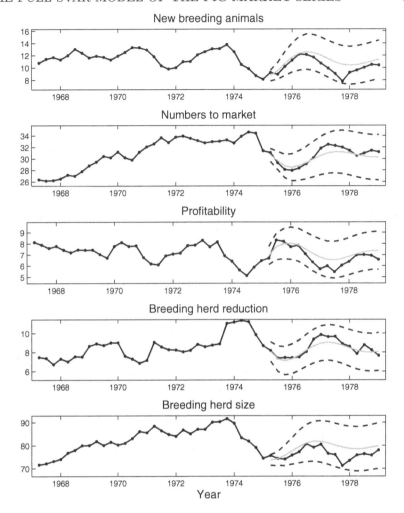

Figure 5.35 *Forecasts of the Pig market series using the SVAR(4) model fitted to the full series. The series are shown by the solid black lines, forecasts of the last 4 years by the solid gray lines and limits by the broken lines.*

The Pig market series were provided to us as official statistics, but with a caution as to their quality because of changes of definition, sampling sizes and other discontinuities in the process of their compilation. Even so, it has been possible to identify a clear structure to the relationships between the series, which can mainly be understood in terms of the known behavior of the market. This has also been of some guidance in the process of model building, and a relevant comment here is that in the later steps of model building a significant dependence of series 2, the Number to market, was found on series 1, the New breeding animals, at lags of 1, 2 and 3. Without this, the Number to market depends only on its immediate past value and the Breeding herd size. But these extra terms can be well interpreted in terms of the breeding cycle, that a breeding pig produces and raises a litter suitable for market typically every two to three quarters. The breeding herd effectively consists of two or three cohorts which produce marketable pigs in successive quarters. The Number to market would be much better explained if the numbers in these breeding herd cohorts were separately known. The Numbers of new breeding animals in past quarters contain information about the cohorts of breeding animals and can therefore contribute to the explanation of the Number to market, implicitly adjusting for the relevant cohort size in the breeding herd.

Forecasting is not an important application of this model, but it may be of value for understanding the cycles which affect the market. Experimentation with the model by changing various coefficients may give insight into how such cycles might be regulated. We present one further example of this type of graphical model building for time series in Section 9.4, but this uses the structural form of the extended VAR models presented in the next chapter.

Searching for significant predictors in a regression context can be automated. Given the directions of dependence between current variables, we could use methods of best subset regression to identify important lagged variables without reference to the links between lagged and current values in the CIG. However, studies have shown that such methods of data mining may be less than reliable at selecting the correct variables, because of the inevitable variability of statistical tests. We have found that using the CIG as a starting point for variable selection, though still subject to some such variability, guides us effectively to an acceptable model. Our graphical modeling approach may also be valuable in the formulation of econometric models based on structural autoregressions. In that context, the relationships between current variables, often expressed in simultaneous equation form, are generally specified on the basis of economic theory. Estimation of the CIG between current variables may challenge, or provide empirical support for, the theoretical specification.

Chapter 6

VZAR: An extension of the VAR model

6.1 Discounting the past

There are several approaches to introducing the models presented in this chapter, and many of the underlying ideas have a long history. We introduced and motivated the VAR(p) model by considering the prediction of x_t based upon past series values up to a finite lag. This is an example of finite memory prediction. One of the main ideas of this extension is to base the prediction upon values extending to much greater lags but still using a limited number of coefficients. One of the earliest forms of predictor based on this idea was the exponentially weighted moving average (EWMA) of Brown (1962), or simply the *exponential smoother*. Past values are given positive weights which sum to one and decay exponentially with discount factor $(1 - \lambda)$. There have been various generalizations of this idea, some of which are based on ideas closely related to the models we propose; see, for example, Cogger (1974), Lütkepohl (1982) and Burman and Shumway (2006). Some of these generalizations use several selected discount factors, but in our models a single one is specified. A prediction may be constructed as a linear combination of variables obtained by applying the exponential smoother repeatedly to past values—the so-called single and double smoothing, etc. However, this leads to multicollinearity in determining the coefficients of the required linear combination.

The formulation we use is closely related, but based upon ideas that may be traced back to the Laguerre filters of Wiener (1949), which were developed in the context of continuous time. These have desirable orthogonality properties which avoid problems of multicollinearity. Discrete time versions have been widely used in systems modeling (see Wahlberg (1991)), and have been applied in univariate time series modeling in Wahlberg and Hannan (1993). In this chapter we present a formulation of these models that is a natural extension of the multivariate autoregressive model presented in Chapter 4. The application of (discounted) weights to values extending to the far past allows a multivariate time series to be predicted and modeled more accurately, and possibly with fewer parameters. Although this does *not* include the kind of long-memory processes that have received much interest in recent years and that are the subject of Beran (1994), a generalization mentioned in the closing section of this book does have this capacity. The models we propose are also extended to give weight to higher lead time prediction. We considered this point at the start of Section 2.6, where we introduced, and later illustrated, the construction of models tuned to predict a combination of exponentially discounted future series values. The prediction of such a combination is the subject of Hansen and Sargent (1981), which, though not specifically concerned with the formulation of new forecasting models, contains some related developments. Within this chapter, we introduce and define the models and present their properties. We describe how they may be used to approximate multivariate time series and how they can be fitted to observed series. This parallels the presentation in Chapter 4; the Yule–Walker, lagged regression, maximum likelihood and order selection methods are described as extensions to the methods for VAR models. The main example used for illustration is that of the Infant monitoring series shown in Figure 1.3.

We have previously presented an account and application of the univariate form of our extended model in Tunnicliffe Wilson and Haywood (2012). Multivariate time series applications are presented in Morton and Tunnicliffe Wilson (2001), Tunnicliffe Wilson et al. (2001) and Tunnicliffe Wilson and Morton (2004), including some continuous time versions of the model. The Ph.D. theses of Morton (2000), Ibañez (2005) and Lo (2008) study, respectively, the multivariate continuous time model, non-linear models and multi-step prediction properties of the model. The findings of the second thesis are summarized in Ibañez and Tunnicliffe Wilson (2007).

6.2 The generalized shift operator

To introduce the new model, recall the idea of the state variables used at time t to represent the VAR model. These are just a finite subset of present and past series values, from lags 0 to $p - 1$, where p is the order of the model, which are used to predict the future of the series. The new model uses, in place of these states, a finite set of linear combinations of present and past series values. For a single series x_t these new states are defined by

$$s_t^{(k)} = Z^k x_t \tag{6.1}$$

where Z is the *generalized shift operator* defined in terms of the shift operator B by

$$Z = \frac{B - \theta}{1 - \theta B} = -\theta + (1 - \theta^2)(B + \theta B^2 + \theta^2 B^3 + \cdots). \tag{6.2}$$

The parameter θ is a specified discount factor, which we will normally assume to lie in $[0, 1)$, although it may in general take any value in $(-1, 1)$. When required, we will indicate the dependence of Z upon θ by writing Z_θ. As we will illustrate below, the effect of the operator Z on a slowly varying time series is similar to applying a lag of $\ell = (1 + \theta)/(1 - \theta)$.

The initial state is always $s_t^{(0)} = Z^0 x_t = x_t$. Following (6.2), the first order state is

$$s_t^{(1)} = Z x_t = -\theta x_t + (1 - \theta^2)(x_{t-1} + \theta x_{t-2} + \theta^2 x_{t-3} + \cdots), \tag{6.3}$$

but is more readily constructed by the recursive relationship deriving from (6.2):

$$s_t^{(1)} = x_{t-1} - \theta x_t + \theta s_{t-1}^{(1)}. \tag{6.4}$$

More generally, the states $s_t^{(k)}$ are constructed for increasing values of k by

$$s_t^{(k+1)} = s_{t-1}^{(k)} - \theta s_t^{(k)} + \theta s_{t-1}^{(k+1)}. \tag{6.5}$$

These equations can be cast in matrix form to show how the vector of states up to any order k is generated as the output from a linear system with x_t as input:

$$
\begin{pmatrix}
I & 0 & \cdots & 0 \\
\theta I & I & 0 & \cdots \\
\ddots & \ddots & \ddots & \ddots \\
\cdots & 0 & \theta I & I
\end{pmatrix}
\begin{pmatrix}
s_t^{(0)} \\
s_t^{(1)} \\
\vdots \\
s_t^{(k)}
\end{pmatrix}
$$

$$
=
\begin{pmatrix}
0 & 0 & \cdots & 0 \\
I & \theta I & 0 & \cdots \\
\ddots & \ddots & \ddots & \ddots \\
\cdots & 0 & I & \theta I
\end{pmatrix}
\begin{pmatrix}
s_{t-1}^{(0)} \\
s_{t-1}^{(1)} \\
\vdots \\
s_{t-1}^{(k)}
\end{pmatrix}
+
\begin{pmatrix}
x_t \\
0 \\
\vdots \\
0
\end{pmatrix}. \tag{6.6}
$$

If $\theta = 0$, then $Z = B$, so the new states are simply the same lagged values as those of the standard autoregressive model. For $\theta > 0$, although the states $s_t^{(k)}$ bear some similarity to the subset of lagged values $x_{t-k\,\ell}$, where $\ell = (1 + \theta)/(1 - \theta)$, the states have the advantage of forming a basis of all present and past series values. No information is lost, therefore, by using them as a set of possible predictors, whereas using a subset of lagged values does lead to some loss of information. Figure 6.1 shows, for $k = 1, 2$ and 5, the weights attached to the present and past values of x_t that are implied by this construction of $s_t^{(k)}$.

Figure 6.2 shows a section of the Respiration rate series sampled every 10 seconds, and the first and second order states obtained by applying the generalized shift operator once then twice, with θ set to 0.5. The effect of the operator on low frequencies in the series is similar to a lag of $(1 + 0.5)/(1 - 0.5) = 3$ time units.

The use of these states is of particular value when the series are sampled at high frequency. For example, the Respiration rate series is actually recorded at intervals of one tenth of a second. We have used subsamples at 5 or 10 second intervals in previous analyses of this and the related series of Pulse rate and Oxygen level measurements. It is, however, quite possible to model the original high frequency series without any subsampling, by using states obtained by applying the generalized shift operator with an appropriate parameter θ. This is a somewhat extreme example in which the high frequency

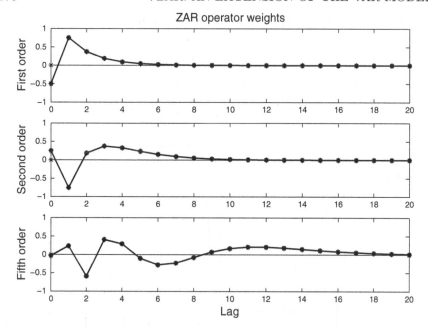

Figure 6.1 *The first plot shows the lag weights of the generalized shift operator with parameter $\theta = 0.5$ and those below show the lag weights of the square and fifth power of the operator.*

sampled series may be more usefully viewed as a continuous record, and it will be treated as such in the next chapter. Even so, most series are sampled at a rate designed to avoid information loss that might possibly occur by sampling at any lower frequency. The auto- and cross-correlations of the series typically extend to quite high lags, and the spectrum tends to be concentrated at lower frequencies. In these cases, predictors using generalized shifts of the series can be expected to have some advantage.

To complete this introduction to the generalized shift operator, we present three further points:

1. We define its inverse:

$$Z^{-1} = \frac{1 - \theta B}{B - \theta} = \frac{B^{-1} - \theta}{1 - \theta B^{-1}}. \tag{6.7}$$

Thus $Z^{-1}x_t$ applies exactly the same weights to present and *future* values of x_t as Zx_t applies to present and past values. The quantities $Z^{-k}x_t$ for $k \geq 0$ similarly form a basis of present and future values of x_t, and we can consider them as states $s_t^{(-k)}$ of the future.

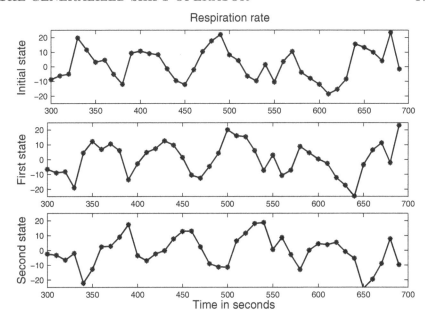

Figure 6.2 *The first plot shows a section of the Respiration rate series and those below show the results of applying the generalized shift operator once then twice, with θ set to 0.5. The effect of the operator on low frequencies in the series is similar to a lag of $(1 + 0.5)/(1 - 0.5) = 3$ time units.*

2. We may express the backward shift operator B in terms of Z as

$$B = \frac{Z + \theta}{1 + \theta Z}, \tag{6.8}$$

from which any combination of lagged values of x_t can also be expressed in terms of the states $Z^k x_t$.

3. Consider substituting numerical values for B in the definition of Z. In particular, take $B = \exp(-2\pi i f)$, a value on the unit circle $|B| = 1$. Then $Z = \exp(-2\pi i g)$, also a value on the unit circle $|Z| = 1$. We say that Z is a *unimodular operator*, or an *all-pass filter*. By application of (4.25), we can show that for any k, $Z^k x_t$ has exactly the same lagged covariances and spectrum as x_t. The dependence of g on f is given by

$$\cos(2\pi g) = \frac{(1 + \theta^2) \cos(2\pi f) - 2\theta}{1 + \theta^2 - 2\theta \cos(2\pi f)} \tag{6.9}$$

and is illustrated in the next section in Figure 6.3(a) for $\theta = 0.6$, where it is used to give insight into the spectrum of models based upon the generalized shift. It is known as a *frequency warp*. Substituting a value of B

outside or inside the unit circle gives a value of Z inside or outside the unit circle, respectively. To summarize, each of the unit circle, its interior and its exterior is mapped onto itself.

We will also find useful the expression for the derivative

$$\frac{dg}{df} = \frac{1 - \theta^2}{1 + \theta^2 - 2\theta \cos(2\pi f)}. \tag{6.10}$$

The inverse map for f in terms of g, and its derivative, are obtained by interchanging f and g in these expressions and replacing θ by $-\theta$.

6.3 The VZAR model

The finite parameter models we introduce in this section are a generalization of the VAR(p) model. They share with that model the property that the co-variances and minimum mean square error linear predictions of any stationary process can be approximated with arbitrary accuracy by a model of this form of sufficiently high order. Their extra value is that for many processes they can achieve a better approximation using a model of lower order.

We introduce our models by considering the prediction of future series values x_{t+k} for $k > 0$ as linear combinations of a finite number of the present states $s_t^{(k)}$, $k \geq 0$. As shown by their definition in the previous section, these are themselves linear combinations of present and past series values x_t, x_{t-1}, ..., the collection of which we refer to as x_{P}. We will write $\mathrm{L}(x_{t+1} | x_{\mathrm{P}})$ for the projection of x_{t+1} on x_{P}, i.e., the minimum mean square error linear predictor of x_{t+1} on all present and past values. To develop our first model, we assume that this can be expressed:

$$\mathrm{L}\left(x_{t+1} | x_{\mathrm{P}}\right) = \xi_1 s_t^{(0)} + \xi_2 s_t^{(1)} + \cdots + \xi_p s_t^{(p-1)} \tag{6.11}$$

$$= \xi_1 x_t + \xi_2 Z x_t + \cdots + \xi_p Z^{p-1} x_t, \tag{6.12}$$

where the ξ_k are matrix coefficients. Our assumption implies that the prediction error $e_{t+1} = x_{t+1} - \mathrm{L}\left(x_{t+1} | x_{\mathrm{P}}\right)$ is the linear innovation in x_{t+1}. In the same manner as for the VAR model in Chapter 2, we use this prediction equation to motivate a model for the series. It is conventional to express this as an equation for x_t in terms of its dependence upon previous values, which introduces a backward shift of one time unit in (6.12), giving

$$x_t = \xi_1 x_{t-1} + \xi_2 Z x_{t-1} + \cdots + \xi_p Z^{p-1} x_{t-1} + e_t \tag{6.13}$$

$$= \xi(Z) B x_t + e_t, \tag{6.14}$$

where

$$\xi(Z) = \xi_1 + \xi_2 Z + \cdots + \xi_p Z^{p-1}. \tag{6.15}$$

From here on we will refer to $s_t^{(k)}$ as the ZAR *state* at time t with ZAR *lag* k, where we are prefixing the standard abbreviation for *autoregressive* with the symbol of the generalized shift operator. We will also call (6.14) the *predictive* form of a multivariate or vector ZAR model of order p with discount factor θ, abbreviated to VZAR(p,θ) model.

Because e_t is assumed to be the linear innovation in x_t, it is a white noise series with

$$\text{Cov}(e_t, e_{t-k}) = 0 \text{ for } k > 0 \tag{6.16}$$
$$\text{Var}(e_t) = V_e. \tag{6.17}$$

For the purpose of model estimation, we will assume further that e_t is multivariate Gaussian.

We will now develop two other forms of this model, the first of which is the *natural* form of this model, introduced by replacing the future series value $B^{-1}x_t = x_{t+1}$ by the future state $Z^{-1}x_t = s_t^{(-1)}$ in the prediction equation (6.12) to give

$$L\left(Z^{-1}x_t \,\big|\, x_{\mathbf{P}}\right) = \varphi_1 s_t^{(0)} + \varphi_2 s_t^{(1)} + \cdots + \varphi_p s_t^{(p-1)} \tag{6.18}$$

$$= \varphi_1 x_t + \varphi_2 Z x_t + \cdots + \varphi_p Z^{p-1} x_t, \tag{6.19}$$

where the φ_k are matrix coefficients. The point of this is that $Z^{-1}x_t$ is similar, for slowly varying series, to the future value $x_{t+\ell}$ where $\ell = (1+\theta)/(1-\theta)$. The predictor therefore acquires some of the robustness of multi-step or higher lead time predictors. The fact that $Z^{-1}x_t$ places a weight of $-\theta$ on the present value x_t does not detract from this; as θ increases, the influence of future values increases.

On applying the operator Z to both sides of (6.19), we derive the model in a conventional form as an expression for x_t:

$$x_t = \varphi_1 s_t^{(1)} + \varphi_2 s_t^{(2)} + \cdots + \varphi_p s_t^{(p)} + n_t$$
$$= \varphi_1 Z x_t + \varphi_2 Z^2 x_t + \cdots + \varphi_p Z^p x_t + n_t \tag{6.20}$$

or

$$\varphi(Z)x_t = n_t, \tag{6.21}$$

where

$$\varphi(Z) = I - \varphi_1 Z - \varphi_2 Z^2 - \cdots - \varphi_p Z^p. \tag{6.22}$$

The error series n_t in these representations is not, however, white noise. Assuming that the error in the predictor $L\left(Z^{-1}x_t \,\big|\, x_{\mathbf{P}}\right)$ in (6.19) is uncorrelated with $x_{\mathbf{P}}$, it can be shown that n_t follows a standard VAR(1) process with *scalar* autoregressive coefficient θ:

$$n_t = \theta n_{t-1} + \varepsilon_t, \tag{6.23}$$

where ε_t is multivariate white noise with variance V_ε related to the variance V_n of n_t by $V_\varepsilon = (1 - \theta^2)V_n$. In taking (6.20) to define a model for the series, we therefore assume this structure for n_t, which we will refer to as the model disturbance term.

Note that the linear combination of states on the RHS of the model (6.20) is *not* a practical predictor of x_t, because x_t must be known to construct these states at time t. However, n_t is the error in the projection of x_t on all the states $s_t^{(k)} = Z^k x_t$ for $k = 1, 2, \ldots$.

We will see that this second model opens up new ways of understanding the VZAR model and fitting it to a time series. However, a most important point is that the two model forms that we have presented are *equivalent*. One form of model can be simply transformed into the other and the coefficients of one derived from those of the other, as we show below: the two forms are interchangeable. The equivalence between them is obtained by algebraic comparison of the natural form $(1 - \theta B)\varphi(Z)x_t = \varepsilon_t$ with the predictive form $\{I - B\xi(Z)\}x_t = e_t$, using (6.8). On multiplying through by $(1 + \theta Z)$, this gives

$$(1 - \theta^2)\varphi(Z) = A\{(I + \theta Z) - (Z + \theta)\xi(Z)\} \tag{6.24}$$

and

$$\varepsilon_t = Ae_t. \tag{6.25}$$

The scaling matrix A may be determined by setting either $Z = -\theta$ or $Z = 0$, as

$$A = \varphi(-\theta) \text{ or } A = (1 - \theta^2)(I - \theta\xi_1)^{-1}, \tag{6.26}$$

the former being useful for deriving the predictive from the natural form and the latter for the reverse. Note that when the model order is $p = 1$, the predictive form reduces to the first order autoregression $x_t = \xi x_{t-1} + e_t$, but the natural form $x_t = \varphi Z x_t + n_t$ still has a different parameterization with $\varphi_1 = (I - \theta\xi)^{-1}(\xi_1 - \theta I)$.

The two forms of model introduced above, predictive and natural, are in fact both special cases of a *general* form of model that is equivalent to both and is also very useful. To derive this, we introduce the operator Z_ρ to which the subscript is attached to emphasize that the discount factor ρ used in its definition is not, in general, the same as that previously used to define Z, which is θ. To keep notation simple, we will continue to use Z to mean Z_θ unless otherwise specified. The model is introduced in a similar manner to the natural model by considering the prediction equation:

$$L\left(Z_\rho^{-1}x_t \,\middle|\, x_P\right) = \zeta_1 s_t^{(0)} + \zeta_2 s_t^{(1)} + \cdots + \zeta_p s_t^{(p-1)} \tag{6.27}$$

$$= \zeta_1 x_t + \zeta_2 Z x_t + \cdots + \zeta_p Z^{p-1} x_t, \tag{6.28}$$

where the ζ_k are again matrix coefficients. The model for x_t that we derive from (6.28) is expressed as

$$x_t = Z_\rho(\zeta_1 x_t + \zeta_2 Z x_t + \cdots + \zeta_p Z^{p-1} x_t) + n_t \tag{6.29}$$

$$= Z_\rho \zeta(Z) x_t + n_t$$

where

$$\zeta(Z) = \zeta_1 + \zeta_2 Z + \cdots + \zeta_p Z^{p-1}. \tag{6.30}$$

The *disturbance term* term n_t in (6.29) now follows a standard VAR(1) process with *scalar* autoregressive coefficient ρ:

$$n_t = \rho n_{t-1} + \varepsilon_t, \tag{6.31}$$

where ε_t is multivariate white noise with variance V_ε related to the variance V_n of n_t by $V_\varepsilon = (1 - \rho^2) V_n$. The argument for this is exactly the same as that for the natural model, being based on the same assumption that the error in the prediction of $Z_\rho^{-1} x_t$ by the RHS of (6.28) is uncorrelated with x_{t-k} for $k \geq 0$.

Note that the predictive form of model corresponds to setting $\rho = 0$ and the natural form to setting $\rho = \theta$. The advantage of the general form is that it provides an extra dimension of robustness. All models are in practice approximations to real data, and to fit them so as to best predict just one step ahead, as in the predictive form, may not provide the best approximation for prediction to higher lead times. Specifying ρ takes future values at higher lead times into account when fitting the model. The effective low frequency lead time of the predicted variable is $(1 + \rho)/(1 - \rho)$. The natural model also achieves this but is constrained to use the same discount factor in constructing the predictor from past series values.

The equivalence between the general form and the predictive form of model is given by

$$(1 - \rho B)I - (B - \rho)\zeta(Z) = M\{I - B\xi(Z)\} \tag{6.32}$$

and

$$\varepsilon_t = M e_t. \tag{6.33}$$

Setting $B = (Z + \theta)/(1 - \theta Z)$, (6.32) becomes a polynomial identity:

$$(1 + \tau Z)I - (Z + \tau)\zeta(Z) = \frac{M}{1 - \rho\theta}\{(1 + \theta Z)I - (Z + \theta)\xi(Z)\}, \tag{6.34}$$

where

$$\tau = \frac{\theta - \rho}{1 - \theta\rho}. \tag{6.35}$$

To derive $\zeta(Z)$ from $\xi(Z)$, we find $M = (1 - \rho^2)\{I - \rho\xi(-\tau)\}^{-1}$ by setting $B = \rho$ and $Z = -\tau$. To derive $\xi(Z)$ from $\zeta(Z)$, we find $M = \{I + \rho\zeta(-\theta)\}$ by setting $B = 0$ and $Z = -\theta$.

6.4   Properties of the VZAR model

We now list several inter-related properties of the VZAR(p, θ) model:

1. The model can be expressed as a restricted VARMA$(p, p - 1)$ of the form

$$\Phi(B)x_t = (I - \Phi_1 B - \Phi_2 B^2 - \cdots - \Phi_p B^p)x_t = (1 - \theta B)^{p-1} e_t. \tag{6.36}$$

This is derived by substituting for $Z = (B - \theta)/(1 - \theta B)$ in $(1 - \theta B)A^{-1}\varphi(Z)x_t = e_t$, multiplying through by $(1 - \theta B)^{p-1}$, expanding and collecting terms in powers of B. Apart from the factor A^{-1}, the coefficients of $\Phi(B)$ are linear in those of $\varphi(B)$. This expression is useful for deriving further properties but is of little practical value because for models of reasonably high order, the VARMA parameterization typically leads to collinearity problems in model fitting.

2. The predictive form (6.14) explicitly defines a process whose parameter space is restricted by the requirement of stationarity. This requirement is, however, more conveniently expressed in terms of the parameter space of the natural form of the VZAR(p, θ) model. This is identical to that of the classical VAR(p) model described in Chapter 4. To show this, we consider the VARMA form (6.36) for which the stationarity condition is that $\Phi(B)$ should be non-singular for $|B| \leq 1$. Now $\Phi(B) = (1 - \theta B)^p \varphi(-\theta)^{-1}\varphi(Z)$ and $Z = (B - \theta)/(1 - \theta B)$ maps each of the boundary and interior of the unit circle onto itself. Therefore $\Phi(B)$ has a singularity for some B in or on the unit circle if and only if $\varphi(Z)$ does so for the corresponding Z. The process is therefore stationary if and only if $\varphi(Z)$ is non-singular for $|Z| \leq 1$.

3. There is a simple state space representation of the predictive form of the model, in terms of the ZAR states. The state transition equation is

$$
\begin{pmatrix} I & 0 & \cdots & 0 \\ \theta I & I & 0 & \cdots \\ \ddots & \ddots & \ddots & \ddots \\ \cdots & 0 & \theta I & I \end{pmatrix}
\begin{pmatrix} s_t^{(0)} \\ s_t^{(1)} \\ \vdots \\ s_t^{(p-1)} \end{pmatrix}
$$
$$
= \begin{pmatrix} \xi_1 & \xi_2 & \cdots & \xi_p \\ I & \theta I & 0 & \cdots \\ \ddots & \ddots & \ddots & \ddots \\ \cdots & 0 & I & \theta I \end{pmatrix}
\begin{pmatrix} s_{t-1}^{(0)} \\ s_{t-1}^{(1)} \\ \vdots \\ s_{t-1}^{(p-1)} \end{pmatrix}
+ \begin{pmatrix} e_t \\ 0 \\ \vdots \\ 0 \end{pmatrix}. \qquad (6.37)
$$

The only difference from (6.6) is that the input is now e_t, with x_t being constructed in the first row, which is simply the predictive model. The remaining rows representing the recursive calculation of the higher order states are exactly the same. Expressing (6.37) as $L\,S_t = R\,S_{t-1} + E_t$, where S_t is the state vector at time t, the conventional form of a state space representation is obtained on dividing through by L. The state transition matrix is then $T = L^{-1}R$ and the observation equation is simply $x_t = s_t^{(0)}$. This representation is useful for calculating model properties, constructing predictions and model estimation. In particular, the covariance matrix V_S of the state vector S_t can be calculated using standard state equation methods, and then $\text{Cov}(S_t, S_{t-k}) = T^k V_s$. The first (block) element in this matrix is $\text{Cov}(x_t, x_{t-k})$.

4. The spectrum of the VZAR model is readily calculated from either the predictive or natural form, but use of the latter gives more insight into the nature of the model. The context for this insight is the statistical properties of the ZAR states. For any fixed t, define a new series in terms of these states:

$$X_k = s_t^{(-k)} = Z^{-k} x_t. \tag{6.38}$$

Then X_k is also a stationary multivariate series, with the terms for $k \leq 0$ providing a basis for $x_s, s \leq t$ and the terms for $k \geq 0$ providing a basis for $x_s, s \geq t$. That X_k is stationary is verified by deriving, by application of (4.25), its lagged covariances in terms of the spectrum $S_x(f)$ of x_t:

$$
\begin{aligned}
\Gamma_{X,v} = \text{Cov}\,(X_k, X_{k-v}) &= \int_{-0.5}^{0.5} Z^{-k}\, \bar{Z}^{-(k-v)}\, S_x(f) df \\
&= \int_{-0.5}^{0.5} Z^{-v} S_x(f) df,
\end{aligned} \tag{6.39}
$$

which depends only upon v. To clarify this derivation, we have set $B = \exp(-2\pi i f)$ in $Z = (B - \theta)/(1 - \theta B)$ within the integral and used the property that $\bar{Z} = Z^{-1}$. Further insight is now obtained by transforming the integral (6.39) by substituting f in terms of g as defined in (6.9). Then, on including the Jacobian of the transformation,

$$\Gamma_{X,v} = \int_{-0.5}^{0.5} \exp(2\pi i v g) S_x\{f(g)\} \frac{1-\theta^2}{1+\theta^2 + 2\theta \cos 2\pi g} dg. \tag{6.40}$$

We deduce (see (4.14)) that the spectrum of X_k is given in terms of that of x_t by

$$S_X(g) = S_x\{f(g)\} \frac{1-\theta^2}{1+\theta^2 + 2\theta \cos 2\pi g}, \tag{6.41}$$

and conversely

$$S_x(f) = S_X\{g(f)\} \frac{1-\theta^2}{1+\theta^2 - 2\theta \cos 2\pi f}. \tag{6.42}$$

Figure 6.3 illustrates the effect of the frequency warp, shown in Figure 6.3(a), on transforming the spectrum of a univariate process x_t, shown in Figure 6.3(b), into the spectrum of the corresponding process X_k, shown in Figure 6.3(c). We say that the spectrum of X_k is displayed on the warped frequency scale g over the warped frequency range or on the warped frequency domain.

The process x_t used in this illustration is constructed as the sum of three independent components, an AR(1) which contributes the spectrum peak at frequency zero, an AR(2) which contributes the spectral peak close to frequency 0.07, and a white noise which adds a uniform component. It may be shown that x_t is an ARMA(3,3) process. The value of $\theta = 0.6$

was used in this illustration. At frequency zero, the gradient (6.10) of the transformation (6.9) is $(1 + \theta)/(1 - \theta) = 4.0$, so the frequency range of the main features of the spectrum in Figure 6.3(b) is expanded by a factor of approximately 4 in Figure 6.3(c).

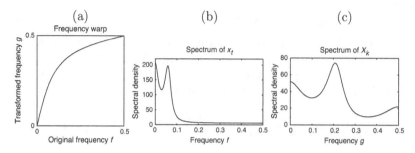

Figure 6.3 *The frequency warp map shown in (a) transforms the spectrum in (b) of the process x_t into the spectrum in (c) of the process X_k.*

We now use the general relationship (6.42) to demonstrate, as stated in (6.23), that the disturbance term n_t of the natural form of the model is an AR(1) process. We define the series $N_k = Z^{-k} n_t$, and recall the model assumption that N_1 is uncorrelated with x_{t-k} for $k \geq 0$. This implies that N_1 is uncorrelated with N_k for $k \leq 0$, and so N_k is multivariate white noise, with variance $V_n = \operatorname{Var} n_t$ and constant spectrum $S_N(g) = V_n$. From (6.42) we have

$$S_n(f) = V_n \frac{1 - \theta^2}{1 + \theta^2 - 2\theta \cos 2\pi f}, \tag{6.43}$$

which is the spectrum of a VAR(1) process with autoregressive coefficient θI, variance V_n and innovation variance $V_\varepsilon = V_n (1 - \theta^2)$. The spectrum of the natural VZAR model (6.21) is derived from $S_n(f)$ as

$$S_x(f) = \varphi(Z)^{-1} \frac{V_n(1 - \theta^2)}{1 + \theta^2 - 2\theta \cos 2\pi f} \varphi(\bar{Z})^{-1}, \tag{6.44}$$

where $Z = \exp(-2\pi i g)$ with $g = g(f)$. Comparing this with (6.42), we conclude that X_k has spectrum

$$S_X(g) = \varphi(Z)^{-1} V_n \varphi(\bar{Z})^{-1}, \tag{6.45}$$

where now $Z = \exp(-2\pi i g)$. This implies that X_k follows the regular VAR(p) model:

$$\varphi(Z) X_k = N_k, \tag{6.46}$$

where Z is the shift operator acting on the index k, i.e., $Z X_k = X_{k-1}$. Methods developed for the VAR(p) model can then be used to calculate the variance and lagged covariances of the states and other properties of a specified VZAR(p, θ) model.

6.5   Approximating a process by the VZAR model

This section serves as a background to the fitting of a VZAR model to an observed time series, which is the subject of the next section. We ask the same question as we asked of the VAR model: how well can the VZAR model approximate a stationary multivariate process with specified covariance function or spectrum? The answer is the same, that the properties and predictions of any stationary process can be approximated with arbitrary accuracy by a VZAR(p) model of sufficiently high order. As we will see, however, the additional flexibility given by the coefficient θ may allow us to approximate a process to the same degree of accuracy as a VAR model but using a much lower order model.

Each of the three forms of the VZAR model, the predictive, the natural and the general form, leads to slightly different approximations when applied to an arbitrary process. For all these models, the approximation is obtained by projecting x_t onto the appropriate set of predictors. We start with the natural form (6.20), for which the approximation of order p is constructed by projecting x_t upon $Z\,x_t$, $Z^2\,x_t$, ..., $Z^p\,x_t$. This gives us a set of multivariate Yule–Walker equations for φ_1, φ_2, ..., φ_p in terms of the covariances defined in (6.39) as $\Gamma_{X,k} = \mathrm{Cov}\,(x_t, Z^k x_t)$. For $k = 1, 2, \ldots, p$ we have

$$\Gamma_{X,k} = \varphi_1\Gamma_{X,k-1} + \varphi_2\Gamma_{X,k-2} + \cdots + \varphi_p\Gamma_{X,k-p}. \qquad (6.47)$$

Provided only that x_t is not a deterministic process, these equations have a unique solution for φ_1, φ_2, ..., φ_p and the approximating VZAR model with these coefficients satisfies the stationarity condition. The variance V_n of the disturbance term n_t in the approximating model is given in terms of the covariances and model coefficients as

$$V_n = \Gamma_{X,0} - \varphi_1\Gamma'_{X,1} - \varphi_2\Gamma'_{X,2} - \cdots - \varphi_p\Gamma'_{X,p}. \qquad (6.48)$$

The theory and application of this approximation is exactly the same as presented for the VAR model in Chapter 2. The recursive method of Whittle (1963) can again be used to derive the approximating models for increasing orders $p = 1, 2, \ldots$. The advantage of using the VZAR model is illustrated by the approximations shown in Figure 6.4 of the spectra presented in Figure 6.3 of a univariate process x_t. Figure 6.4(a) shows the approximation by an AR(4) model, which fails to distinguish the two peaks. Figure 6.4(b) shows the approximation by a ZAR(4,0.62) model, which very well approximates the two peaks. Figure 6.4(c) shows the same as Figure 6.4(b), but on the warped frequency domain. The point of this third plot is that the approximation is exactly that which would be obtained by using a *standard* AR(4) model to approximate the spectrum of the process X_k. That spectrum, shown by the solid line in Figure 6.4(c), occupies the full frequency range, unlike that of the original spectrum in Figures 6.4(a) and (b), which is confined to the lower quarter of the range. A standard AR(4) model can adequately approximate the warped spectrum, but not the original, which requires an AR(16) to give

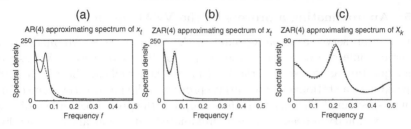

Figure 6.4 *(a) The spectrum of a specified process x_t (solid line) with the spectrum of an approximating AR(4) model (broken line); (b) the spectrum of x_t but with the approximation given by a ZAR(4,0.62) model; (c) the same as (b) but on the warped frequency domain, which corresponds to the approximation of X_k by an AR(4) model.*

an approximation comparable to that of the ZAR(4,0.62). This example has been selected to illustrate the main advantage of the ZAR model, that it can approximate such spectra with a much lower order model. It is a reasonably typical example of the processes for which these models are well suited: those with a spectrum confined to a lower range of frequencies. In fact, the parameter θ in this example was set at 0.62 in order that the approximating ZAR spectrum appear distinct from that of the true spectrum. With θ set to 0.6, they were indistinguishable.

The natural form of model has a useful property that is not shared by the predictive or general forms. If the model is used to approximate an arbitrary process, it will *capture* the lagged covariances $\Gamma_{X,k}$. A process generated from the approximating model will have the same values of $\mathrm{Cov}(x_t, Z^k x_t)$ as those used to derive the approximation in (6.47) and (6.48).

We should, properly, have further indexed the approximating model coefficients by the order p of the approximation, because if p is increased, the set of coefficients will change and be extended. In the limit, as p increases, the variance of the disturbance series will reduce to a limit V_n which, it is very reasonable to assume, will be a strictly positive matrix. The disturbance series will itself converge to a limit n_t, which follows a VAR(1) process with scalar coefficient θ, excited by a multivariate white noise process ε_t, as in the model equation (6.23). The model coefficients will also converge to limits φ_k and, in particular, there is a non-singular limit of $A = \varphi(-\theta)$. The corresponding predictive forms of the approximating natural VZAR models will therefore also converge with a limiting white noise innovation series $e_t = A^{-1}\varepsilon_t$. The important point of this discussion is that, by taking a sufficiently high order p, the approximating model will be indistinguishable, for all practical purposes, from a true VZAR model as defined by (6.20) and (6.23).

We now consider the predictive form (6.14) of the VZAR model, for which the approximation of order p is constructed by projecting x_t upon x_{t-1}, $Z x_{t-1}$, $Z^2 x_{t-1}$, ..., $Z^{p-1} x_{t-1}$. The resulting set of equations for the coefficients $\xi_1, \xi_2, \ldots, \xi_p$ is similar to the multivariate Yule–Walker equations

(6.47), and we will also refer to them as such. The matrix of the new equations is exactly the same, because the covariances between the predicting variables are again $\Gamma_{X,k} = \text{Cov}(x_{t-1}, Z^k x_{t-1})$. However, the variables on the left hand side are changed as follows. For $k = 1, 2, \ldots, p$ we have

$$\text{Cov}(x_t, Z^{k-1} x_{t-1}) = \xi_1 \Gamma_{X,k-1} + \xi_2 \Gamma_{X,k-2} + \cdots + \xi_p \Gamma_{X,k-p}, \qquad (6.49)$$

and the variance of the model error term is

$$\begin{aligned} V_e \;=\; & \text{Var}(x_t) - \xi_1 \text{Cov}(x_t, x_{t-1})' - \xi_2 \text{Cov}(x_t, Z\, x_{t-1})' - \cdots \\ & \cdots - \xi_p \text{Cov}(x_t, Z^{p-1} x_{t-1})'. \end{aligned} \qquad (6.50)$$

Because the matrix of the equations is the same, there is again a unique solution. We can also establish that the resulting model represents a stationary process.

For the general model (6.29), the approximation is obtained from projecting $Z_\rho^{-1} x_t$ upon x_t, $Z\, x_t$, $\ldots$, $Z^{p-1} x_t$, or equivalently, x_t upon $Z_\rho x_t$, $Z_\rho Z\, x_t$, $\ldots$, $Z_\rho Z^{p-1} x_t$, giving, for $k = 1, 2, \ldots, p$, the equations

$$\text{Cov}(x_t, Z^\rho Z^{k-1} x_t) = \zeta_1 \Gamma_{X,k-1} + \zeta_2 \Gamma_{X,k-2} + \cdots + \zeta_p \Gamma_{X,k-p}. \qquad (6.51)$$

The variance of the projection error term is

$$\begin{aligned} V_n \;=\; & \text{Var}(x_t) - \zeta_1 \text{Cov}(x_t, Z_\rho x_t)' - \zeta_2 \text{Cov}(x_t, Z_\rho Z\, x_t)' - \cdots \\ & \cdots - \zeta_p \text{Cov}(x_t, Z_\rho Z^{p-1} x_t)'. \end{aligned} \qquad (6.52)$$

The matrix of these equations is, once more, the same as for the other methods, and we can also confirm, once more, that the model resulting from this projection is stationary.

The advantage of the general form of model lies in the flexibility of its two discount factors ρ and θ. We can show that the approximating model obtained by solving (6.51) minimizes the divergence (4.60) of the model spectrum $S_\zeta(f)$ from the true spectrum $S_x(f)$ of x_t, but with the integral in (4.60) *weighted* by a function $w(f)$ of frequency equal to the spectrum of a scalar AR(1) process with unit variance and parameter ρ, given by

$$w(f) = \frac{1 - \rho^2}{1 + \rho^2 - 2\rho \cos 2\pi f} \qquad (6.53)$$

and illustrated in Figure 6.5 for $\rho = 0.5$.

This places greater weight, shown by the shaded portion in Figure 6.5, upon the lower fraction $(1 - \rho)$ of the frequencies from zero up to the Nyquist frequency. If the spectrum of the process is confined to a range of low frequencies, we may then choose ρ so that preference is given to the fit of the model over this range. We have already shown by the illustration in Figure 6.4 that the parameter θ can result in closer approximation of a spectrum concentrated at low frequencies, and this is by virtue of the *functional form* of the ZAR model spectrum.

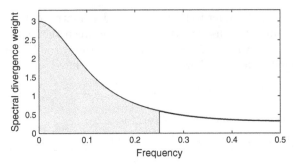

Figure 6.5 *The weight function applied to the divergence of the spectrum of the approximating model from that of the true process when fitting the general form of model with future discount factor* $\rho = 0.5$.

If $\rho = \theta$, we have the natural form of model. We gain insight into the approximation then obtained as the solution of (6.47) by considering the spectrum divergence (4.60) transformed to the *warped* spectrum space. This is

$$\int_{-0.5}^{0.5} \text{trace} \left\{ S_X(g) S_{X,\beta}(g)^{-1} \right\} - \log \det \left\{ S_X(g) S_{X,\beta}(g)^{-1} \right\} dg - m, \quad (6.54)$$

where, for clarity, we also attach the subscript X to the transformed model spectrum $S_{X,\beta}(g)$. The approximation of the natural model, as illustrated in Figure 6.4(c), minimizes the divergence (6.54) with *uniform* weighting applied across the transformed frequency domain. Furthermore, approximating the general model corresponds to minimizing (6.54) with a weighting within the integral given by

$$W(g) = \frac{(1 - \tau^2)}{1 + \tau^2 + 2\tau \cos(2\pi g)}, \quad (6.55)$$

where τ is given by (6.35). This is uniform if $\rho = \theta$, whence $\tau = 0$. If $\rho > \theta$, then $\tau < 0$, and $W(g)$ has the same shape as that shown in Figure 6.5, being concentrated towards low frequencies, but on the transformed frequency domain. If $\rho < \theta$, including the predictive form of model with $\rho = 0$, then $\tau > 0$ and the weight is concentrated towards the higher frequencies in this domain, the shape being as that in Figure 6.5 reversed from right to left. A corresponding weighting is illustrated in the context of continuous time models in Section 7.8. Depending upon the particular example, either of these possibilities might be appropriate.

Some loss of statistical efficiency is incurred by fitting the general form of model with positive value of ρ to real data. This is in comparison to the efficiency of the predictive form, which is the special case of the general form with $\rho = 0$. However, this loss of efficiency need not be large. Consider the simplest case, where the process is a univariate AR(1). The variance of the fitted value of the parameter ϕ is increased by a factor $(1 - \rho^2\phi^2)/(1 - \rho^2)$, which is not much greater than 1, provided ϕ is much closer to 1 than ρ. For example, the increase is only 8.5% if $\phi = 0.99$ and $\rho = 0.9$. In some applications of the general form of model, we may take $\rho < \theta$, which would reduce the loss of efficiency, but in others, we may take $\rho > \theta$ so that the approximation is more robust and appropriate for higher lead time prediction.

If x_t actually followed a true VZAR(p, θ) model, then the approximations obtained from all three approaches would exactly recover models equivalent to that. In other words, when transformed to a common, say predictive, form, the fitted model would be the same whatever value of ρ was used for fitting, including the cases of $\rho = 0$ (predictive) and $\rho = \theta$ (natural). In practice, however, the models fitted using different values of ρ will not be equivalent. The approximation furnished by fitting the predictive model form will, by construction, have a smaller innovation variance V_e than that furnished by the natural or general model. Our experience shows that when the order is chosen to ensure a good approximation, fitting the different model forms will give results that are very nearly equivalent. However, an example in the next section will show that noticeable differences can arise when comparing the spectra after fitting the different model forms. The greatest difference between the models is for very high frequency sampled data, which will be considered in the context of continuous time models in Chapter 7. The robustness and flexibility of the general model form is then particularly useful.

6.6   Yule–Walker fitting of the VZAR

In this and the following two sections we consider how the coefficients of a VZAR(p, θ) model can be estimated for specified values of the model order p and discount factor θ. We will then use the information obtained from fitting the model, particularly the estimate of the model prediction error variance, to assist in selecting the order p and discount factor θ. The methods we use are natural extensions of those presented for the VAR model in Chapter 4; in particular, we develop a generalization, ZIC, of the AIC for determining p and θ. However, there is a new consideration. We have a further choice of model structure, in that the general form of model involves a new parameter ρ which is fixed not by an objective criterion, but by a judgment regarding model robustness. The natural and predictive forms of model are special cases and judgment is implicit in the selection of either of these. Our example will illustrate how this judgment can be made by building models for a small number of prescribed values of ρ.

Following from the previous section, the method we describe in this section is based on solving the Yule–Walker equations (6.47), (6.49) and (6.51) for, respectively, the natural, predictive and general forms of model, using sample values for the constituent covariance matrices of these equations. This implicitly requires generation of the series of states $s_t^{(k)} = Z^k x_t$ from the observed time series x_t, $t = 1, 2, \ldots, n$.

We again encounter an end-effect problem, that transient errors are introduced into the states through lack of knowledge of x_t for $t < 1$, or equivalently, knowledge of the states at time $t = 0$. Typically, these will all be set to zero, and the consequent transient errors will generally decrease with increasing t at the discount rate θ, so that they can be neglected for very long series. One practical solution is to omit a number of values of the states up to some time point t_0 before computing the sample covariances, the partial loss of information being preferable to the bias introduced by the errors. We will consider other solutions when we come to estimation methods based upon regression and maximum likelihood. The solution that we recommend for the Yule–Walker method is first to apply a taper to the series x_t before computing the states. This was previously recommended for reducing bias when estimating the VAR model by the Yule–Walker method. It leads to some loss of information, but that is preferable to the bias.

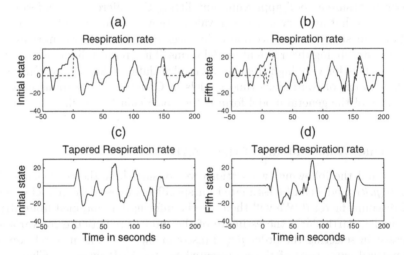

Figure 6.6 *The ZAR states of a section of the Respiration rate: (a) Solid line shows the series (the initial state) and the broken line the truncated series with truncated values set to zero; (b) the fifth state, after applying the Z operator five times with $\theta = 0.5$ to the series shown in (a); (c) the series after applying 10% tapering and (d) the fifth state after applying the Z operator five times to the tapered series.*

Figure 6.6(a) shows a section of the Respiration rate series in which the solid line is the original series values over the time range from -50 to 200 seconds, and the broken line the series truncated by setting to zero the values outside the time range from 1 to 150 seconds. Figure 6.6(b) shows the results of applying the Z operator 5 times to these series. In fact, the solid line is generated using original series values from time origin -100, so it shows the true ZAR state series over the whole range displayed. The broken line shows the results of calculating the states after setting all of them to zero at time zero, which is equivalent to setting all original series values to zero up to time zero. The transient errors at the start of the series are seen, by comparing the solid and broken lines, to extend up to about time 20. The earlier values contain sharp oscillations due to the large discontinuity in the original truncated series. The discontinuity at the end of the truncated series causes a similar oscillation in the corresponding state series, but then does follow the true ZAR state fairly closely for about 15 seconds before dying out. Figure 6.6(c) shows the original series after applying 10% tapering over the time range from 1 to 150 seconds, and Figure 6.6(d) the state series calculated from this. The point to note is the smooth transition at each end of the series.

When the ZAR states are constructed from a tapered series of length n, the sample covariances between them are formed over a time range of length N that extends beyond the end of the original series to a point where all the states have effectively decayed to zero. This ensures that the sample value $C_{X,v}$ of $\mathrm{Cov}(Z^k x_t, Z^{k-v} x_t)$ does indeed depend only on the difference v in ZAR lag. The length n of the original series is retained as the divisor in the definition, which ensures that each state series has the same sample variance as the original series from which it was generated. The matrix of the Yule–Walker equations constructed from these sample covariances is then positive definite, and the VZAR models estimated from their solution is stationary. The sample covariance $C_{X,v}$ can also be obtained from the sample spectrum of the tapered series x_t (see (4.25)) as

$$C_{X,v} = \int_{-0.5}^{0.5} Z^{-v} S_x^*(f) df, \qquad (6.56)$$

and for the estimated coefficients of the predictive form of model, the sample values of the covariances on the LHS of (6.49) can be obtained as

$$\widehat{\mathrm{Cov}}(x_t, Z^{k-1} x_{t-1}) = \int_{-0.5}^{0.5} Z^{-(k-1)} B^{-1} S_x^*(f) df. \qquad (6.57)$$

For the general form, we use

$$\widehat{\mathrm{Cov}}(Z_\rho^{-1} x_t, Z^{k-1} x_t) = \widehat{\mathrm{Cov}}(x_t, Z_\rho Z^{k-1} x_t)$$
$$= \int_{-0.5}^{0.5} Z_\rho^{-1} Z^{-(k-1)} S_x^*(f) df. \qquad (6.58)$$

In these integrals, we set $B = \exp(-2\pi i f)$ and $Z = \exp(-2\pi i g)$ with $g = g(f)$. This results in exactly the same statistics as the time domain sample covariances, provided the integrals are evaluated as approximating sums over a division of the frequency range into N intervals, where, as above, N is the extended length of the ZAR state series. The normalization of the taper avoids bias in the estimate of the variance of the tapered series, but, as for the VAR model, tapering leads to some loss of efficiency that is measured by the tapering factor γ defined in (4.35). This loss is equivalent to having a series of reduced length γn.

We illustrate the Yule–Walker estimation methods by applying them to the three series of Oxygen level, Pulse and Respiration rate series. The methods based on the predictive and natural forms of model correspond, respectively, to the general method with $\rho = 0$ and 0.5, and we will also apply the method based on the general model with $\rho = 0.9$. For comparison of the results, we will set the order $p = 9$ and parameter $\theta = 0.5$ to be the same for all three methods, these values being determined as optimal for the general method with $\rho = 0.9$, as explained shortly. We will again take records over a period of 1000 seconds, but now with a sampling interval of 0.5 seconds.

The resulting three models can all be cast in the same form, for example the natural form, for the purpose of comparing their parameters. We do not do this, but instead plot, in Figure 6.7, selected spectral properties of the three fitted models (solid lines), with the smoothed sample spectra superimposed (broken lines). This is appropriate because fitting the models now corresponds to minimizing the Whittle deviance (4.58) weighted by the function defined in (6.53) and illustrated in Figure 6.5. The bandwidth of 0.02 cycles per second (cps) was used for the smoothed sample spectra. Note that the Nyquist frequency is 1 cps, but we have set the maximum of the frequency range in the plot to 0.25 cps because there is little spectral power above this limit.

The spectra are similar for all three different methods of fitting the model, but using the general model appears to resolve better the lower frequency pattern of the spectra. The choice of the parameter $\rho = 0.9$ applies about 80% of the weight to frequencies up to 0.1 cps and corresponds to an effective (low frequency) lead time of 19 sampling intervals, close to 10 seconds. The results are not particularly sensitive to this choice; they are very similar if one chooses $\rho = 0.8$ with an effective lead time of 5 seconds. Fitting the natural model corresponds to a weight function with $\rho = \theta = 0.5$, with 80% of the weight now applied to frequencies up to 0.5 cps. Fitting the predictive model, the weight is uniformly spread up to 1 cps.

We now describe how to determine the order p and parameter θ to use in the model, but this is conditional upon the specification of the parameter ρ, which must be set by subjective consideration of the particular application. More than one value of ρ might be considered, including $\rho = 0$ for the predictive model form, with p and θ determined for each choice of ρ. Plots such as those in Figure 6.7 can then be compared in order to make a final choice. For the example illustrated in Figure 6.7, we first set the value of $\rho = 0.9$. This was determined according to the guidance given at the end of the previous section. The univariate spectra of the series are essentially confined to the frequencies lower than 0.1 cps, 10% of the Nyquist frequency, which is 1 cps. This suggests using $\rho = 1 - 0.1 = 0.9$.

To determine θ, we consider a selection of possible values: in this example we took 0.0, 0.1, 0.2, ..., 0.9. For each of these, we select an optimal order p based on an information criterion (ZIC) derived for the VZAR model. We then compare the optimal ZIC for each value of θ and select the model with the overall minimum.

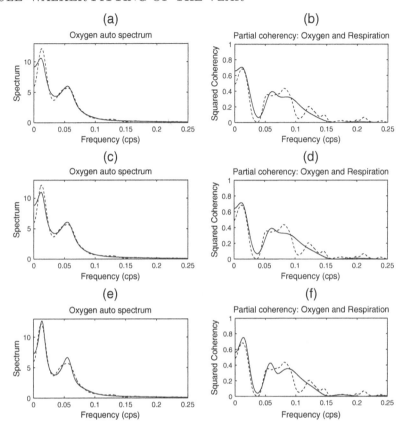

Figure 6.7 *Plots of the model spectra (solid lines) and smoothed sample spectra (broken lines) for the VZAR(9,0.5) model fitted by the Yule–Walker method to the Oxygen level and Pulse and Respiration rates sampled at intervals of 0.5 seconds, using in (a) and (b) the predictive form of the model, in (c) and (d) the natural form of the model and in (e) and (f) the general form of the model with $\rho = 0.9$. The bandwidth used for the smoothed spectral estimation is 0.02 cps.*

We shortly motivate and present an expression for the ZIC following its application to the Infant monitoring series. In the next section, we give a further explanation for this expression based upon the large sample model estimation properties that can be derived using the regression method. We continue here with our example based upon $\rho = 0.9$ by showing in Table 6.1 the orders selected for each θ and the corresponding ZIC, with the overall minimum ZIC subtracted to better reveal the variation.

We therefore select $p = 9$ and $\theta = 0.5$ as giving the minimum ZIC. We note that, for the lowest values of θ, the optimal orders are very much higher and the order of the selected model is much lower than these, though this will not necessarily be the case. The ZIC criterion is based upon the estimated

Table 6.1 *The selected order p and corresponding value of the ZIC for a range of discount factors θ used to fit the VZAR(p, θ) model to the Oxygen level and Pulse and Respiration rate series using the Yule–Walker method.*

θ	0.0	0.1	0.2	0.3	0.4	0.5	0.6	0.7	0.8	0.9
p	26	22	18	15	11	9	14	11	8	5
ZIC	7.6	5.7	5.6	8.1	4.6	0	9.0	8.4	178.0	563.1

variance matrix $\hat{V}_n$ of the prediction error, as defined in (6.52). Recall that, for the standard VAR(p) model, the AIC is expressed as $n\gamma \log \det \hat{V}_e + 2\,m^2\,p$. The factor γ is the correction applied when the model is fitted by solving Yule–Walker equations generated from the tapered series. The second term in the expression for the AIC is a penalty derived from the combined effects of estimation bias in V_e and the increase in prediction error variance due to parameter estimation. For the VZAR(p, θ), fitted in the general model form, the penalty term is only the same as for the AIC if $\rho = 0$, and more generally, it depends upon the parameters of the true model. Nevertheless, we can apply the following generalized information criterion to select the model order for a given value of θ:

$$\text{ZIC} = \text{ZDev} + 2\,m^2\,p\,\frac{(1 + \rho\theta)}{(1 - \rho\theta)}. \tag{6.59}$$

Here, ZDev is the measure minimized to fit the model, analogous to the deviance for the predictive model, and given by

$$\text{ZDev} = n\gamma \log \det \hat{V}_n. \tag{6.60}$$

The ZIC as defined in (6.59) is a simple and effective expression for selecting the order for a given value of θ, but it cannot then be used to select between different values of θ. For that purpose we replace (6.59) by ZIC = ZDev + P where we use the full expression of the penalty term P as a function of the model transformed to its natural form:

$$P = 2\,m^2\,p\,\frac{(1 + \rho\theta)}{(1 - \rho\theta)} - 2\,m\,\text{trace}\left\{\varphi(-\tau)^{-1}\varphi'(-\tau)\right\}\rho\frac{(1 - \theta^2)}{(1 - \rho\theta)^2}. \tag{6.61}$$

Here, $\varphi(Z)$ is the operator of the natural model form as defined in (6.22), $\varphi'(Z)$ is the derivative of $\varphi(Z)$ with respect to Z and τ is defined in (6.35). This expression for the penalty can be consistently estimated from the fitted VZAR(p, θ) model. It is applied to form the ZIC of that model for each value of θ and its corresponding selected value of p. The model that has the lowest ZIC amongst these determines the selected value of θ, as in Table 6.1.

The reason why we can use the simpler penalty term in (6.59) to select the model order for a given value of θ is that the second term in (6.61) will remain constant for any model order p equal to or greater than that of the true model. We need then only use the first term of (6.61) as the penalty in (6.59).

Using the full expression (6.61) at that stage would introduce unnecessary extra sampling variability in the second term as the order increased.

Figure 6.8 shows the result of fitting the VZAR model having set $\rho = 0$, which specifies the predictive model form. The selected value of θ was zero, and the order $p = 21$, though there was a local minimum in the table of ZIC—which becomes the AIC in this special case—at $\theta = 0.5$ and $p = 9$. In comparison with Figure 6.7(c) and (d), we see that the fit to the smoothed sample spectra is much less close at low frequencies, less close even than fitting the predictive form with $\theta = 0.5$ and $p = 9$, as shown in 6.7(a) and (b). This underlines the fact that for models to be robust to higher sampling frequencies, we may be wise to avoid fitting the model so as to minimize the prediction error variance one sampling period ahead.

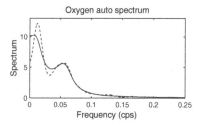 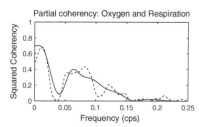

Figure 6.8 *Plots of the model spectra (solid lines) and smoothed sample spectra (broken lines) for the VZAR(21,0.0) model fitted by the Yule–Walker method to the Oxygen level and Pulse and Respiration rates sampled at intervals of 0.5 seconds, using the predictive form of model. The bandwidth used for the smoothed spectral estimation is 0.02 cps.*

6.7   Regression fitting of the VZAR

We now extend the method of OLS regression which we used for VAR models to fit the VZAR model. We will present the treatment for the general model by regressing the response variables $Z_\rho^{-1}x_t$ upon the explanatory variables $Z^k x_t$ for $k = 0,\ldots,p-1$. The estimates for the predictive form of model can then be found by taking $\rho = 0$, and for the natural form of model by taking $\rho = \theta$. The regression approach provides a way of dealing with the end-effect problem by including constructed sequences which compensate for the transient errors introduced into the response and explanatory variables through lack of knowledge of x_t for, respectively, $t > n$ and $t < 1$.

The response vectors are constructed to have elements $y_t = Z_\rho^{-1}x_t$ for $t = 1,\ldots,n$ using

$$y_t = x_{t+1} - \rho x_t + \rho y_{t+1} \quad \text{for} \quad t = n, n-1, \ldots, 1, \tag{6.62}$$

and setting x_{n+1} and y_{n+1} to zero to start the recursions. In general, y_t is multivariate, but the regression will be carried out for each component in turn as the response, as for VAR models. The elements of the vectors of explanatory variables in each of these regressions are the values of the states $s_t^{(k)}$ for $t = 1, \ldots, n$. These are generated using (6.4) and (6.5), or equivalently the systems equation (6.6), setting the vector of states $s_{t-1}^{(k)}$ to zero to start the recursions at $t = 1$.

The transient error in y_t is $\rho^{n-t}(x_{n+1} + \rho y_{n+1})$, so we introduce a new explanatory vector with elements ρ^{n-t} for $t = 1, \ldots, n$ into each regression to compensate for this. The transient errors in the regression vectors constructed from the states $s_t^{(k)}$ depend linearly on $s_0^{(j)}$ for $j \leq k$. They span a space of dimension k which is the same for all of the component series. Moreover, the space for $k = p-1$ contains the space for all lower values of k, so to compensate for the errors we need only include in the regression a further set of $p-1$ basis vectors of this space. These are readily generated as the impulse responses from any one component of the unknown series value x_0 on the corresponding component of $s_t^{(k)}$ for $k = 1, \ldots, p-1$. These can be generated by applying the system equation (6.6) but with the dimension m set to 1. Take all $x_t = 0$ for $t = 1, \ldots, n$ and set all $s_0^{(k)} = 0$ except $s_0^{(0)} = 1$. The series $s_t^{(0)}$ so generated is identically zero; the remaining $p - 1$ series constitute the required basis vectors.

Note that if we set $\rho = 0$ and $\theta = 0$, this procedure reduces to the standard lagged regression for the VAR(p) model. The response vector is the series x_{t+1} with the unknown last element at time $t = n$ set to zero. The related compensatory vector is the indicator variable for this last element so that its effect is the same as removing the last element of all the variables, response and explanatory. The explanatory variables are the vectors of lagged values x_{t-k} for $k = 0, 1, \ldots, p - 1$ with values for $t - k < 1$ set to zero, and the compensatory vectors associated with these are the indicator variables for the first $p-1$ elements. The effect of these is the same as removing the first $p-1$ elements of all the vectors. The response vector then contains $x_{1+p}, \ldots, x_n$, and the explanatory vectors of greatest lag contain $x_1, \ldots, x_{n-p}$.

We have selected a VZAR(p, θ) model using this regression estimation procedure in the same manner as that using the Yule–Walker procedure. We used the same series and the same value of ρ. However, for comparing models of increasing orders k for a fixed value of θ, we have used the strategy of including in each regression the set of compensating vectors appropriate to the model of maximum order p. The reason for this is best explained in the simpler context of ordinary lagged regression. Our strategy corresponds to taking as response series the vector containing $x_{1+p}, \ldots, x_n$; however, many lagged regressors (up to the maximum of p) are used, rather than taking, for any $k < p$, the longer response vector containing $x_{1+k}, \ldots, x_n$. The only change as the order is increased is then the set of lagged regressors. Unfortunately, this means that the choice of maximum order does influence the estimated value of the measure of fit of the model, $\hat{V}_n$. We fixed upon a maximum order of $p = 20$, and Table 6.2 shows that the selected model has order 11 with discount factor 0.7, both

somewhat higher than those selected by the Yule–Walker method. We will
shortly discuss reasons for the differences between this table and Tables 6.1
and 6.3 corresponding to Yule–Walker and maximum likelihood estimation
methods, but move on now to compare the selected models.

Table 6.2 *The selected order p and corresponding value of the ZIC for a range of
discount factors θ used to fit the VZAR(p, θ) model to the Oxygen level and Pulse
and Respiration rate series, using the regression method.*

θ	0.0	0.1	0.2	0.3	0.4	0.5	0.6	0.7	0.8	0.9
p	20	20	18	14	11	9	15	11	8	20
ZIC	408	383	372	372	359	292	138	0	107	252

Figure 6.9 shows plots of the same properties of the fitted model as were
shown for the Yule–Walker procedure in Figure 6.7(e) and (f). The main
difference is that the peak in the Oxygen auto-spectrum is now much sharper,
and the peak in the squared coherency spectrum is higher. The bandwidth of
0.01 cps was used for the smoothed sample spectra superimposed on the plots
to better match the model spectra. Reducing the bandwidth to 0.005 gave an
even closer match of the low frequency peaks but gave very variable smoothed
spectral estimates at higher frequencies. However, for a given value of ρ, the
spectral properties of the model fitted by the regression approach are almost
identical to those of the model fitted by the Yule–Walker approach if the model
parameters, (p, θ) are the same. Thus, if we fit the VZAR(9,0.5) model by the
regression approach, the properties, though not shown separately, are the same
as those shown in Figure 6.7 using the Yule–Walker approach, and if we fit the
VZAR(11,0.7) model by the Yule–Walker approach, the properties are again
the same as those shown in Figure 6.9 using the regression approach. This
arises because the series is quite long ($n = 2000$ points) and the estimation
is relatively insensitive to the method of treating the end-effects. The least
squares estimation equations from the regression method, when divided by n,
will therefore be close to the Yule–Walker equations. The difference between
the two methods in this example lies solely in the selection of θ; they select
the same order of 9 for $\theta = 0.5$ and of 11 for $\theta = 0.7$.

For relatively short series, Yule–Walker estimation of univariate autoregressions is known to be biased, with the result that peaks in the model
spectrum are broadened and reciprocal roots of the autoregressive operator
shrunk away from the unit circle. The use of tapering can reduce both this
bias and the broadening effect on the spectral peaks, but lagged regression
based estimation of the models is not prone to such bias and in general is better able to fit very sharp spectral peaks. The regression approach can suffer
a different drawback, however: the fitted model may not satisfy the stationarity condition. This extends to the regression procedure we have described
for VZAR models; the reciprocal roots of $\det \varphi(Z) = 0$ may lie outside the

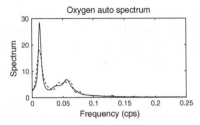
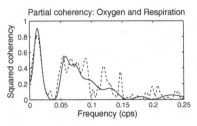

Figure 6.9 *Plots of the model spectra (solid lines) and smoothed sample spectra (broken lines) for the VZAR(11,0.7) model fitted by the regression method to the Oxygen level and Pulse and Respiration rates sampled at intervals of 0.5 seconds, using the general form of model with $\rho = 0.9$. The bandwidth used for the smoothed spectral estimation is 0.01 cps.*

unit circle. This is most likely to occur when the series may be described by a model with reciprocal roots very close to the unit circle, which are typically associated with sharp peaks in the model spectrum. We have checked that this problem does not occur for the model estimates used in the examples we have presented, but it does for some of the models when rather higher orders are used. The Yule–Walker procedure always estimates a stationary model. In the next section we describe another approach to model estimation using a pseudo-likelihood criterion, which treats the end-effects in a more sophisticated manner. This approach overcomes the drawbacks of the Yule–Walker and regression approaches, but at the cost of computational complexity.

We conclude this section with a consideration of the large sample properties of the model parameters estimated by the regression method. Our three methods of estimation will be equivalent in large samples, but the regression approach is convenient for developing these properties. The results we now present apply, however, only to estimation of the saturated VZAR(p, θ) model in which the full set of $m^2 p$ parameters is estimated. As we showed for the VAR model, least squares estimation of the dependence of each component series of the response *separately* on the explanatory regressors gives the same result as minimizing the general estimation criterion, which is the determinant of the prediction error covariance matrix, *simultaneously* with respect to the regression coefficients for all the responses. The large sample properties of estimated model parameters may be useful for testing whether individual parameters might be removed to obtain a subset regression model. However, they are also used to derive the model selection criterion (6.61), and we present an application where they are used to determine error limits on model properties, such as the responses of one of the series variables to another.

Let us assume that we are fitting the predictive form (6.14) of the VZAR(p, θ) model, with p and θ given, to a multivariate Gaussian time series x_t of length n which can be represented exactly by this model. The variance matrix of the parameter estimates obtained from the application of ordinary least squares theory is then valid for inference in large samples. This matrix is given, for the set of coefficients relating to the response series $x_{i,t}$, by the standard formula from least squares:

$$\hat{\sigma}_{e,i}^2 (X'X)^{-1} \tag{6.63}$$

where $\hat{\sigma}_{e,i}^2$ is the estimated prediction error variance and X is the regression matrix with the row for time t composed of elements of the state vector S_{t-1} as defined in (6.37). We omit the dummy vectors for the transient state errors as being negligible in sufficiently large samples. Now $n^{-1}X'X$ converges to the variance matrix V_S of S_t, which leads to the large sample property of the vector $\hat{\xi}_{(i)}$ of the subset of coefficients $\xi_{i,j,k}$ relating to the response $x_{i,t}$:

$$n^{-1/2}(\hat{\xi}_{(i)} - \xi_{(i)}) \sim N(0, \sigma_{e,i}^2 V_S^{-1}). \tag{6.64}$$

This is readily extended to the large sample property for the whole vector of coefficients:

$$n^{-1/2}(\hat{\xi} - \xi) \sim N(0, V_S^{-1} \otimes \Sigma_e), \tag{6.65}$$

where the variance matrix of the normal distribution is the Kronecker product (see (4.37)) of V_S^{-1} with the variance matrix Σ_e of the error vector e_t. We now state the further extension to the large sample properties for estimates of the vector of coefficients ζ in the general model form. The regressors are exactly the same, but allowance must be made for the correlation of the errors n_t in this model, giving

$$n^{-1/2}(\hat{\zeta} - \zeta) \sim N(0, (V_S^{-1} \Omega V_S^{-1}) \otimes \Sigma_n), \tag{6.66}$$

where Σ_n is the variance matrix of the error vector n_t and

$$\begin{aligned} \Omega &= \sum_{k=-\infty}^{\infty} \rho^{|k|} \mathrm{Cov}(S_t, S_{t-k}) \\ &= V_S + \rho(I - \rho T)^{-1} V_S + \rho V_s (I - \rho T')^{-1}. \end{aligned} \tag{6.67}$$

Here, T is the state transition matrix as defined following (6.37), where the expression for $\mathrm{Cov}(S_t, S_{t-k}) = T^k V_S$ is also given. The matrix Ω is readily and consistently estimated from the fitted model, but one practical point is that the computation of V_S and Ω requires that the fitted model is stationary, so this needs to be checked if the regression method of estimation is used. One application of (6.66) is that the penalty term in the ZIC may be derived as

$$2m\,\mathrm{trace}\left\{V_S^{-1}\Omega\right\} = 2m\,\mathrm{trace}\left\{(I - \rho T)^{-1}(I + \rho T)\right\}. \tag{6.68}$$

The eigenvalues of T are the reciprocal roots of $\det \varphi(Z) = 0$, the operator in the natural form of the model, and this relationship can be manipulated to give the expression (6.61).

6.8   Maximum likelihood fitting of the VZAR

We first present the procedure for deriving the exact likelihood of the predictive form of the VZAR(p, θ) model. The approach we use may be viewed as overcoming the transient effects of the unknown states S_t at time $t = 0$ by estimating these states from the observations. Given a set of model parameters ξ and V_e, the first step is to present the joint density of the set of observations x, consisting of x_t, $t = 1, 2, \ldots, n$, and the initial states S_0, which we will write simply as S for convenience of notation in the following development:

$$
\begin{aligned}
p(x, S) &= p(x|S) \times p(S) \\
&= |V_e|^{-n/2} \exp\left(-\sum_{t=1}^{n} e_t' V_e^{-1} e_t / 2\right) \\
&\quad \times |V_S|^{-1/2} \exp(-S' V_S^{-1} S / 2) \\
&= |V_e|^{-n/2} |V_S|^{-1/2} \exp\left\{-Q(S)/2\right\}.
\end{aligned} \tag{6.69}
$$

The first factor in the second line is the density of the observations given the initial states. It is expressed in terms of the prediction error vectors e_t, $t = 1, 2, \ldots, n$, that can be regenerated from the observations for a proposed value S. The second term is the marginal density of the initial states. In the third line the quantity $Q(S)$ in the exponential part of the product represents a quadratic in S. To derive this explicitly, we set $e_t = e_t^{(0)} + a_t S$ where e_t^0 is the error at time t regenerated from the value $S = 0$ and a_t is the (matrix) coefficient in the linear dependence of e_t upon S. The elements of a_t can be derived by regenerating e_t with all $x_t = 0$ and indicator variables substituted for S. Then

$$
\begin{aligned}
Q(S) &= \sum_{t=1}^{n} e_t^{(0)\,'} V_e^{-1} e_t^{(0)} + 2S' \sum_{t=1}^{n} a_t' V_e^{-1} e_t^{(0)} \\
&\quad + S' \left\{ \sum_{t=1}^{n} a_t' V_e^{-1} a_t + V_S^{-1} \right\} S \\
&= a + 2S'b + S'BS.
\end{aligned} \tag{6.70}
$$

The minimum of $Q(S)$ occurs at the solution $\tilde{S}$ of $b + BS = 0$. Write $\tilde{e}_t$ for e_t regenerated using $\tilde{S}$ so that

$$
Q(\tilde{S}) = \sum_{t=1}^{n} \tilde{e}_t' V_e^{-1} \tilde{e}_t + \tilde{S}' V_S^{-1} \tilde{S} \tag{6.71}
$$

and factorize the joint density as

$$
\begin{aligned}
p(x, S) &= |V_e|^{-n/2} |V_S|^{-1/2} |B|^{-1/2} \exp{-Q(\tilde{S})/2} \\
&\quad \times |B|^{1/2} \exp{-(S - \tilde{S})' B (S - \tilde{S})/2} \\
&= p(x) \times p(S|x).
\end{aligned} \tag{6.72}
$$

The quantity of minus twice the log likelihood of the model parameters, which we call the deviance, is then

$$\text{Dev}(\xi, V_e) = n \log |V_e| + \log |V_S| + \log |B| + Q(\tilde{S}). \tag{6.73}$$

Maximum likelihood estimation of ξ and V_e is then conveniently carried out by numerical minimization of the deviance. Initial parameter estimates obtained using the Yule–Walker or regression method reduce the computation time.

We remark that the prediction error decomposition form of the likelihood of the predictive form of model can be derived using the state space approach. We have implemented it and find that it is not quite as rapid computationally, though it does of course give exactly the same value.

To fit the general form of model, we modify the likelihood method as presented above to generate a quasi-likelihood and corresponding quasi-deviance. Following (6.60), we will also call this ZDev, though, because it takes into account the end-effect of estimating initial states, it has a somewhat more complex form. We will retain the parameterization in terms of ξ and V_e and also retain the step of estimating the initial state S_0. However, we replace the first term of $Q(\tilde{S})$ in (6.71) by an objective function relating to the error of prediction of $Z_\rho^{-1} x_t$, which is

$$Z_\rho^{-1} n_t = (1 - \rho B^{-1})^{-1} M e_t = \nu_t. \tag{6.74}$$

Here, n_t is the autoregressive disturbance term (6.31) in the general form of model (6.29) and M is the matrix coefficient relating the error terms ε_t and e_t in (6.33). The variance of ν_t is then

$$\text{Var}(\nu_t) = V_\nu = M V_e M' / (1 - \rho^2). \tag{6.75}$$

To implement this approach we generate

$$\nu_t = \rho \nu_{t+1} + M \tilde{e}_t \quad \text{for} \quad t = n, n-1, \ldots, 1 \tag{6.76}$$

taking the unknown $\nu_{n+1} = 0$ because it is completely uncorrelated with the observations. The variance of ν_t will be reduced, for t close to n, by the factor $f_t^2 = 1 - \rho^{2(n-t+1)}$ and in the following we will take it that ν_t has been corrected to constant variance by dividing by f_t. The quasi-deviance is then

$$\begin{aligned} \text{ZDev}(\xi, V_e) &= n \log |V_\nu| + \log |V_S| + \log |B| \\ &\quad + \sum_{t=1}^{n} \nu_t' V_\nu^{-1} \nu_t + \tilde{S}' V_S^{-1} \tilde{S}. \end{aligned} \tag{6.77}$$

Again, this may be minimized numerically using starting values of ξ and V_e transformed from the estimates of ζ and V_ζ obtained by the Yule–Walker or regression approaches. We call this the quasi-maximum likelihood (QML) estimation procedure.

We applied QML using the same illustrative data set used for the Yule–Walker and regression approaches and using the same formulation of the ZIC to select the model orders. These are shown in Table 6.3. We note that the selected values of p and θ are the same as for the regression method. The table is also generally more similar to Table 6.2 for the regression method than Table 6.1 for the Yule–Walker method, though there are appreciable patterns of differences, particularly in the ZIC for lower values of θ, which are much higher for the regression method. We have not examined closely the reasons for the differences between the three tables, but believe that they arise because the fitted models are close to the boundary of the stationarity region. Indeed, this is why we limited the regression method to a maximum order of 20, because some of the fitted models of higher order were not stationary. The penalty terms in either of the forms (6.68) and (6.61) are sensitive to small changes in the model coefficients close to the stationarity boundary. The Yule–Walker estimates tend to have coefficients constrained to be further from this boundary than those of the maximum likelihood estimates, and the regression method suffers the least constraint so that it may transgress the boundary. We have carried out simulations (Tunnicliffe Wilson and Haywood (2012, p. 520)), which confirm that the QML method has good order selection properties when applied to a univariate series following a ZAR(4, 0.9) model estimated in the general form with $\rho = 0.7$. This model also had coefficients close to the stationarity border.

For the present example we limited the orders to 25 when using the QML method, but only for the practical reason of avoiding excessive computing time. The order of 25 was only selected for $\theta = 0$. The spectral properties of the selected VZAR(11, 0.7) model were indistinguishable from those for the regression method shown in Figure 6.9. Selection of models using either the Yule–Walker or regression approaches is very fast compared with the maximum likelihood method, though it should be possible, given some technical effort, to improve this last method substantially by calculating the derivatives of the terms in (6.77). It therefore seems sensible for long series of observations to use the regression method, say, for selecting the model orders, though finally estimating the model parameters by QML. For short series the three approaches might not be so consistent in their selection of orders, and the computational cost of using maximum likelihood may not be so great as to deter its use for order selection. Furthermore, if subset regression models were to be estimated after eliminating some parameters by statistical testing, the Yule–Walker and regression methods may not be fully efficient and QML estimation is strongly recommended.

The AIC has been criticized because of a tendency to overestimate the order of a model. Other criteria, such as that due to Hannan and Quinn (1979), consistently select the correct model order. Logically, this criticism only applies if the true model lies in the set of models being fitted. The more practical situation is that the data generating model can only be approximated by the set of models fitted, though approximated to arbitrary accuracy by a

Table 6.3 *The selected order p and corresponding value of the ZIC for a range of discount factors θ used to fit the VZAR(p, θ) model to the Oxygen level and Pulse and Respiration rate series, using the QML method.*

θ	0.0	0.1	0.2	0.3	0.4	0.5	0.6	0.7	0.8	0.9
p	25	22	18	14	11	9	15	11	8	21
ZIC	38	32	30	31	26	22	11	0	171	313

sufficiently high order model. In that case the choice of criterion represents a compromise between the disadvantages of over-fitting and the accuracy of model approximation. Our criterion ZIC was motivated by the concept of the final prediction error (FPE) criterion which developed into the AIC. However, it is easily modified to become a generalization of the Hannan–Quinn criterion by replacing the factor 2 premultiplying the second term in the ZIC by $2 \log(\log n)$. The simulation study of Tunnicliffe Wilson and Haywood (2012, p. 520) compares the effect on the selection of model order of using this modification. For our sample length of $n = 2000$ the value of the factor $2 \log(\log n)$ is 4.0565, so the penalty term is doubled. Table 6.4 shows the effect this has on the selection of model orders, which are much reduced. Interestingly, the order of the selected model is now the highest, contrasting with Table 6.3, in which the order of the selected model is among the lower ones.

Table 6.4 *The selected order p and corresponding value of the Hannan–Quinn modification of the ZIC for a range of discount factors θ used to fit the VZAR(p, θ) model to the Oxygen level and Pulse and Respiration rate series, using the QML method.*

θ	0.0	0.1	0.2	0.3	0.4	0.5	0.6	0.7	0.8	0.9
p	5	5	4	4	5	7	6	6	4	5
ZIC	136	133	148	160	114	0	11	88	405	506

The spectral properties of the VZAR(7,0.5) model selected by this Hannan–Quinn modification of the ZIC when fitted by QML to the general model form are not shown. This is because they are very similar to those in Figure 6.7(a) and (b) using the predictive form of the VZAR(9,0.5) model. In comparison with the properties, shown in Figure 6.9, of the model selected by the unmodified (or standard) form of ZIC and fitted by regression, the main consequence is the much lower resolution at low frequencies, as evidenced by the much broader spectral peak of the Oxygen autospectrum in Figure 6.7(a). We consider that the higher resolution of this spectrum, shown in Figure 6.9, corresponds to the better model because of its closer match to the low bandwidth spectrum estimate shown on the figures. This is achieved by using the general form of model, either the regression or QML estimation, and the standard form of the ZIC.

6.9   VZAR model assessment

In the previous sections of this chapter we have compared the spectral properties of the VZAR model with those of the smoothed sample spectra of the series. We gave an expression for these properties in terms of the natural form of the model in (6.42). The spectrum $S_X(g)$ used in that expression is given in (6.45). Other properties of the models that may be derived are the auto- and cross-covariances, which may again be compared with sample values in order to assess how well the model captures these statistics. The calculation of the covariance properties of the VZAR model was described using the state space formulation in Section 6.2. Figure 6.10 shows the sample lagged cross-correlations between the Oxygen level and Respiration rate series superimposed on those derived from the VZAR(11,0.7) model fitted by QML with $\rho = 0.9$. The correspondence is good.

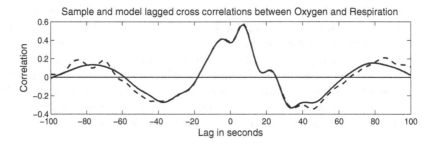

Figure 6.10 *The sample lagged cross-covariances between the Oxygen level and Respiration rate series (broken line) superimposed on those of the VZAR(11,0.7) model fitted by QML with $\rho = 0.9$ (solid line).*

For the example we used, the Oxygen level and Pulse and Respiration rate series, further model properties of interest are the impulse and step responses from one variable to another in the absence of any random disturbance. For example, the step response of Oxygen to Respiration is the observed increase in Oxygen level in response to raising the Respiration level by one unit. In the infant this cannot be observed experimentally, but a model of the system can be used to infer this response, which may be a useful characteristic of the respiratory system. Figure 6.11 shows the responses inferred using the VZAR(11,0.7) model. The system is conveniently simulated by using the model in state space form and setting all initial states to zero at time $t = 0$, except the states corresponding to the value of Respiration, which is set to one. The transition matrix is then applied repeatedly, and with no disturbances added, to generate successive states from which the resulting sequence of values of Oxygen can be extracted. This is the impulse response of Oxygen to Respiration, and the step response is found by cumulative summation of this series. The eventual level of the step response is difficult to estimate precisely in systems where disturbance to one variable feeds back to all remaining variables. It is a property of the low frequency characteristics of the system, which is why it is important to select a model which represents the spectrum well, with good resolution at low frequencies.

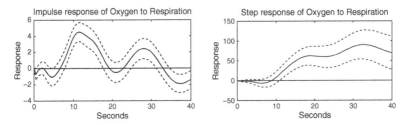

Figure 6.11 *Properties of the VZAR(11,0.7) model fitted to the Oxygen level and Pulse and Respiration rate series, sampled at intervals of 0.5 seconds, using the QML method with $\rho = 0.9$: (a) the impulse response of Oxygen level to a unit increase in Respiration rate and (b) the step response of Oxygen level to a unit increase in Respiration rate. The solid lines are the responses and the dashed lines show plus or minus two standard deviations.*

We show error limits for the responses in Figure 6.11. The classical approach to constructing such intervals is based on the profile likelihood ratio of the property of interest, but this can be computationally very demanding so we again used the simulated delta method presented in Section 4.4. Although the limits derived by this method are approximate, they do give a useful indication of sampling variability, as illustrated in the figure. We believe that the value of the VZAR models for describing the dynamic interdependence of time series is well illustrated by this example.

portable. It is a property of the low-frequency characteristics of the system, which is why it is important to select a model within this region that performs well with post-estimation use in question.

Figure 6.9: Response of the VNAR (VAR) model fitted to the 3-year bond and 10-year bond yields. The top left segment at intervals of 0.5 shows the response of the 3-month bond. The response (the 3-year bond) to a step increase at 10-year bond yields. The bottom row shows the step response of 10-year bond to a shift increase in the 3-year bond yield. The solid lines are the responses and the dashed lines show plus-or-minus two standard errors.

We show error limits for the responses in Figure 6.9. The classical approach to constructing such intervals is based on the profile likelihood ratio of the property of interest, but this can be computationally very demanding so we again used the simulated delta method presented in Section 4.3. Although the limits derived by this method are approximate, they do give insight. In illustration of sampling variability is illustrated in the figure. We believe that the value of the VNAR model for describing the dynamic interdependencies of time series is well illustrated by this example.

Chapter 7

Continuous time VZAR models

7.1 Continuous time series

The early time series work of Wiener was developed in the context of electrical engineering when the nature of the series was of a continuous analog signal. In the present digital age, variables which are naturally continuous, such as the Oxygen levels of the infant monitoring system, are not recorded continuously, but are discretized by sampling at high frequency and then recorded digitally. The recorded music industry has long since made this transition from analog vinyl records to the digitized compact disc format. Before analyzing high frequency sampled discretized signals, it is usual to smooth them to reduce observation noise and then to sub-sample them at a reduced rate, chosen so that little information is lost. This is done by confirming that there is negligible power in the spectrum beyond the Nyquist frequency associated with the sampling rate of the final recorded signal. In the context of sound recording, this enables the continuous signal to be accurately reconstructed as an electrical voltage to be fed into an amplifier and speaker system.

Whereas the discrete time models presented in previous chapters used summation of weighted values of lagged variables to construct predictors, the natural operations applied to continuous times series, as presented in Wiener's early work, were differential and integral operators. Continuous time systems are still very important in simulation and control applications, and the software used in these systems still naturally uses models expressed in terms of differential and integral operations. However, these are applied to discretized data using quadrature and related numerical techniques. The reason for using models expressed in this form is not simply that it is natural to do so, but because models expressed in terms of discrete lags, such as the standard autoregressive predictor, can be very ill-conditioned with severe parameter collinearity when applied to high frequency sampled data.

The ZAR models introduced in the previous chapter were motivated in part by this same limitation on the standard autoregressive model, when applied to time series with spectra in which power tends to be concentrated at low frequencies. The continuous time models that we now propose are the natural limiting form of the ZAR model, as the sampling interval reduces to zero. The formulation and application of the continuous time ZAR model, or

CZAR model, is therefore very similar to that of the discrete time version. In fact, our own formulation of CZAR models evolved from the continuous time models of Belcher et al. (1994) and thence to the discrete ZAR models. Continuous time process models based upon the Laguerre filter methods of Wiener (1949) have previously been formulated and applied, for example, to point processes by Ogata and Akaike (1982). More recently, Solo and Pasha (2012) have reformulated these models for the joint modeling of analog and point process signals. These models are formally identical to those of Belcher et al. (1994) and our present CZAR models.

However, before presenting this model, we remark on another important context for the use of continuous time models. This is in the representation of irregularly sampled time series data where the instances of the observations do not lie on any grid of regularly spaced time points. A continuous time formulation of the model is then obligatory, and in Chapter 8 we show how it can be applied in this way. It was the univariate modeling of such irregularly sampled data by Belcher et al. (1994) which led to the present CZAR model formulation. This data was recorded at equal intervals of depth in a core of ocean sediment, but varying sedimentation rates meant that the measurements corresponded to unequally spaced intervals on a time scale inferred from geophysical theory. The example we use in Chapter 8 to illustrate our models is of data measured from exposed geophysical strata, where the time can again be determined scientifically but does not fall onto any regular scale. There has also been ongoing interest in applying continuous time models to regularly sampled econometric data (Bergstrom (1990)) because the parameters of the continuous time model may have more meaningful interpretation. In Belcher et al. (1994) an application is given to measurements taken on an individual every two hours, except when they were asleep; a continuous time model can also be useful for such data, which is regular except for long periods of missing values. Sometimes the timing of measurements that should be regular are *jittered*, i.e., delayed by some small irregular amount, and a continuous time model may then again be useful.

For continuous data sampled at very high frequency, a practical approach is to fit an approximating discrete ZAR model. However, an understanding of the estimation issues involved in that approach is only gained by a proper formulation of the continuous time models and an investigation of their properties. The main issue relates to estimation of the model using criteria related to the accuracy of prediction over an infinitesimal lead time for continuous series, or a very short lead time for high frequency sampled series. For irregularly sampled series the prediction intervals are invariably finite, from one observation of the series to the next, so this issue does not arise directly, but the process model must be integrated over the time spans between observations. Important new concepts and issues arise, relating to the stochastic calculus of differential and integral operations, as we extend our ZAR models to the continuous time context. We will present an informal introduction to these concepts as we meet them, beginning with the most simple continuous time first order autoregressive model, the CAR(1).

7.2   Continuous time autoregression and the CAR(1)

There are some important distinctions to be made between discrete (regularly sampled) and continuous time series (measured in continuous time). It is usual to express the time dependence of such a series using the notation $x(t)$. Assuming second order stationarity, the (multivariate) autocovariance function is defined for the mean corrected series very much as in the discrete case by

$$\Gamma_x(h) = \mathrm{E}\left\{ x(t)\,x(t - h)' \right\}, \tag{7.1}$$

where h is a time interval of any real value. It becomes essential to state the scale of time measurement since it cannot default to that of a discrete sampling index. The value of $x(t)$ need not be continuous in time; for example, the size of a simple stationary queue is integer valued and has jumps at the arrival or departure of a customer. However, the autocovariance function is continuous, and if $x(t)$ is Gaussian, it must also be continuous. We must therefore face the fact there is no continuous time version of white noise, that most simple of discrete time processes and the essential building block of discrete time models. The autocovariance of such a process would necessarily not be continuous at lag zero, where it would be positive, with zero values elsewhere. The theoretical exclusion of such a process is necessary to avoid a range of inconsistencies, such as the fact that it would, with certainty, be unbounded in any finite interval. The general requirement to ensure consistency in the definition of stochastic processes is that of separability, and this leads to the continuity of the autocovariance function for stationary continuous processes.

Our replacement for white noise as a building block in continuous time models is derived from the continuous time analog of the random walk, which is well defined. This is the Wiener process $W(t)$, which is characterized by the property that any change, or increment, $W(t + \delta) - W(t)$ is Gaussian with mean zero and variance $\delta\,\sigma_W^2$ and independent of all previous increments $W(s + \lambda) - W(s)$ for $s + \lambda \leq t$. Fixing the process at some point in time, such as $W(t) = 0$ at $t = 0$, is necessary but inconsequential. The covariance properties of the process are then

$$\mathrm{Cov}\{W(s), W(t)\} = s\sigma_W^2 \quad \text{for } 0 \leq s \leq t. \tag{7.2}$$

The Wiener process is also known as a Brownian process $B(t)$. Sampling this process at regular intervals of δ gives a discrete random walk $w_k = W(k\,\delta)$ from which discrete white noise can be extracted as the first differences $a_k = w_k - w_{k-1}$. In general, an increment $dW(t)$ defined as $W(t + dt) - W(t)$ is used as the replacement for a white noise term in our models. A somewhat counterintuitive but important result is that σ_W^2 can be determined precisely from a record of $W(t)$ over any finite time range $0 < t \leq T$. Sampling this record at intervals $\delta = T/n$, the estimate

$$\hat{\sigma}_W^2 = T^{-1} \sum_{k=1}^{n} a_k^2 \tag{7.3}$$

has the property that $\mathrm{E}\,\hat{\sigma}_W^2 = \sigma_W^2$ and $\mathrm{Var}\,\hat{\sigma}_W^2 = 2\sigma_W^4/n$. By increasing n the variance can be reduced towards zero. The formal statement of this result is that, with probability one,

$$\sigma_W^2 = T^{-1} \int_0^T dW(t)^2. \tag{7.4}$$

This is the central result of stochastic calculus (see Øksendal (1992)) and is markedly different from the corresponding result for classical calculus. If $W(t)$ in (7.4) is replaced by any differentiable function, the integral is zero. To evaluate an integral involving a function of a Wiener process, the formula of Itô can generally be used to transform it into an integral to which classical analytic methods may be applied.

We can now move on to consider the univariate first order continuous time Gaussian autoregression or CAR(1) model, the most simple consistently defined continuous time stationary process. This is also generally known as the Ornstein–Uhlenbeck or O–U process. It is conveniently specified by its autocorrelation function,

$$\rho(h) = \exp(-\alpha h) \quad \text{for } h \geq 0, \tag{7.5}$$

where $\alpha > 0$. Regularly sampling at intervals δ gives the discrete series $x_k = x(k\,\delta)$, which has autocorrelation function ϕ^k where $\phi = \exp(-\delta)$, so that x_k is an AR(1) process. More generally, it may be checked directly that $x(t+\delta) - \exp(-\alpha\delta)\,x(t)$ is uncorrelated with $x(s)$ for $s \leq t$. We may therefore obtain the prediction equation

$$x(t + \delta) = \exp(-\alpha\delta)\,x(t) + e(t, \delta) \tag{7.6}$$

where $\exp(-\alpha\delta)\,x(t)$ is the predictor of $x(t + \delta)$ made at time t and $e(t, \delta)$ is the prediction error uncorrelated with all the past values $x(s)$. For small δ it is better to express (7.6) in terms of the change in $x(t)$ as

$$\begin{aligned}
x(t + \delta) - x(t) &= -\{1 - \exp(-\alpha\delta)\}\,x(t) + e(t, \delta) \\
&\simeq -\alpha\,\delta\,x(t) + e(t, \delta)
\end{aligned} \tag{7.7}$$

where the error in approximating $(1 - \exp(-\alpha\delta))$ by $\alpha\delta$ has a variance of $O(\delta^2)$, an order less than that of $e(t, \delta)$, which is $\sigma_x^2\{1 - \exp(-2\alpha\delta)\} \simeq \sigma_x^2\,2\alpha\delta$. Finally, the conventional way of expressing (7.7) is in terms of differential quantities

$$dx(t) = -\alpha\,x(t)\,dt + dW(t). \tag{7.8}$$

The term on the left of this equation and the first term on the right are just shorthand for the same terms in (7.7). To the same level of approximation as applied to the first two terms in (7.7), the error term in (7.7) can be shown to have the properties of an increment $dW(t)$ in a Wiener process with $\sigma_W^2 = \sigma_x^2 \, 2\,\alpha$, by which it is replaced in (7.8).

Thus (7.8) is the standard formulation of the CAR(1) model. It should be stressed, however, that the differentials $dx(t)$ and $dW(t)$ are associated with a *positive* differential in time dt, which the continuous time notation does not make explicit. In particular, $dW(t)$ is the differential innovation of the process, independent of present and past values of $x(t)$. The autocorrelation will decay rapidly with increasing lag for a CAR(1) process with a high value of α, which may therefore approximate continuous white noise.

The spectrum of a stationary continuous time process is defined in terms of its autocovariance function as

$$S(f) = \int_{h=-\infty}^{\infty} \Gamma(h)\cos(2\pi hf)dh \quad \text{for} \quad -\infty < f < \infty. \tag{7.9}$$

It is useful to relate this to the spectrum of the discrete process obtained by high frequency sampling at intervals of δ. This has autocovariance function $\Gamma_k = \Gamma(h)$ at $h = k\,\delta$ and spectrum, from (3.15), of

$$S(f) = \sum_{k=-\infty}^{\infty} \Gamma(k\delta)\cos(2\pi k\delta f). \tag{7.10}$$

We have, however, re-scaled f to range between the Nyquist limits of $\pm 0.5/\delta$. It will be seen that the sum in (7.10) is just a numerical quadrature approximation to the integral in (7.9). Thus, for sufficiently high sampling frequency, the spectrum of the discrete process will be a close approximation to that of the continuous process. In practice, the sample spectrum of a continuous process is just that computed as for a discrete time series using the discretized data. The same is true for the sample autocovariances. However, some texts will present definitions of these sample quantities in which the sums over the discrete time points are replaced by integrals over the time range. For graphical presentation of the spectrum, the frequency range needs to be chosen to include the main features, up to a point where the power becomes negligible.

The spectrum of the CAR(1) model in (7.8) can be shown to be

$$S(f) = \frac{\sigma_W^2}{\alpha^2 + (2\pi f)^2} \quad \text{for} \quad -\infty < f < \infty. \tag{7.11}$$

The total integral of $S(f)$ is the process variance, $\sigma_x^2 = \sigma_W^2/(2\,\alpha)$. Note that the spectrum decays fairly slowly with increasing frequency. The spectrum decays more rapidly for smoother processes which have derivatives of higher order. The CAR(1) process is continuous but not even first order differentiable. This may be deduced from the fact that its autocorrelation function,

$\exp(-\alpha|h|)$, is not differentiable at the origin. If the autocorrelation function is differentiable to order $2k$ at the origin, then the process is differentiable to order k. The fact that the smoothness of the process is sensitive to the behavior of the autocorrelation function at a single point raises concerns about how this smoothness property may, in practice, be inferred from discretized data. Our modeling procedures will take account of this point.

For irregularly sampled data, it may appear that smoothness of the series cannot be determined, but much depends on the pattern of sampling. Conditions can be specified under which the complete autocovariance function and spectrum can be determined from a sufficiently long record, for example, if the sampling time points are those of events from a Poisson process; see Shapiro and Silverman (1960), Beutler (1970) and Masry (1978). For other sampling patterns, such as jittered data, that is not possible. The estimation of the autocovariance function for irregularly sampled data is also a challenge (see Masry (1983)); one approach is by means of the variogram. Direct estimation of the spectrum for irregularly sampled data is also difficult, which is one reason why it is so useful to fit a parametric model from which the model spectrum and autocovariance function can be derived. It is quite possible, however, to extrapolate the properties of a parametric model to regions where there is no data, such as when distributions are fitted to censored random samples. For irregularly sampled data, this caution applies to the estimation of the spectrum at high frequencies.

7.3   The CAR(p) model

The notation for discrete time models is based on the backward shift operator B. In a similar manner, the notation for discrete time operators is built on the differential operator s. This notation derives from the formal use of the Laplace transformation used to solve differential equations. In this methodology, s is a real or complex valued variable and for some purposes we will set it to such values. However, in practice, it has come to be used as convenient shorthand notation for expressing and manipulating linear differential equations. When t is the continuous time argument of a function, $s\,x(t)$ represents $d/dt\,x(t)$. As a first example, we take the CAR(1) model (7.8) and divide through by dt, giving

$$\frac{d}{dt}x(t) = -\alpha\,x(t) + e(t). \tag{7.12}$$

We have replaced $dW(t)/dt$ by $e(t)$, although, as we have only recently pointed out, $W(t)$ does not have a derivative. This notation is convenient and is acceptable provided that we replace $e(t)dt$ by $dW(t)$ in the integrals that arise in the solution of this equation. It is usual to refer to $e(t)$ as the formal derivative of $W(t)$. The model may then be written using s as the differential operator:

$$(s + \alpha)x(t) = e(t). \tag{7.13}$$

The classical solution for $x(t)$ in terms of $e(t)$ is given by

$$
x(t) = \frac{1}{s+\alpha}e(t) = \int_0^\infty \exp(-\alpha h)e(t-h)dh
$$

$$
= \int_{h=0}^\infty \exp(-\alpha h)dW(t-h). \tag{7.14}
$$

The higher order univariate CAR(p) model is written as $\alpha(s)x(t) = e(t)$ by replacing $s+\alpha$ with the polynomial operator

$$
\alpha(s) = s^p + \alpha_{p-1}s^{p-1} + \cdots + \alpha_1 s + \alpha_0. \tag{7.15}
$$

The solution of this model can be expressed in terms of the roots r_k of $\alpha(s) = 0$, provided these roots are distinct, by the formal expansion in partial fractions of the inverse of $\alpha(s)$:

$$
x(t) = \frac{1}{\alpha(s)}e(t) = \sum_{k=1}^p \frac{A_k}{s-r_k}e(t)
$$

$$
= \sum_{k=1}^p A_k \int_0^\infty \exp(-r_k h)dW(t-h). \tag{7.16}
$$

The CAR(p) model can be fitted to a continuous time series by solving Yule–Walker equations involving the derivatives of the autocovariance function at the origin, i.e., at lag zero; see Hyndman (1993). Of course, an estimate of the parameter α of the CAR(1) model (7.8) can be determined from (7.5) without the use of derivatives, by substituting a sample value for the lagged correlation $\rho(h)$ at any lag h, and this idea can be extended to the CAR(p). A somewhat different idea is to fit the CAR(p) model to a discretely sampled continuous time series. This has been proposed by Jones (1981) as an empirical modeling procedure, but it has a major limitation that the models of order $p-1$ are not strictly nested in the models of order p. A series that follows the CAR(p) model has derivatives to order $p-1$, so is necessarily smoother than a series that follows a lower order model. The CAR(p) model can only approach a CAR($p-1$) model in the limit as one of the roots r_k of $\alpha(s) = 0$ tends to minus infinity, which results in divergence of all the coefficients α_k. Unlike the discrete AR(p) model, it is not therefore practicable to approximate an arbitrary stationary continuous time series by a sufficiently high order CAR(p) model. This limitation is overcome by the class of CZAR(p) models that we will define.

7.4   The continuous time generalized shift

We extend the definition (6.2) of the discrete time generalized shift operator Z to its counterpart for application to continuous time series. We will not use a different symbol to distinguish the continuous time from the discrete time operator. The main reason for this is that the continuous time operator is the natural limit of the discrete time operator as the discrete sampling interval approaches zero. There will also be no confusion caused by use of the same symbol, because of the two distinct contexts in which they are applied.

We define the continuous time generalized shift in terms of the differential operator s as

$$Z = \frac{\kappa - s}{\kappa + s}, \tag{7.17}$$

where κ is a counterpart of θ in the discrete operator. It is a reciprocal time constant or *rate parameter* that defines the rate of decay of the weight which Z applies to lagged values of $x(t)$. This is apparent in the explicit form of

$$y(t) = Zx(t) = -x(t) + 2\kappa \int_{h=0}^{\infty} \exp(-\kappa h)x(t - h)dh. \tag{7.18}$$

The average value of the lag in this operation is $2/\kappa$, the counterpart of the average lag $(1 + \theta)/(1 - \theta)$ of the discrete operator. Thus $Z\,x(t)$ is similar in appearance, at low frequencies, to the lagged value $x(t-2/\kappa)$. If the continuous time series is actually supplied at high frequency sampled discretized time intervals of δ, the continuous time operator Z may be closely approximated by the discrete time operator. One choice for this approximation is to take the weight obtained by integrating $\exp(-\kappa h)$ in (7.18) over the interval from $k\delta$ to $(k+1)\delta$ and divide it equally between the values of $x(t)$ at each endpoint of the interval. Then the continuous time operator Z with parameter κ becomes the discrete time operator Z with $\theta = \exp(-\kappa\,\delta)$. The approximation is very close if $\kappa\,\delta$ is small, and under this condition an alternative and very similar choice is to set $\theta = (2-\kappa\,\delta)/(2+\kappa\,\delta)$. The advantage of this alternative approximation is that it has the same average lag value of $2/\kappa$ as the continuous time operator. The approximating ZAR operator can be applied to the discretized series using (6.4), but for small δ, to reduce loss of numerical accuracy, this equation may be re-arranged in terms of the parameter $\nu = 1-\theta$, which may be set to either $\nu = 1 - \exp(-\kappa\delta)$ or $\nu = 2\kappa\delta/(2 + 2\kappa\delta)$, or simply $\nu = \kappa\delta$ if δ is extremely small. Setting the discretized series $y_k = y(k\delta)$ and $x_k = x(k\delta)$, the operator is then implemented as

$$y_k = y_{k-1} - (x_k - x_{k-1}) + \nu x_{k-1} - \nu y_{k-1}. \tag{7.19}$$

Note that, although Z is formally defined in (7.17) using the differential operator s, the explicit form in (7.18) is an integral which does not require that $x(t)$ is differentiable.

Applying Z repeatedly generates the series $s^{(0)}(t) = x(t)$, $s^{(1)}(t) = Zx(t)$, $\ldots$, $s^{(k)}(t) = Z^k x(t)$. The vector of *states* $S(t)$ with these as elements may be considered as the output of a linear differential system with $x(t)$ as input, which has the solution

$$S(t) = Hx(t) + \int_0^{\infty} T\exp(hT)\,H\,x(t - h)\,dh \tag{7.20}$$

where $T = A^{-1}B$ and $H = A^{-1}G$ are defined in terms of the vector $G = (1, 0, \ldots, 0)'$ and matrices

$$
A = \begin{pmatrix} 1 & 0 & 0 & \cdots \\ 1 & 1 & 0 & \cdots \\ \ddots & \ddots & \ddots & \ddots \\ \cdots & 0 & 1 & 1 \end{pmatrix}, \quad B = \kappa \begin{pmatrix} 0 & 0 & \cdots & \cdots \\ 1 & -1 & 0 & \cdots \\ \ddots & \ddots & \ddots & \ddots \\ \cdots & 0 & 1 & -1 \end{pmatrix}. \quad (7.21)
$$

The elements of H are alternately $+1$ and -1 so that $Z^k x(t)$ always applies weight $(-1)^k$ to $x(t)$. Figure 7.1 shows the weight function applied to present and past values of $x(t)$ by Z^k for $k = 1, 2, 3$, taking, without loss of generality, $\kappa = 1$, because κ is simply a time scale parameter.

Further properties of the continuous time operator Z that parallel those of the discrete time operator are as follows.

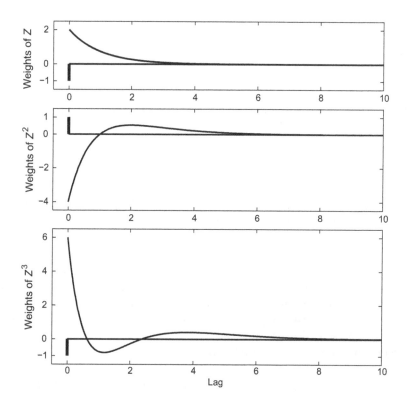

Figure 7.1 *The weights applied to lagged values of the time series by the continuous time generalized shift operators Z, Z^2 and Z^3. The thick line at lag zero indicates the discrete weight of -1 or $+1$ that is always applied at that lag by all powers of Z.*

1. The inverse of Z is an operator on the future given by

$$Z^{-1}x(t) = \frac{\kappa + s}{\kappa - s}\,x(t) = -x(t) + 2\kappa \int_{h=0}^{\infty} \exp(-\kappa h)x(t+h)dh. \quad (7.22)$$

2. $Z^k\,x(t)$ for $k = 0, 1, \ldots$ form a basis for the present and past values $x(t-h)$ for $h \geq 0$, and $Z^{-k}\,x(t)$ for $k = 0, 1, \ldots$ form a basis for the present and future values $x(t+h)$ for $h \geq 0$.

3. The continuous time operator Z shares the property of the discrete time version that it is unimodular, so that for a stationary time series, $Z\,x(t)$, and in general $Z^k\,x(t)$, has exactly the same autocovariances and spectrum as $x(t)$. The unimodular property of Z is derived from the frequency response of the differential operator. If $x(t) = \exp(2\pi i f t)$, then $s\,x(t) = d/dt\,x(t) = 2\pi i f x(t)$, so s has frequency response $2\pi i f$. Substituting for s, we obtain the frequency response of Z, which has modulus $|Z(f)| = 1$ and can therefore be written

$$Z(f) = \frac{\kappa - 2\pi i f}{\kappa + 2\pi i f} = \exp(-2\pi i g), \quad (7.23)$$

where

$$g = \frac{1}{\pi}\arctan\frac{2\pi f}{\kappa}. \quad (7.24)$$

For a stationary series $x(t)$ with spectrum $S_x(f)$, the spectrum of $Z\,x(t)$ is then $|Z(f)|^2 S_x(f) = S_x(f)$.

4. Define the sequence X_k by setting, for any fixed t,

$$X_k = Z^{-k}x(t). \quad (7.25)$$

Then X_k is a discrete stationary time series with spectrum $S_X(g)$ for $0 \leq g \leq 0.5$, given in terms of the spectrum $S_x(f)$ of $x(t)$ by

$$S_X(g) = S_x\{f(g)\}\frac{\kappa}{2}\sec^2(\pi g), \quad (7.26)$$

where $f(g) = (\kappa/2\pi)\tan(\pi g)$ from inverting (7.24). The transformation (7.25) from a continuous to a discrete stationary time series is implicit in the prediction theory presented by Wiener for continuous series and is formally presented in Doob (1953, p. 583).

To demonstrate (7.26), we first derive the autocovariance function of X_k as

$$\Gamma_{X,v} = \mathrm{Cov}(X_k, X_{k-v}) = \int_{-\infty}^{\infty} Z^{-k}\,\overline{Z}^{-(k-v)}\,S_x(f)\,df$$

$$= \int_{-\infty}^{\infty} Z^{-v}\,S_x(f)\,df, \quad (7.27)$$

which depends only on the lag v, confirming that X_k is stationary. In (7.27) now write $Z = \exp(-2\pi i g)$ and substitute for f in terms of g. Then (7.27) becomes

$$\text{Cov}(X_k, X_{k-v}) = \int_{-\infty}^{\infty} \exp(2\pi i v g)\, S_x\{f(g)\} \frac{\kappa}{2} \sec^2(\pi g)\, dg \qquad (7.28)$$

from which (7.26) follows. The transformation (7.24) is a *warp* from the infinite frequency range of the continuous process to the finite frequency range of the discrete process which is the counterpart to that which arises for the discrete ZAR models. Note for later use that the Jacobian df/dg of the warp transformation, which appears as the factor $(\kappa/2) \sec^2(\pi g)$ in (7.28), has inverse $dg/df = 2\kappa/\{\kappa^2 + (2\pi f)^2\}$.

Figure 7.2(a) shows the frequency warp transformation using $\kappa = 2/1.5$, over a limited part of the infinite frequency range to which it applies. The spectrum of the Oxygen level series, sampled at intervals of 0.1 seconds, is shown in Figure 7.2(b). At this sampling rate the series appears continuous. There is very little power in the spectrum beyond frequency 0.2 cycles per second, though the range of frequencies extends to 25 times this point, up to the Nyquist frequency of 5 cycles per second. Although still finite, this is well beyond the frequency range shown in this figure. The choice of κ is made to give the same low frequency delay of 1.5 seconds for Z_κ as that given by the discrete operator Z_θ with $\theta = 0.5$, determined in the previous chapter when selecting the discrete VZAR model for data sampled every 0.5 seconds. Thus $2/\kappa = 0.5 \times (1 + \theta)/(1 - \theta) = 1.5$. The warp maps the true frequency $\kappa/(2\pi)$ to the mid point, 0.25, of the warped frequency range. This can be useful for choosing κ from inspection of the spectrum. In this example the true frequency of $1/(1.5\pi) \approx 0.2$ is mapped to 0.25. The warped spectrum obtained with this transformation is shown in Figure 7.2(c). The warp is close to linear over the lower frequency range containing the greatest power in the spectrum, but the power in the remaining 95% of the frequency range is compressed by the warp into the same range on the warped scale. Because the warp transformation includes the Jacobian, it preserves the power, so it is possible to see the magnitude of this remaining power. We will comment later on the fact that there seems to be a slight rise in the warped spectrum towards the upper limit of its range. This range does not quite reach 0.5 because the true frequency range is in practice finite.

7.5   The continuous time VZAR model, VCZAR

We first present the continuous time counterpart of the predictive form (6.14) of the VZAR model as a multivariate extension of the CAR(1) model (7.8):

$$dx(t) = \{\xi_1\, x(t) + \xi_2\, Z\, x(t) + \cdots + \xi_p\, Z^{p-1}\, x(t)\}\, dt + dW(t). \qquad (7.29)$$

This is the VCZAR(p) model for the multivariate process $x(t)$ with predictors $Z^{k-1}x(t)$ and matrix coefficients ξ_k for $k = 1, \ldots, p$. We will sometimes write

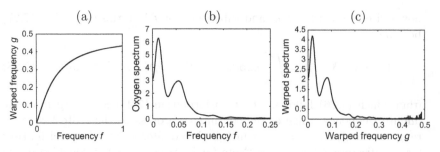

Figure 7.2 *Illustration of the warped spectrum (a) the continuous time frequency warp with $\kappa=2/1.5$, (b) the spectrum of the high frequency sampled Oxygen level series, and (c) the warped spectrum of this series.*

the model as VCZAR(p,κ) to emphasize the dependence of the operator Z on κ. The innovation term $dW(t)$ in (7.29) is the differential of a multivariate Wiener process $W(t)$ with $\text{Var}\{dW(t)\} = V_e\,dt$. The operator form of the model, written as an extension of (7.13), is

$$s\,x(t) = \xi(Z)\,x(t) + e(t) \tag{7.30}$$

where

$$\xi(Z) = \xi_1 + \xi_2\,Z + \cdots + \xi_p\,Z^{p-1} \tag{7.31}$$

and we now refer to $e(t)$, the formal derivative of $W(t)$, as the innovation term.

The counterpart of the natural form (6.21) of the VZAR model is

$$\varphi(Z)x(t) = n(t), \tag{7.32}$$

where

$$\varphi(Z) = I - \varphi_1 Z - \varphi_2 Z^2 - \cdots - \varphi_p Z^p, \tag{7.33}$$

and the *disturbance* term $n(t)$ follows a CAR(1) process with scalar coefficient κ:

$$(s + \kappa)n(t) = \varepsilon(t). \tag{7.34}$$

The innovation term $\varepsilon(t)$ is the formal derivative of a multivariate Wiener process with differential variance $V_\varepsilon dt$ related to the variance V_n of $n(t)$ by $V_\varepsilon = 2\,\kappa\,V_n$.

Finally, the counterpart of the general form (6.29) of the VZAR model is motivated by considering the prediction of the future value Z_λ^{-1} by the same set of past predictors $Z^{k-1}x(t)$ with matrix coefficients ζ_k for $k = 1,\ldots,p$. Here, Z_λ is defined with the reciprocal time constant λ replacing κ, which is implicitly understood to be the constant in the operator Z defining the predictors. Strictly, Z_λ^{-1} is a function of the future *and* present because it places a weight of -1 on the current value $x(t)$, but at low frequencies it is similar to the future value $x(t+2/\lambda)$. We will show later that, as for the general VZAR model, the choice of λ allows us to introduce a more robust estimation procedure for the VCZAR(p) model. The important point, however, is that all three models are equivalent; given only that κ is the same throughout, any one form of model can be algebraically transformed into any other. The general VCZAR model form is then

$$
\begin{aligned}
x(t) &= Z_\lambda\{\zeta_1 x(t) + \zeta_2 Z x(t) + \cdots + \zeta_p Z^{p-1} x(t)\} + n(t) \quad (7.35)\\
&= Z_\lambda \zeta(Z) x(t) + n(t) \quad (7.36)
\end{aligned}
$$

where

$$
\zeta(Z) = \zeta_1 + \zeta_2 Z + \cdots + \zeta_p Z^{p-1}. \quad (7.37)
$$

We retain the same notation $n(t)$ for the *disturbance* term in (7.36) as for the natural model in (7.32), although $n(t)$ now follows a CAR(1) process with scalar coefficient λ:

$$
(s + \lambda) n(t) = \varepsilon(t). \quad (7.38)
$$

This notation is consistent because the natural form of model is a special case of the general form with λ set equal to κ. As for the general VZAR model, the disturbance term $n(t)$ is uncorrelated with the predictors $Z_\lambda\, Z^k\, x(t)$ in (7.36), which is true also for the special case of the natural model (7.32).

The equivalence between the predictive and general forms of the model follows from the proportionality of the innovation series: $\varepsilon(t) = M e(t)$ gives

$$
(s + \lambda)\{I - Z_\lambda \zeta(Z)\} = M \{s - \xi(Z)\}. \quad (7.39)
$$

Multiplying through by $(1 + Z)$ and simplifying by using $(1 + Z) s = \kappa(1 - Z)$ gives an equivalence between polynomials in Z:

$$
(1 + \mu Z)I - (\mu + Z)\zeta(Z) = \frac{M}{\lambda + \kappa}\{\kappa(1 - Z) - (1 + Z)\xi(Z)\}, \quad (7.40)
$$

where

$$
\mu = \frac{\lambda - \kappa}{\lambda + \kappa}. \quad (7.41)
$$

By setting, respectively, $Z = -\mu$ or $Z = -1$, we obtain

$$
M = 2\lambda\{\lambda I - \xi(-\mu)\}^{-1} \quad \text{or} \quad M = I + \zeta(-1) \quad (7.42)
$$

which are needed for deriving $\zeta(Z)$ from $\xi(Z)$ or the reverse. The equivalence between the natural model and the predictive model form is a special case obtained by setting $\lambda = \kappa$ and $\mu = 0$, leading to some simplification of (7.40) and (7.42). However, the VCZAR model differs from its discrete counterpart in that the predictive form of model cannot be expressed directly as a special case of the general form. We need to be aware of the limiting behavior:

$$
\lim_{\lambda \to \infty} \frac{\lambda}{2}\left(Z_\lambda^{-1} - 1\right) = s, \quad (7.43)
$$

from which

$$\xi_1 = \lim_{\lambda \to \infty} \frac{\lambda}{2} (\zeta_1 - I) \quad \text{and} \quad \xi_k = \lim_{\lambda \to \infty} \frac{\lambda}{2} \zeta_k \quad \text{for} \quad k = 2, \dots, p. \tag{7.44}$$

7.6 Properties of the VCZAR model

The VCZAR model extends to the second order stationary continuous time process $x(t)$ the main property noted previously for VAR and VZAR models of a discrete time process. Based upon observations $x(t - h)$ for $h \geq 0$, the minimum mean square error linear prediction of a future value $x(t + \ell)$ can be constructed with arbitrary accuracy using a VCZAR model of sufficiently high order. Thus, for all practical purposes, the VCZAR model can represent any stationary continuous time series. This is a strong argument in favor of using the model; the standard CAR model does not share this property because of the smoothness requirements of the higher order models. Further properties of the VCZAR model are now presented.

1. The predictive model form (7.29) explicitly defines a linear stochastic process which is stationary if and only if the operator $\varphi(Z)$ in the natural model form (7.32) satisfies the stationarity condition of a discrete VAR model. This is explained in point 2 below.

2. The predictive model can be presented in continuous time state space form with state vector $S(t)$ having elements $s^{(k)}(t) = Z^k x(t)$ for $k = 0, \dots, p-1$ and the innovation process $e(t)$ as the input:

$$s \begin{pmatrix} I & 0 & 0 & \cdots \\ I & I & 0 & \cdots \\ \ddots & \ddots & \ddots & \ddots \\ \cdots & 0 & I & I \end{pmatrix} \begin{pmatrix} s^{(0)}(t) \\ s^{(1)}(t) \\ \vdots \\ s^{(p-1)}(t) \end{pmatrix} =$$

$$\begin{pmatrix} \xi_1 & \xi_2 & \cdots & \xi_p \\ \kappa I & -\kappa I & 0 & \cdots \\ \ddots & \ddots & \ddots & \ddots \\ \cdots & 0 & \kappa I & -\kappa I \end{pmatrix} \begin{pmatrix} s^{(0)}(t) \\ s^{(1)}(t) \\ \vdots \\ s^{(p-1)}(t) \end{pmatrix} + \begin{pmatrix} e(t) \\ 0 \\ \vdots \\ 0 \end{pmatrix} \tag{7.45}$$

or $s\, LS(t) = RS(t) + ue(t)$, where $u = (1, 0, \dots, 0)'$. The first row of (7.45) is simply the predictive model and the remaining rows generate the higher order states. The usual form of state space representation is obtained by dividing through by L, leading to the matrix differential equation

$$s\, S(t) = L^{-1} R\, S(t) + L^{-1} u e(t) = T\, S(t) + v e(t). \tag{7.46}$$

The observation equation is simply $x(t) = s^{(0)}(t)$. This representation is particularly useful for application to irregularly sampled series. It has the solution

$$S(t) = \int_0^\infty \exp(T\,h)\, v\, e(t - h) dh = \int_0^\infty \exp(T\,h)\, v\, dW(t - h). \tag{7.47}$$

From this solution the model is seen to be stationary if and only if all the eigenvalues of T have a negative real part. These eigenvalues can be shown to be equal to the zeros, at $s = r_k$, of $\varphi(Z(s))$, considered as a function of s. But a value of s with a negative real part implies a value of $Z(s)$ outside the unit circle, which gives the required stationarity condition.

3. The lagged covariances of the VCZAR model can be conveniently calculated using the state space representation. There are standard algorithms for calculating the stationary distribution of the states in (7.46), given T, v and V_e. However, the covariance $\mathrm{Cov}\{s^{(i)}(t), s^{(j)}(t)\}$ of the states at any given time t is the same as that, $\mathrm{Cov}(X_{k-i}, X_{k-j})$, of the process $X_k = Z^{-k}x(t)$. Now the natural model form (7.32) implies that X_k follows a standard VAR(p) model

$$\varphi(Z)X_k = N_k, \tag{7.48}$$

where Z is the shift operator acting upon k and N_k is a white noise sequence; see below for further explanation. The lagged covariances $\Gamma_{X,k}$ of X_k and hence the variance matrix V_S of $S(t)$ can be evaluated by the same methods as presented in Section 2.5 for the covariances of a standard VAR model with coefficients φ_j and noise variance $V_N = V_e/(2\kappa)$. The lagged covariances of $S(t)$ can be derived as follows. The solution for $S(t)$ of the state equation (7.46) can also be expressed conditional upon the state $S(t - r)$ at some earlier time point as

$$
\begin{aligned}
S(t) &= \exp(T\,r)S(t - r) + \int_0^r \exp(T\,h)\,v\,dW(t - h) \\
&= \exp(T\,r)S(t - r) + w(r),
\end{aligned} \tag{7.49}
$$

where $\exp(T\,r)S(t - r)$ is the conditional expectation, or prediction, of $S(t)$ at time $t - r$ and $w(r)$ is the integrated or prediction error which is uncorrelated with $S(t - r)$, having variance

$$\mathrm{Var}\{w(r)\} = \int_0^r \exp(T\,h)\,v\,V_e v'\,\exp(T'\,h)dh. \tag{7.50}$$

From (7.49) we now obtain

$$\Gamma_S(r) = \mathrm{Cov}\{S(t), S(t - r)\} = \exp(T\,r)V_s. \tag{7.51}$$

The lag covariance $\Gamma_x(r)$ of $x(t)$ is the upper left $m \times m$ block of $\Gamma_S(r)$.

4. The spectrum of the VCZAR model can be derived directly from (7.32) as

$$S_x(f) = \{\varphi(Z)\}^{-1}\frac{V_e}{\kappa^2 + (2\pi f)^2}\{\varphi(\overline{Z})'\}^{-1}, \tag{7.52}$$

in which Z is substituted by its frequency response function $Z(f)$, given in (7.23). Applying the warp transformation to (7.52) gives the spectrum of the discrete process X_k as

$$S_X(g) = \{\varphi(Z)\}^{-1}V_n\{\varphi(\overline{Z})'\}^{-1}, \tag{7.53}$$

where now we have $Z = \exp(-2\pi i g)$. This confirms that X_k follows the standard VAR(p) model (7.48). Note that the central term in (7.52), which is the spectrum of $n(t)$, is transformed to the warped spectrum of N_k, which becomes, after taking into account the Jacobian of the warp, the constant white noise spectrum V_n.

Quite generally, the warped spectrum $S_X(g)$ will have a strictly positive value at its upper limit $g = 0.5$. Consequently, the spectrum $S_x(f)$ will decay, as $f \to \infty$, at the same rate as that of the CAR(1) spectrum of $n(t)$. In consequence, any process $x(t)$ following a VCZAR model will have the same smoothness limitations as a CAR(1) process: its derivative does not exist. This does not, however, prevent the VCZAR model from approximating any stationary continuous process, however smooth, in the sense set out at the start of this section.

7.7 Approximating a continuous process by a VCZAR

We first consider the approximation of a stationary continuous process with known covariance structure by the general form of model. The coefficients are determined by projection of $x(t)$ upon $Z_\lambda Z^k x(t)$ for $k = 0, \ldots, p-1$, leading to normal equations, which, following our practice for the VAR and VZAR models, we will also refer to as Yule–Walker equations:

$$\text{Cov}\{x(t), Z^\lambda Z^{k-1} x(t)\} = \zeta_1 \Gamma_{X,k-1} + \zeta_2 \Gamma_{X,k-2} + \cdots + \zeta_p \Gamma_{X,k-p} \quad (7.54)$$

for $k = 1, 2, \ldots, p$. The variance of the projection error term is

$$\begin{aligned} V_n &= \text{Var}\{x(t)\} - \zeta_1 \text{Cov}\{x(t), Z_\lambda x(t)\}' - \zeta_2 \text{Cov}\{x(t), Z_\lambda Z x(t)\}' \\ &\quad \cdots - \zeta_p \text{Cov}\{x(t), Z_\lambda Z^{p-1} x(t)\}'. \end{aligned} \quad (7.55)$$

These equations are the same in structure as (6.51) and (6.52) for the general VZAR model, and the VCZAR model resulting from this projection is stationary. If $x(t)$ truly follows a VCZAR(p, κ) model, then the solution of (7.54) and (7.55) will always exactly recover this model, whatever value of λ is used. An approximation by the natural form of model is obtained if $\lambda = \kappa$ and this has the property that it captures the lagged covariances $\Gamma_{X,k}$ for $k = 0, 1, \ldots, p$ from which the coefficients are derived. A process generated from the model will have exactly these covariance properties. The advantage of the general form of VCZAR model, as for the VZAR model, is that it allows a more robust approach to the approximation of an arbitrary process. By construction, it seeks, at time t, to predict $Z_\lambda^{-1} x(t)$, which is similar at low frequencies to the future value $x(t+2/\lambda)$, at the finite lead time $2/\lambda$. It can also be shown to attach greater weight to the lower frequencies in the data in constructing the approximation: 90% of the weight is attached to frequencies below λ. This is in contrast to the approximation that is derived from the predictive form of model which seeks to predict only an infinitesimal step dt into the future and places equal weight upon all frequencies.

The normal equations used to derive the approximating predictive model (7.29) are found by projecting $dx(t)$ upon $Z^k x(t)$ for $k = 0, 1, \ldots, p-1$ and consequently depend on derivatives of covariances. We illustrate this for the univariate first order model which reduces to the CAR(1) model (7.8). The normal equations are found from taking the covariance of the model equation with the predictor $x(t)$:

$$\text{Cov}\{dx(t), x(t)\} = -\alpha \,\text{Cov}\{x(t), x(t)\}\, dt + \text{Cov}\{dW(t), x(t)\} \qquad (7.56)$$

and setting the last term, the covariance between the error and the predictor, to zero. To investigate this equation, on the right set the differential dt to the *positive* value δ and on the left replace the differential in $x(t)$ with the finite difference $x(t+\delta) - x(t)$. In terms of the autocovariance function $\gamma_x(v)$ of $x(t)$, after dividing through by δ, we obtain

$$\{\gamma_x(\delta) - \gamma_x(0)\}/\delta = -\alpha \,\gamma_x(0) \qquad (7.57)$$

and in the limit as $\delta \to 0$ we have the normal equation

$$\frac{d}{dh}\gamma_x(h)|_{h=0+} = -\alpha \,\gamma_x(0). \qquad (7.58)$$

Note that the derivative of $\gamma_x(h)$ is on the right at the origin. By symmetry, the derivative on the left is its negative.

For the general VCZAR(p) model, let

$$\Delta_{X,k} = \frac{d}{dh}\text{Cov}\{x(t+h), Z^k x(t)\}|_{h=0+} \quad \text{for } k \geq 0, \qquad (7.59)$$

so for the univariate CAR(1) model the term on the left of (7.58) is $\Delta_{X,0}$. For the general VCZAR(p) model, the normal equations are then

$$\Delta_{X,k-1} = \xi_1 \Gamma_{X,k-1} + \xi_2 \Gamma_{X,k-2} + \cdots + \xi_p \Gamma_{X,k-p}, \qquad (7.60)$$

for $k = 1, 2, \ldots, p$. By considering $\text{Var}\{x(t) + dx(t)\}$, we derive the equation for the error variance:

$$V_e = -\{\Delta_{X,0} + \Delta'_{X,0}\}. \qquad (7.61)$$

Using (7.59) with $k = 1$ to replace $\Delta_{X,0}$ in (7.61), we obtain the error variance for the CAR(1) as

$$\sigma_W^2 = 2\alpha \,\sigma_x^2. \qquad (7.62)$$

We can show that the coefficients ξ_k determined by (7.60) are those obtained from the coefficients ζ_k of the general model in the limit as $\lambda \to \infty$, using (7.44). Consequently, the model so determined is either stationary or on the borderline of the stationarity region. An example of borderline stationarity is readily presented by considering the approximation of a CAR(2) process by a CAR(1) model. The smoothness of the CAR(2) process results in the derivative of the covariance function being zero at lag zero, so the coefficient in the approximating CAR(1) model is found to be zero. But this gives $dx(t) = dW(t)$, i.e., the approximating model is a Wiener process which is not stationary. The approximating model is therefore sensitive to the smoothness property of the process covariances at lag zero, which relate to the high frequency properties of the spectrum. This issue will arise again as we consider estimating VCZAR models from observations of a continuous time series in the following sections. As for the VAR and VZAR models, we will consider the three estimation methods: Yule–Walker, regression and maximum likelihood.

7.8 Yule–Walker fitting of the VCZAR model

For the general model form (7.36), the Yule–Walker estimates are obtained by substituting sample values for the covariances which arise in the projection equations (7.54) and (7.55). As for the discrete models, these sample covariances can be conveniently constructed from the sample spectra. We use the Oxygen level and Respiration rate series sampled at intervals of 0.1 second, the highest frequency available, for our example. These are shown at this sampling frequency in Figure 7.3. In fact, the Oxygen level series only changes every second, i.e., 10 sampling points. It therefore appears less smooth than the Respiration rate series.

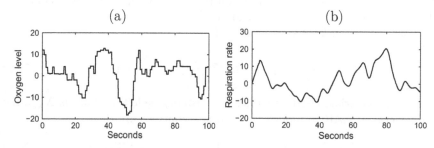

Figure 7.3 *The Oxygen level and Respiration rate series at sampling interval 0.1 second.*

Given series that are, in practice, discrete, though sampled at a high frequency, we have used the discrete ZAR operator to approximate the continuous time operator and fitted the model using the same method described for discrete models. To approximate Z_κ we used the discrete ZAR operator with equivalent lag of 1.5 seconds and to approximate Z_λ we used the equivalent lag of 9.5 seconds. These were the lags of the model we identified in Section 6.6 using the same series with a sampling interval of 0.5 second. With the sampling interval now being 0.1 second, we approximate the continuous operators Z_κ and Z_λ using the respective discrete operators Z_θ and Z_ρ with $\theta = 0.875$ and $\rho = 0.9792$. We will also use the same model order of $p = 9$. The fitted model is again displayed by comparing the sample and model spectra, and the sample and model lagged covariances. However, it is useful to compare the spectra on the warped frequency scale, which enables us to see how the whole of the power in the spectrum is approximated. If we were to use the true frequency scale as in Figure 7.2(b), we would be omitting a small but not totally negligible fraction of the power residing in the upper 95% of the frequency range. Figure 7.4 shows the auto-spectra of the Oxygen level and Respiration rate series and their squared coherency displayed on the warped frequency scale.

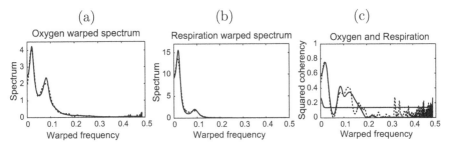

Figure 7.4 *The warped spectra (dashed lines) of the Oxygen level and Respiration rate series with sampling interval 0.1 second, with the model spectra (solid lines) superimposed: (a) the Oxygen spectrum, (b) the Respiration spectrum and (c) their squared coherency.*

The Oxygen auto-spectrum rises to a strictly positive limit at the upper end of the frequency range. Such a limit is a property of the model upon which we remarked following (7.53) at the end of Section 7.6. The Respiration auto-spectrum does appear to reach a limit of zero at this frequency, but in fact it is strictly positive, though small. The model is therefore able to approximate a wide variety of processes with varying degrees of smoothness. This is further illustrated by the autocorrelation properties shown in Figure 7.5. Although symmetric, these are shown for both positive and negative lags to highlight that for the Oxygen series the autocorrelations shown in Figure 7.5(a), with a magnified view in Figure 7.5(d), the pattern is not smooth at the origin. This is the characteristic pattern of any CZAR model, but is not at all evident in the corresponding Figures 7.5(b) and (e) for the Respiration series autocorrelations, because the effect is so small. Instead, the Respiration series autocorrelations appear smooth and locally quadratic at the origin.

For both series, the model provides very good approximations to the spectra and lagged correlations, including the squared coherency and lagged cross-correlations shown in Figures 7.4(c) and 7.5(c). Note that the Jacobian of the warp does *not* affect the squared coherencies. What might appear to be a rise in the level of smoothed squared coherencies towards the upper end of the frequency range is due only to the increasing density of the sampling fluctuations. In Chapter 9 we will illustrate smoothing of spectra on the warped frequency space, which avoids this appearance.

The natural form of VCZAR model is a special case of the general form taking $\lambda = \kappa$. As for the discrete case, the approximation of a true spectrum,

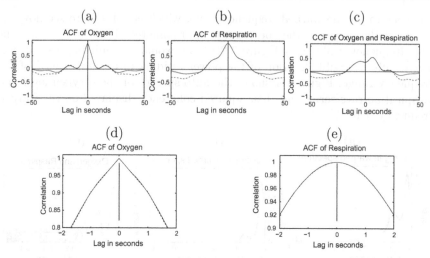

Figure 7.5 *The sample lagged correlations (dashed lines) between Oxygen level and Respiration rate at sampling interval 0.1 second with the model correlations (solid lines) superimposed: (a) the autocorrelations of Oxygen, (b) the autocorrelations of Respiration, (c) the cross-correlations of Oxygen and Respiration, (d) the autocorrelations of Oxygen close to lag zero and (e) the autocorrelations of Respiration close to lag zero. For (d) and (e) the sample and model correlations are barely distinguishable.*

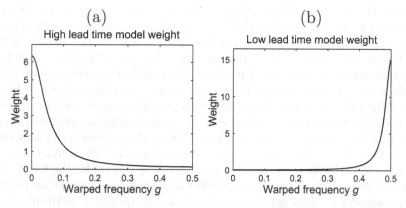

Figure 7.6 *The weights implicitly applied to the sample spectra, on the warped frequency domain, when fitting the general model with CZAR equivalent lag of 1.5 seconds and (a) an equivalent lead time of 9.5 seconds and (b) an equivalent lead time of 0.1 second, which approximates the predictive model. For the natural model the weighting is uniform and is not shown.*

or the fitting of a sample spectrum, by the predictive form of model, implicitly minimizes the information divergence or Whittle deviance weighted by a function of λ. The natural frequency space is infinite, and it is more helpful to display this weighting on the warped frequency scale. Figure 7.6(a) shows the weighting that is implicitly applied by using the general model with the values of κ and λ we have just used to fit the infant monitoring series. The formula for the weighting is

$$W(g) = \frac{(1 - \mu^2)}{1 + \mu^2 + 2\mu \cos(2\pi g)},$$ (7.63)

where μ is given by (7.41). The integral of $W(g)$, from -0.5 to 0.5, is one. By reference to Figure 7.4, we note that most of the weight is given to the frequencies where power is most concentrated and the squared coherency significant. For the natural model $\lambda = \kappa$ and $\mu = 0$, so the weighting is uniform on the warped frequency range and is not shown in the figure. For both values of λ, for the general and natural models, respectively, the fitted spectra are very similar at low frequencies to those shown in Figures 6.7(c), (d), (e) and (f) for the VZAR model of the series sampled at 0.5 second intervals.

The Yule–Walker equations for approximation of a process by the predictive model involve the right derivatives, at the origin, of the lagged covariances between $x(t)$ and $Z^k x(t)$ as defined in (7.59). These may be difficult to estimate numerically. Following (7.43), we may obtain approximations to the coefficients in the predictive model by fitting the general model with a large value of λ, after scaling the response variable as

$$y(t) = \frac{\lambda}{2} \left(Z_\lambda^{-1} - 1\right) x(t).$$ (7.64)

Then, as λ increases,

$$\text{Cov}\left\{y(t), Z^k x(t)\right\} \to \Delta_{X,k}.$$ (7.65)

Figure 7.6(b) shows, on the warped frequency scale, the weights implicitly applied to the Whittle deviance of our example series, by using the general model with a large value of λ corresponding to the lead time of the sampling interval, 0.1 second. This concentration of weight at the upper frequencies on the warped scale corresponds to a weighting which approaches uniformity across the infinite scale of the true frequencies as $\lambda \to \infty$ and the general form of model converges to the predictive form. As a result of this weighting, the model coefficients may be greatly influenced by the pattern of the spectrum at high frequencies well away from the low frequency region where most of the power is concentrated. This can lead to increased precision of prediction at very short lead times but loss of accuracy in prediction at high lead times.

In practice, to fit the predictive form of model, we actually recommend fitting the general model using a suitably large value of λ and then transforming the coefficients ζ_k of the general form to the coefficients ξ_k of the predictive form using (7.40). For our example series, this results in fitted spectra similar to those in Figures 6.7(a) and (b) for the VZAR predictive model of the series sampled at 0.5 second intervals. The fitted spectra shown in this figure do still follow the general pattern of the sample spectra at low frequencies, despite the much increased weight attached to higher frequencies. The fit is not, however, nearly as close as for the general model fitted using the small value of λ

with the equivalent lead time of 9.5 seconds. The reasonably good fit for the predictive model reminds us that the strength of the VZAR model lies not only in the choice of equivalent lead time. The fit that we have observed very much relies on the choice of the reciprocal time constant κ of the model, and if this is set to an equivalent lag equal to the sampling interval of 0.1 second the fit does become very poor. A good model depends on suitable choices of both κ and λ.

To complete this section, we investigate how the covariance derivatives $\Delta_{X,k}$, which appear in the formal equations (7.60) for estimation of the predictive model, are determined from the series spectra. We gain insight into this by expressing the covariance quantity in (7.65) in terms of the warped spectrum of $x(t)$:

$$\text{Cov}\left\{y(t), Z^k x(t)\right\} = \int_{-0.5}^{0.5} H(g) Z^{-k} S_X(g) dg. \tag{7.66}$$

Here $Z = \exp(-2i\pi g)$ and

$$H(g) = D(g) + i B(g) = \frac{\lambda}{2}\left(Z_\lambda^{-1} - 1\right) \tag{7.67}$$

is the frequency response function of $(\lambda/2)(Z_\lambda^{-1} - 1)$, expressed in terms of the warped frequency g and the parameter μ defined in (7.41), as

$$D(g) = \frac{\kappa(1-\mu)^2(1 - \cos 2\pi g)}{2(1 + \mu^2 + 2\mu\cos 2\pi g)} \tag{7.68}$$

and

$$B(g) = \frac{\kappa(1+\mu)^2 \sin 2\pi g}{2(1 + \mu^2 + 2\mu\cos 2\pi g)}. \tag{7.69}$$

Setting also the spectrum terms associated with the predictors as

$$Z^{-k} S_X(g) = R_X(g) + i I_X(g), \tag{7.70}$$

we get

$$\text{Cov}\left\{y(t), Z^k x(t)\right\} = \int_{-0.5}^{0.5} D(g) R_X(g) - B(g) I_X(g) dg. \tag{7.71}$$

Figure 7.7(a) shows the weight of the kernel $D(g)$ applied to the real part $R_X(g)$ of the spectrum quantity in this expression. Both of these are even functions of g. Figure 7.7(b) shows the weight of the kernel $B(g)$ applied to the imaginary part $I_X(g)$ of the spectrum quantity. Both of these are odd functions of g.

The covariance derivative $\Delta_{X,k}$ is obtained from (7.71) in the limit as $\lambda \to \infty$. To investigate this limit, we assume that $S_X(g)$ is continuous in its value and first derivative at $g = 0.5$. As λ increases, the total weight of $D(g)$ approaches κ and all the weight moves (symmetrically) towards $g = \pm 0.5$, so that the first term in the integral of (7.71) becomes $\kappa R_X(0.5)$. This is of

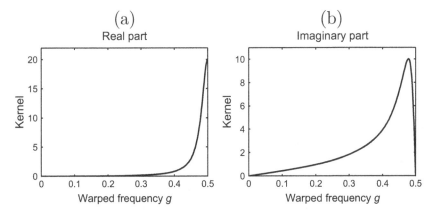

Figure 7.7 *The kernel weights implicitly applied to the sample spectra in the construction of the sample covariance between a low lead time (0.1 second) future value and the predictors in the Yule–Walker equations: (a) the real part of the kernel weighting and (b) the imaginary part of the kernel weighting.*

interest for the case $k = 0$, as it then gives the value of the symmetric part of $\Delta_{X,0}$ for use in (7.61), resulting in

$$V_e = 2\kappa S_X(0.5). \tag{7.72}$$

However, for large λ, $B(g)$ approaches the fixed function $\kappa \tan \pi g$ for all values of g between ± 0.5, and its total absolute weight diverges. The second term in the integral of (7.71) converges because $I_X(g)$ will behave like $\pm 0.5 - g$ close to ± 0.5, in consequence of being a smooth odd function of g.

For both terms in the integral, the value in the limit as $\lambda \to \infty$ is very sensitive to the behavior of $S_X(g)$ at $g = 0.5$. There are difficulties if the warped sample spectrum is used in place of $S_X(g)$, when, as in our example, the series is in fact discrete though sampled at high frequency. As is evident in Figure 7.4, the frequency range of the warped spectrum falls just short of $g = 0.5$ because the true frequency range is not in fact infinite. However, it is also evident, in particular in Figures 7.4(a) and (c), that, as g approaches 0.5, the warped spectra exhibit considerable sampling variability, leading to uncertainty of the value of their limits. Using the general model with a large, but finite, value of λ provides a compromise solution to the problem this raises for estimation of the predictive model form.

To reiterate, the strategy we finally recommend for series which appear continuous, but are actually high frequency sampled discrete series, is to fit the discrete VZAR model with the values of ρ and θ corresponding to the chosen continuous time parameters λ and κ. The coefficients of $\zeta_1, \ldots, \zeta_p$ and disturbance variance V_n for the continuous time model can then be taken to be those of the fitted general form VZAR model. The innovation variance will be given by $V_\varepsilon = 2\lambda V_n$. The continuous time model so determined can then be used to derive those properties of the series which are best presented in the continuous time context. This strategy can also be applied when the model is fitted to the same type of data using regression or maximum likelihood methods, as described in the next section.

7.9   Regression and ML estimation of the VCZAR

As indicated in the introduction to this chapter, there is an issue relating to the estimation of continuous time series models which does not arise with discrete series. It was circumvented in the Yule–Walker approach to estimation because that was based on projection principles which could be implemented without difficulty in continuous time. The issue arises in the estimation of the models by regression because the response variables in the regression are not finite in number, but follow a continuous record. The formal treatment of this issue should be to express the least squares criterion in terms of an integral over this record. However, the resulting criterion requires modification to make it workable for the predictive form of the VCZAR model. We can explain this if we consider such an integral as the limit of approximating sums obtained by discretizing the series, and it is instructive to apply this approach to the simple example of the CAR(1) model (7.8). Suppose we have a record $x(t)$ of this series for $0 \leq t \leq T$, sampled at intervals of $\delta = T/n$ for large n, giving the discrete series $x_1, x_2, \ldots, x_{n+1}$. The natural discretization of the model (7.8) is

$$\Delta x_k = x_{k+1} - x_k = -\alpha x_k \, \delta + e_k \tag{7.73}$$

where it is important that the forward difference and not the backward difference is used. Now on the RHS of (7.73) the predictor $-\alpha \delta x_k$ has variance proportional to δ^2, whereas the error term e_k has variance $\delta \sigma_W^2$. The expected sum of squares from the regression is $\delta \alpha^2 T \sigma_x^2$ and, as δ decreases, this becomes a vanishingly small proportion of the expected sum of squares of residuals, which is $T \sigma_W^2$. In the continuous limit, the sum of squares of the response becomes equal to the sum of squares of the residuals:

$$\int_0^T dx(t)^2 = \int_0^T dW(t)^2 = T \sigma_W^2. \tag{7.74}$$

Consequently, it is not possible to explain any of this variation by the predictor.

The least squares equations of the discretized regression do not, however, suffer any such problem in the limit. The required modification to the least squares criterion is found by examining the deviance quantity (minus twice the log likelihood), which is obtained by scaling the sum of squares by the error variance:

$$\mathrm{Dev}(\alpha) \;=\; \sum_{k=1}^{n} e_k^2 / (\delta \sigma_W^2)$$

$$= \sum_{k=1}^{n} (\Delta x_k + \alpha x_k \delta)^2 / (\delta \sigma_W^2)$$

$$= \sum_{k=1}^{n} \{ (\Delta x_k)^2 + 2\alpha x_k \Delta x_k \delta + \alpha^2 x_k^2 \delta^2 \} / (\delta \sigma_W^2). \quad (7.75)$$

With this scaling, the first term in the sum diverges as δ decreases. However, the relative likelihood converges, or equivalently the deviance difference $\text{Dev}(\alpha) - \text{Dev}(\alpha_0)$ converges for any given α_0. Taking $\alpha_0 = 0$ in this example, the terms of the deviance difference have the finite limit

$$\text{Dev}(\alpha) = \left\{ 2\alpha \int_0^T x(t) dx(t) + \alpha^2 \int_0^T x(t)^2 \, dt \right\} / \sigma_W^2. \quad (7.76)$$

Minimizing this criterion gives the least squares equation:

$$\hat{\alpha} \int_0^T x(t)^2 dt = - \int_0^T x(t) dx(t). \quad (7.77)$$

The criterion (7.76) is a special case of the general likelihood for diffusion models. For an overview of this topic, see Phillips and Yu (2009). For the predictive form of the CZAR(p) model (7.30), the deviance criterion becomes

$$\begin{aligned} \text{Dev}(\xi) &= \int_0^T \{\xi(Z)x(t)\}' V_e^{-1} \{\xi(Z)x(t)\} dt \\ &\quad - 2 \int_0^T \{\xi(Z) V_e^{-1} x(t)\}' dx(t). \end{aligned} \quad (7.78)$$

The model variance matrix V_e which appears in this expression is taken as known because, in theory, generalizing (7.74), it can be determined with probability one as

$$V_e = T^{-1} \int_0^T dx(t) dx(t)'. \quad (7.79)$$

However, if the CZAR model is saturated, i.e., all elements of the coefficients ξ_k are to be estimated without constraint, the estimation equations obtained by minimizing (7.78) do not depend on V_e:

$$\sum_{k=1}^{p} \hat{\xi}_k \int_0^T \{Z^{k-1} x(t)\} \{Z^{j-1} x(t)\}' dt = \int_0^T dx(t) \{Z^{j-1} x(t)\}' \quad (7.80)$$

for $j = 1, \ldots, p$.

The issue we have now resolved does not, however, arise with the general form (7.36) of the VCZAR model, because the response variable $Z_\lambda^{-1} x(t)$ is not a differential, and we can use the integrated prediction error as our regression criterion:

$$\text{Dev}(\zeta) = \int_0^T \{Z_\lambda^{-1} x(t) - \zeta(Z)x(t)\}' V_n^{-1} \{Z_\lambda^{-1} x(t) - \zeta(Z)x(t)\} dt. \quad (7.81)$$

Minimizing this gives equations similar to (7.80):

$$\sum_{k=1}^{p} \hat{\zeta}_k \int_0^T \{Z^{k-1}x(t)\}\{Z^{j-1}x(t)\}'dt = \int_0^T \{Z_\lambda^{-1}x(t)\}\{Z^{j-1}x(t)\}'dt \quad (7.82)$$

for $j = 1, \ldots, p$. An estimate of V_n can also be obtained as

$$\hat{V}_n = \int_0^T x(t)x(t)'dt - \sum_{j=1}^{p} \hat{\zeta}_j \int_0^T \{Z^{j-1}x(t)\}\{Z_\lambda^{-1}x(t)\}'dt \quad (7.83)$$

and using (7.42) we can derive

$$\hat{V}_\epsilon = 2\lambda\{I + \hat{\zeta}(-1)\}^{-1}\hat{V}_n\{I + \hat{\zeta}(-1)'\}^{-1}. \quad (7.84)$$

It is again instructive to illustrate these estimation procedures with reference to the CAR(1) model of a process recorded over the interval $0 \le t \le T$. We simulate samples of this process at intervals of $\delta = T/n$ for large n, from which we approximate the integrals used to construct the estimator $\hat{\alpha}$ in (7.77). The integral on the left, when divided by T, we write as

$$\hat{\sigma}_x^2 = \frac{1}{T}\int_0^T x(t)^2\, dt \approx \left(\sum_{k=1}^{n} x_k^2\right)\frac{\delta}{T} = \left(\sum_{k=1}^{n} x_k^2\right)\frac{1}{n}. \quad (7.85)$$

This approximating sum could be extended to $(\sum_{k=1}^{n+1} x_k^2)/(n+1)$, including the last observation, but with negligible effect for large n. On the right we set

$$\frac{1}{T}\int_0^T x(t)dx(t) \approx \frac{1}{T}\sum_{k=1}^{n} x_k\,\Delta x_k. \quad (7.86)$$

Figure 7.8(a) shows one sample of a series simulated from this model taking $\alpha = 0.2$, $\sigma_W^2 = 1.0$ and $T = 100$. The series was generated at intervals $\delta = 0.001$, giving $n = 100,000$, using a discrete autoregression with autoregressive parameter $\phi = \exp(-\alpha\delta)$ and innovation variance $(1 - \phi^2)\sigma_W^2/(2\alpha)$. Figure 7.8(c) shows the histogram of the estimates of α from 10,000 samples. This estimate of α does, however, suffer from the defect that it is not necessarily positive. Estimates from a shorter record with $T = 10$ have a small proportion of negative values. This defect can be understood and avoided by examining the integral in (7.85) as follows. The Itô formula, applied to this simple example, gives

$$d\{x(t)^2\} = 2x(t)dx(t) + dx(t)^2 \quad (7.87)$$

from which we can transform the integral in (7.85), giving

$$\int_0^T x(t)dx(t) = \left[\frac{1}{2}x(t)^2\right]_0^T - \frac{1}{2}\int_0^T dx(t)^2$$

$$= \{x(T)^2 - x(0)^2\}/2 - T\sigma_W^2/2. \quad (7.88)$$

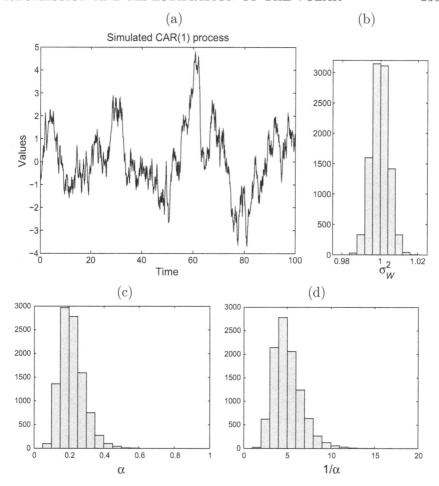

Figure 7.8 *Simulation results for estimation of the CAR(1) model: (a) a typical realization, (b) histogram of 10,000 sample values of $\hat{\sigma}_W^2$, (c) histogram of $\hat{\alpha}$ and (d) histogram of $1/\hat{\alpha}$.*

The last term in this result arises from the stochastic nature of the integral and is the dominant term for large T. The discretized approximation to the integral can also be arranged to give the corresponding terms in the result:

$$\sum_{k=1}^{n} x_k \, \Delta x_k \;=\; \sum_{k=1}^{n} x_k \, (x_{k+1} - x_k) \tag{7.89}$$

$$=\; (x_{n+1}^2 - x_1^2)/2 - T \, \hat{\sigma}_W^2 /2 \tag{7.90}$$

where

$$\hat{\sigma}_W^2 = \sum_{k=1}^n (\Delta x_k)^2 / T. \tag{7.91}$$

Then we have

$$\hat{\alpha} = \frac{\hat{\sigma}_W^2 - \{x(T)^2 - x(0)^2\}/T}{2\hat{\sigma}_x^2}. \tag{7.92}$$

By removing the term $\{x(T)^2 - x(0)^2\}/T$, which becomes negligible for large T, we obtain a modified estimate of α which avoids negative values. The effect on our simulated values is quite small, though the mean square error of $\hat{\alpha}$ is slightly reduced, by almost 2%. This modified estimate is the same as the moment estimate obtained from the model property $\sigma_W^2 = 2\alpha\sigma_x^2$. The value of $\hat{\sigma}_W^2$ is equal to σ_W^2 with probability one, in the limit as n increases. Figure 7.8(b) shows the distribution of $\hat{\sigma}_W^2$ from our simulations, supporting this result. The sampling variability of the modified estimator therefore comes almost wholly from the denominator $\hat{\sigma}_x^2$. The distribution of $\hat{\alpha}$ shown in Figure 7.8(c) has substantial bias of 0.0189 with standard error 0.0007, but $1/\hat{\alpha}$, being proportional to $\hat{\sigma}_x^2$, has only a marginally significant bias of 0.033 with standard error 0.016, from the 10,000 simulations. Its distribution is shown in Figure 7.8(d).

This modification of the estimate is special to the simple CAR(1) model. There is another approach which can be more generally applied. This is to incorporate information into the likelihood from the initial stationary distribution of the series, which is normal with mean zero and variance σ_X^2. This leads to an additional term in the deviance (7.76):

$$\text{Dev}(\alpha) = \left\{ 2\alpha \int_0^T x(t)dx(t) + \alpha^2 \int_0^T x(t)^2\, dt \right\} / \sigma_W^2$$
$$+ \log(\sigma_W^2) - \log(2\alpha) + 2\alpha x(0)^2 / \sigma_W^2. \tag{7.93}$$

For $\alpha > 0$, $\text{Dev}(\alpha)$ has positive second derivative and diverges to plus infinity at both zero and infinity, so it has a unique minimum which satisfies the equation

$$\hat{\alpha} \int_0^T x(t)^2 dt = - \int_0^T x(t)dx(t) - x(0)^2 + \sigma_W^2/(2\hat{\alpha}). \tag{7.94}$$

Replacing the integrals as before gives

$$\hat{\alpha} = \frac{\hat{\sigma}_W^2 - \{x(T)^2 + x(0)^2\}/T + \hat{\sigma}_W^2/(\hat{\alpha}T)}{2\hat{\sigma}_x^2}. \tag{7.95}$$

We have expressed this in a form which shows how, for large T, this is close to the estimate (7.93). However, $\hat{\alpha}$ in (7.95) can be determined as the positive solution of a quadratic equation. Our simulations of this value resulted in a population of estimates $\hat{\alpha}$ with a further 2% reduction in mean square error.

The main point, however, is that including the information from the stationary distribution of initial states will guarantee that a stationary process model is estimated. For these simulations we also fitted the discrete AR(1) model with parameter ϕ to points sampled at intervals of $\delta = 0.01$, using the maximum likelihood methods we described for the VZAR model, and generated estimates of $\alpha = (1 - \phi)/\delta$. These were identical to those generated from the solution of (7.95). This did not work, however, when using the smaller interval $\delta = 0.001$ because of the numerical problems relating to the issues described at the start of this chapter. It is necessary to move to numerical procedures constructed for the continuous time models if the predictive form of the model is to be used.

The picture is different if we fit the general form (7.36) of the VCZAR model by solving the equations (7.82). For discretized series this may be implemented using the least squares methods described for the discrete VZAR models. The values of ρ and θ are set to correspond to the chosen continuous time parameters λ and κ as in our implementation of the Yule–Walker estimation procedure. We have experienced no numerical difficulty with this method, even using a very small sampling interval in our simulations. The operators may be implemented using (7.19) to reduce numerical error, but we have not found this to be necessary. The criterion (7.81) from which the least squares equations (7.82) were derived may again be extended to include the information from the initial stationary distribution of the series. Similarly, this may be implemented using the likelihood estimation procedure developed for the discrete VZAR model, with the advantage of ensuring that the fitted model is stationary.

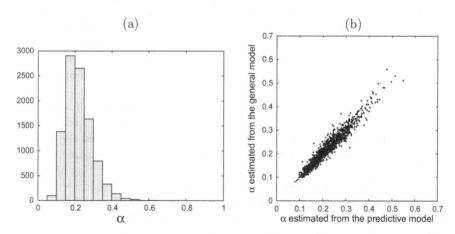

Figure 7.9 *Simulation results for estimation of the general form of CAR(1) model: (a) histogram of 10,000 sample values of $\hat{\alpha}$, (b) scatter plot of estimates obtained from fitting the predictive and general forms of model.*

We have again used simulations to illustrate these estimation procedures with reference to the CAR(1) model with the same parameter $\alpha = 0.2$, sampling period $0 \leq t \leq T = 100$ and sampling interval $\delta = 0.001$. To fit the model we set $\lambda = 4.0$ in (7.36). This corresponds to a time delay of $2/\lambda = 0.5$ in comparison with the time constant of $1/\alpha = 5.0$ over which the series autocorrelation decays to $1/e$. The setting of $\rho = \exp(-\lambda\delta)$ was then used to fit the general ZAR(1) model. For the first order model the ZAR parameter θ has no influence, so it was set to zero. The estimated coefficient $\hat{\zeta}$ of the general ZAR(1) model was transformed to the estimate $\tilde{\alpha} = \lambda(1 - \hat{\zeta})/(1 + \hat{\zeta})$ of the CAR(1) model parameter. Figure 7.9(a) shows the histogram of this estimate and Figure 7.9(b) the scatter plot between this estimate and the estimate $\hat{\alpha} = \hat{\sigma}_W^2/(2\hat{\sigma}_x^2)$ from fitting the predictive form of the CAR(1) model. The ratio of the mean square error of $\hat{\alpha}$ to $\tilde{\alpha}$ was 0.908. The theoretical efficiency for large T is $\lambda/(\lambda + \alpha) = 0.95$, and this is attained using a longer sample with $T = 500$. The loss of efficiency is more than outweighed by the robustness of this procedure. We illustrate this by repeating the simulations but applying a small amount of smoothing to the discretized series, replacing x_k by $0.2x_{k-1} + 0.6x_k + 0.2x_{k+1}$. With the sampling interval of $\delta = 0.001$ there is no noticeable change at all in the visual appearance of the plot of the series such as is shown in Figure 7.8(a), unless a very high magnification is used. However, the estimates of α from fitting the predictive model become badly biased towards zero, with a mean of 0.097. Figure 7.10(a) shows the histogram of the estimates from fitting the general form of model, which have a very similar distribution to that in Figure 7.9(a) and the same root mean square error of 0.076. The parameter σ_W^2 is well estimated from fitting the general form of model, as shown by the histogram in Figure 7.10(b), though not highly concentrated close to the true value, as in Figure 7.8(b). Of course, the model fitted in this way does not correctly describe the properties of the smoothed series at very short time lags, but will give optimal predictions at higher lead times. These would be very poor if the predictive form of model were fitted.

We close this section with an application to the estimation of a model of the Infant monitoring series with the highest sampling rate at intervals of 0.1 second. Our strategy, as for the Yule–Walker method, is to fit the discrete VZAR model, but now using the regression method. To approximate Z_λ we use the discrete operator Z_ρ with $\rho = 0.9792$ for the equivalent lag of 9.5 seconds. We will determine the model order and the rate parameter κ in terms of the discrete discount factor θ using the generalized information criterion ZIC and penalty term P defined in (6.59) and (6.61). The procedure selected values of $\theta = 0.94$ and $p = 10$. Using the QML estimation method selected $\theta = 0.92$ and $p = 13$, though the ZIC showed a local minimum for the same values as the regression method. The lag corresponding to $\theta = 0.92$ is 2.4 seconds. The corresponding continuous time parameter is therefore $\kappa = 0.0833$. Figure 7.11 shows the impulse and steps responses derived from the model estimated by QML. These are very close to those shown in Figure 6.11 for

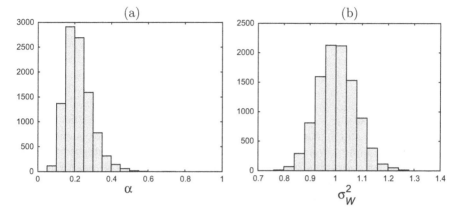

Figure 7.10 *Simulation results for robust estimation of the CAR(1) model: (a) histogram of 10,000 sample values of $\hat{\alpha}$, (b) histogram of 10,000 sample values of $\hat{\sigma}_W^2$.*

the model fitted using the lower sampling rate with interval 0.5 second. It is reassuring that increasing the sampling rate by a factor of 5 gives essentially the same model properties. The model can therefore be fitted to the highest frequency sampled data, avoiding the need for sub-sampling at a lower frequency.

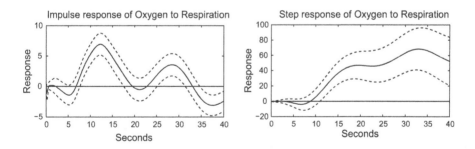

Figure 7.11 *Properties of the VZAR(13,0.92) model fitted to the Oxygen level and Pulse and Respiration rate series sampled at intervals of 0.1 second, using the QML method with the value of ρ equivalent to a lag of 9.5 seconds: (a) the impulse response of Oxygen to a unit increase in Respiration and (b) the step response of Oxygen to a unit increase in Respiration. The solid lines are the responses and the dashed lines show plus or minus two standard deviations.*

A final point concerning this model is that the estimated innovation variance V_e has correlation matrix

$$C_e = \begin{pmatrix} 1.0000 & 0.0856 & 0.0365 \\ 0.0856 & 1.0000 & -0.0531 \\ 0.0365 & -0.0531 & 1.0000 \end{pmatrix}. \tag{7.96}$$

The low values of the correlations indicate that the disturbances affecting the three series are independent. For a continuous or very high frequency sampled series, contemporaneous causality is not a convincing interpretation of any significant correlation between the innovation series; the most plausible explanation would be that the series include a common component variable that is not separately recorded. In this example we have no need to postulate such explanations. In conclusion, the strategy we have employed in this example has enabled us to reliably model a system with very high frequency sampling rate.

Chapter 8

Irregularly sampled series

8.1 Modeling of irregularly sampled series

The interest in this topic is evidenced by the dedication of a whole conference to its consideration, the collection of papers appearing in Parzen (1984). At the close of Section 7.2 we raised the theoretical problem of whether the whole autocovariance function or spectrum of the process could be determined from irregular sampling. The papers in Parzen (1984) consider, inter alia, both the theoretical and practical aspects of this problem. The approach of parametric model fitting that we present in this chapter was developed in part from ideas found among those papers. A successful demonstration of our approach in the univariate case has appeared in Belcher et al. (1994).

Non-parametric approaches to estimation of the autocorrelation function and spectrum of irregularly sampled data are presented by Kirchner and Weil (2000), who supplied us with the bivariate series that we use as our main example in this chapter, and which we displayed in Figure 1.1. They estimated auto- and cross-covariances of these irregularly sampled extinction and origination rate series by a binning procedure: the covariance at a given lag is formed from all the points with a separation within 5 Myr of this lag. This gives similar estimates to those formed by applying some interpolation scheme, such as kernel smoothing, to the series (Wand and Jones (1994)). Further, Kirchner and Weil (2000) present similar, but alternative, estimates of the cross-correlations obtained from direct spectrum estimates computed using the Lomb–Scargle algorithm (Scargle (1982)). In order to provide insight into our initial experiences of fitting univariate parametric models to these series, which we recount in Section 8.3, we will present direct spectrum estimates for the same series in Section 8.4. Our method of spectrum estimation does, however, require some prior assumption relating to how the power in the spectrum is spread over the frequency range and for this we use a simple parametric model. We explain how this is equivalent to using this model to interpolate the series onto a regular grid of time points, so that the usual sample spectra (3.36) may be formed. This example opens up a range of issues on inference for models of irregularly sampled noisy series, and it is to shed light on these that we include this digression into the estimation of sample spectra for such data. We believe that the experiences we relate will prove of value in similar modeling exercises.

The parametric models that we use to represent irregularly sampled data are the same VCZAR models described in Chapter 7. However, if such data appear to be affected by substantial observational noise or measurement errors, it is reasonable to assume that these errors will be independent at the sampling points because of the finite time intervals between them. To represent this, a white noise term can be included as an additive component in VCZAR models for such data. If such a term is not included, the selected VCZAR model operator will contain a zero of large magnitude which approximates this white noise term.

The time series from Kirchner and Weil (2000) are the rates of extinctions and originations of species determined from fossil records. The numbers are accumulated over stratigraphic intervals of varying lengths, there being 106 intervals over a period of some 540 Myr (millions of years). Interest lies in determining whether extinctions are associated with subsequent originations. The observation time assigned to an origination is that of the boundary that begins the interval, and the time assigned to an extinction is that of the boundary that ends it. The authors describe this as a conservative approach to determining any association between extinctions and subsequent originations. The point, we believe, is that any effect of sampling from a given stratigraphic interval that is common to the two series cannot then contribute to a correlation with positive time delay from extinction to origination. The two series are again shown in Figure 8.1 together with the histogram of the sampling intervals. We would expect a white observational noise term to be appropriate in a model for such series because each observation is obtained from a finite sample at a limited number of sites. This contrasts with the infant monitoring series where, although uncertainty does exist in the measured variables, the

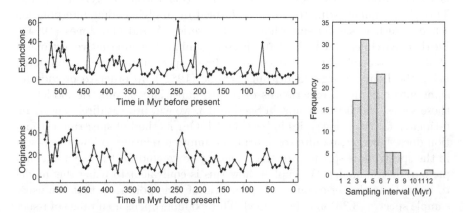

Figure 8.1 *Irregularly observed time series of rates of extinctions and originations of species, with the histogram of intervals between sampling time points of the series.*

record is essentially continuous and any component of this uncertainty cannot be readily distinguished in the series or the model used to represent them.

We have decided to model both series on a logarithmic scale. This transformation helps to symmetrize the distribution of the observations, moderating the peaks that appear in the extinctions series. It also reveals more clearly downward trends in the series, particularly for the extinctions series. The analysis by Kirchner and Weil (2000) removed low frequency variations in the series before estimating the lagged cross-correlations, which would otherwise have been less pronounced. We will just fit linear trends in time as part of our modeling of the transformed series.

We will also assign to each observation the time point that is the center of the stratigraphic interval to which it relates. We believe that this gives a fairer indication of the proximity of the information in successive intervals. Figure 8.2 shows the transformed series on this time scale and the histogram of the intervals between sampling points, which are more compactly distributed. From here on, for convenience, we will refer to these transformed series with these sampling points as the Extinctions and Originations series.

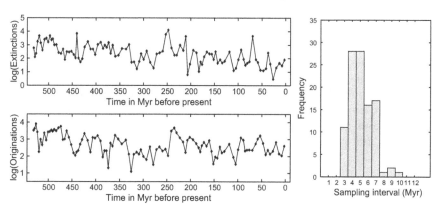

Figure 8.2 *Irregularly observed time series of transformed extinction and origination rates of species with sampling time points set as the centers of stratigraphic intervals, with the histogram of intervals between the sampling points.*

One of the first considerations when fitting the VCZAR model to such series is the choice of the rate parameter κ, or rather the range of values of κ that might be considered. As a guide to this choice, suppose that the sampling time points were *equally spaced* at intervals of 5 Myr, which is close to the mean of the actual sampling intervals. The Nyquist frequency would then be 0.1 Myr^{-1}. Now it is generally desirable to choose a model for which the warp transformation (7.24) maps the central region of this frequency range, around $f = 0.05$ Myr^{-1}, to the central region of the warped frequency space, around $g = 0.25$. Substituting these values into (7.24) gives

$$\kappa = 2\pi f / \tan(\pi g) = 2\pi f = \pi \times 0.1 \approx 0.3 \text{ Myr}^{-1}. \qquad (8.1)$$

A lower value of κ may be useful if the fitted model spectrum is concentrated at much lower frequencies. The warped spectrum of the model can be inspected and ideally should be well spread across the lower and central regions of the range $0 \leq g \leq 0.5$. Because the histogram of Figure 8.2 shows a good proportion of sampling intervals less than the mean, there should be some information in the series about frequencies higher than the Nyquist frequency of 0.1 based on the supposition of equal spacing.

8.2   The likelihood from irregularly sampled data

We will fit our models by maximum likelihood assuming the joint Gaussian distribution of the observations. For any set of model parameters, the VCZAR determines the covariance matrix of the observations and hence the likelihood. However, practical computation of the likelihood is carried out by applying the Kalman filter to obtain the prediction error decomposition form of the likelihood. The state space form of the model (7.46) is integrated forward as in (7.49) from each observation time point to the next. Let the total number of observations in the series taken together be N. Let these observations be ordered by time so that $y_j = x_i(t_j)$ is the i-th element of the series $x(t)$ at time t_j for $j = 1, \ldots, N$. Whenever two or more series are observed at the same time, they can be ordered by the series index. The sequence of observations y_j so constructed is used to obtain the prediction error decomposition of the likelihood. The Kalman filter gives the prediction error e_j of y_j and its standard error σ_j, from the model-based prediction of each observation y_j given previous observations. Maximum likelihood estimates are obtained by minimizing the deviance:

$$\mathrm{Dev}(\xi, V_e) = \sum_{j=1}^{N} 2 \log \sigma_j + (e_j/\sigma_j)^2. \qquad (8.2)$$

A further consideration is the choice of parameterization of the model. The coefficients of the predictive representation appear in the state space form (7.45), but it is a simple matter to transform between different parameterizations. The deviance in (8.2) is generally non-linear in its arguments. It is minimized using an extension of non-linear least squares, and the choice of parameterization can influence the speed and success of this procedure. From our experience we have found that using the natural parameters φ and V_ϵ results in more reliable minimization. Note that the likelihood is the same whichever parameterization is used. Although different criteria were presented in the motivation for the predictive, natural and general forms of model, there appears to us only the one possible criterion in the context of irregularly sampled data, and that is the deviance in (8.2) based on the prediction of each successive observation from the time point of the most recent previous observation. We have also found it useful to have an alternative parameterization in structural form of the error variance matrices V_e of the disturbance term $e(t)$ in (7.30) or V_ϵ of the disturbance term $\varepsilon(t)$ in (7.34).

For the natural form of model we modify the disturbance process model (7.34) by introducing a coefficient φ_0 and error vector ω_j with diagonal variance matrix V_ω, so that

$$\varphi_0 \varepsilon(t) = \omega(t). \tag{8.3}$$

By default φ_0 is taken to be upper triangular with all diagonal elements equal to one, representing a regression of successive elements of $\varepsilon(t)$ on elements with a higher index (or lower in the ordering), with regression errors $\omega(t)$. However, a choice can be made of which other off-diagonal elements of φ_0 are set as free parameters and which constrained to zero. We will call this a *structural error* parameterization. A similar structural model parameterization, but with a somewhat different interpretation, is to allow the natural model to be fully structural. The coefficient φ_0 then takes the place of the identity matrix in the natural model operator (7.33). Then φ_0, in the default form, represents the regression of successive elements of $x(t)$ upon elements with higher index, in the model form (7.32). In the disturbance process model (7.34), the error vector $\varepsilon(t)$ is *replaced* by $\omega(t)$, so that the disturbance $n(t)$ also has diagonal variance matrix. The same possibilities are allowed for in the parameterization by the predictive model form by the introduction of a coefficient ξ_0, so that in the structural error form

$$\xi_0 e(t) = f(t), \tag{8.4}$$

with $f(t)$ having diagonal variance matrix V_f. In the structural model form the coefficient ξ_0 premultiplies $x(t)$ on the left hand side of (7.30), and on the right hand side the disturbance term becomes $f(t)$.

These structural forms have some advantages when interpreting the model. They can also improve the parameterization of the error structure by simplifying the positive definite constraint on the variance matrices.

Our models for irregularly sampled data will also include the possibility of observation noise, so that

$$y_j = x_i(t_j) + w_{i,j} \tag{8.5}$$

where $w_{i,j}$ is a white noise error with variance dependent upon the series index i. We will call $x_i(t)$ the *noise-free* series. It is a latent process that is not directly observed and is represented by a VCZAR model. We will call $w_{i,j}$ the *observation noise* term, which is a discrete process. We allow no correlation between any of the $w_{i,j}$ except that, if observations on different series are made at the same time, it is reasonable to consider the possibility that they are correlated. This may be appropriate for our example where we associate the mid time point of each stratigraphic interval with the observations of both series associated with that interval. The state space observation equation is then extended to $y_j = x_i(t_j) + w_{i,j}$, and the deviance defined by (8.2) should then also include amongst its arguments the variance and correlation coefficients of the observation noise. Mean and trend correction of the series may be applied by ordinary least squares prior to estimating the VCZAR model, or the coefficients of such fixed components may also be estimated by maximum likelihood. The use of maximum likelihood may not lead to very different estimates of these coefficients, but will give more reliable standard errors.

8.3 Irregularly sampled univariate series models

There are several statistical issues of time series interest which arise in selecting the models for the Extinctions and Originations series. These occur because of (a) the fairly small number of observations, just over 100 in each series, (b) the presence of observation noise, particularly in the extinctions series, and the inclusion in the model of terms to represent this, (c) the effect of trend estimation on depleting low-frequency variation of the series and (d) the non-linearity of the models. The issues are more evident when fitting higher order models and can be described as due to over-fitting. The deviance function may have one local minimum corresponding to what appears to be a good model and a further, generally lower minimum very close to the boundary of the parameter space, corresponding to a model with some extreme properties. We will now demonstrate and explain these issues in the context of modeling the univariate series, first the Originations, then the Extinctions series. Examination of the sample spectrum of the series in Section 8.4 will give further insight and guidance on how a good model may reliably be found. We apply this when developing the bivariate model for the series.

Table 8.1 *Deviance and AIC values for models with and without an observation noise term fitted to the univariate Originations series.*

Model	With obs. noise term			Model	Without obs. noise term		
Order	κ	Deviance	AIC	Order	κ	Deviance	AIC
1		−65.06	−61.06	1		−64.99	−62.99
2	0.5	−65.44	−59.44	2	0.05	−65.11	−61.11
3	0.25	−67.52	−59.52	3	0.10	−65.01	−59.01
4	0.0625	−68.58	−58.58	4	0.315	−67.53	−59.53

We fitted models up to order 4 to the Originations for a range of values from $\kappa = 0.05$ to $\kappa = 0.5$, at ten intervals with approximate geometric spacing. The deviance values for the first order models were constant over the range of κ because these all represent the CAR(1) model, though with different parametrization. Table 8.1 shows the minimum deviance and AIC values of the fitted models and the value of κ at which the minimum was found for model orders up to 4. The number of parameters counted for the AIC was the order plus one if an observation noise term was included. The constant term and linear trend term were included in all models so are left out of this count. The t value for the trend term was close to -2.3 for all the models so this term was retained. The minimum AIC is achieved by the simplest CAR(1) model. The properties of the fitted models for $p = 1$ and 2, including the observations noise term, were similar to those of the fitted models for $p =$

2 and 3, respectively, with no observation noise term. The autocorrelations generally decayed in a roughly exponential pattern by around lag 30, and the spectrum peaked at frequency zero, falling to low values above frequency 0.1. Moreover, the t values for the observation noise term were 0.25 and 1.24 for model orders $p = 1$ and 2, respectively. All this supports the selection of the CAR(1) model, but we draw attention to the fact that the AIC *does decrease* as the order increases from 2 to 3 in the model with observation noise and from 3 to 4 in the model without this noise term.

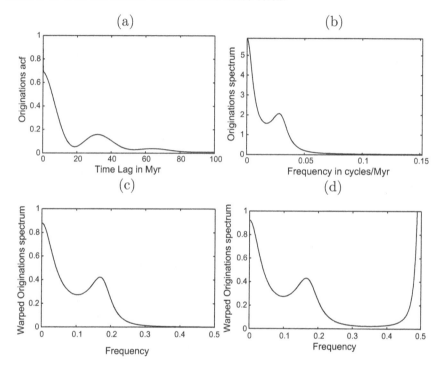

Figure 8.3 *Properties of models fitted to the Originations series: (a) the autocorre-lation function and (b) the spectrum of the order 4 model; (c) and (d) the warped spectra of the order 3 model with an observation noise term and the order 4 model, respectively.*

It is instructive to look closer at these models. When the observation noise term is included in the model, the autocovariance necessarily has a disconti-nuity at the origin, with a jump equal to the error variance. Figures 8.3(a) and (b) show the autocorrelation function and spectrum for the fitted model of or-der 4 without observation noise. Note particularly the apparent discontinuity in the autocorrelations which appear to jump from the limiting value of 0.69 as the lag reduces to zero to the value of 1.0 at the origin. There is in fact no discontinuity, the autocorrelation function being theoretically continuous, but dropping away extremely rapidly from zero. The autocorrelation for the fitted model of order 3 with observation noise is not shown because it is very similar in appearance to Figure 8.3(a), but it does have a true discontinuous jump at lag zero. It is as well to be aware of the capacity of the CZAR(4) model, without observation noise to approximate this discontinuity. We gain insight by comparing the warped spectrum of the CZAR(3) model with observation noise with that of the CZAR(4) model without observation noise in Figures 8.3(c) and (d), respectively. Figure 8.3(c) shows the warped spectrum of the CZAR(3) model with operator $\varphi(Z)$. This covers, as desired, the lower to central region of the warped frequency space. Although the warp is based upon a slightly different value of κ, Figure 8.3(d) shows the CZAR(4) spectrum, which is quite similar except that it has a peak at the upper frequency limit, arising from a zero of $\varphi(Z)$ close to $Z = -1$. It is this peak which transforms back to a very low level spectrum above frequency $f = 0.08$, which is barely evident in Figure 8.3(b). This extends to very high frequency, approximating continuous white noise as a proxy for the observation noise and is *not* present in the spectrum of the order 3 model, which includes the observation noise term. Close inspection reveals that the continuous acf of Figure 8.3(a) falls away from 1.0 at lag zero to 0.69 at lag 0.5 Myr to give the appearance of discontinuity. The shortest sampling interval of the series is 2.5 Myr, so this pattern is fully compatible with the presence of an uncorrelated observation noise component.

Although the AIC does not select for either of these models, the t values of all the model coefficients φ_k exceed 2.0, and the t value for the observation noise term in the order 3 model is 3.76. The reduction in deviance from the models with orders reduced by one is only a little over 2.0. This is not significant when referred to the chi-squared distribution on one degree of freedom, which would be expected to be appropriate in a large sample test of the acceptability of the lower order model. The contradictory inferences gained from the t values and the deviance reduction are explained by the non-linearity of the model and low sample size. There is some hint that these higher order models are revealing a valid feature of the data, but a simulation study would be necessary to determine the true significance. We move on, however, to consider modeling the univariate Extinctions series, which is also an instructive exercise.

Table 8.2 shows results of fitting low order models to the Extinctions series. Note, however, that under the models with observation noise there are two models listed of order 2, and under the models without observation noise there are two models listed of order 3. Constant and trend terms were included as components in all models, and the t value for the trend coefficient was typically around -4.5. As for the Originations models, the deviance for a model of order p with observation noise is similar to that of the model of order $p + 1$ without observation noise, and the model autocorrelations and spectra are also similar. We will therefore just consider the models with observation noise and in particular the two models of order 2 with respective κ values of 0.08 and 0.5, which we will refer to as models I and II.

Table 8.2 *Deviance and AIC values for models with and without an observation noise term fitted to the univariate Extinctions series.*

Model	With obs. noise term			Model	Without obs. noise term		
Order	κ	Deviance	AIC	Order	κ	Deviance	AIC
1		−5.07	−1.07	1		−3.92	−1.92
2	0.08	−9.01	−3.01	2	0.5	−4.98	−0.98
2	0.5	−6.06	−0.06	3	0.05	−8.97	−2.97
3	0.5	−8.62	−0.62	3	0.5	−5.857	0.143

Model I has the lowest AIC in Table 8.2. It was found when the estimation over the range of κ from 0.05 to 0.5 was started from the default values of zero coefficients for $\varphi(Z)$. Model II was found when starting the estimation from the coefficients of the first order model, with the coefficient of φ_2 set to zero. The autocorrelation and spectrum properties of these two models are shown in Figure 8.4.

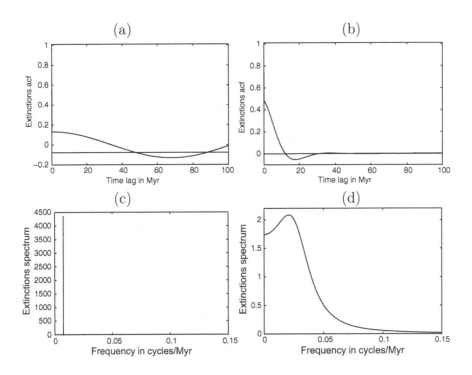

Figure 8.4 *Properties of models of order 2 with observation noise fitted to the Extinctions series: (a) and (b) the autocorrelation functions of models I and II; (c) and (d) the spectra of models I and II.*

Model II shows a decaying autocorrelation function and smooth spectrum, whereas model I is seen to be degenerate with a spectral line and an autocorrelation function that is a persistent cycle. The coefficients of model I are extremely close to the boundary of the stationarity region, with $\varphi_2 = -0.99997$. The CZAR model then represents a process almost indistinguishable from a pure sinusoid, to which is added the observation noise. Exactly the same degeneracy can occur when fitting a standard AR(2) model with observation noise to a discrete time series, or more generally when fitting an ARMA(2,2) model. Such a model suggests that the process contains a sinusoid of the observed frequency as a regression component. To follow up this suggestion, we fitted to the Extinctions series cosine and sine regression terms at the frequency 0.00735 of the spectral line, along with the constant term and trend, taking a simple CAR(1) to model the variations about these terms. The deviance of this fitted model (the same for all κ) was -16.57 and the AIC for comparison with the values in Table 8.2 was -10.57. Figure 8.5(a) shows the Extinctions series with the line of the fitted trend and cycle. The evidence of the AIC and the visual appearance of this fit seem to provide convincing support for this model. However, we question the statistical validity of this conclusion and argue that a model with a smooth spectrum such as model II is to be preferred.

To justify our questioning of the validity of model I, we simulated several continuous time series from the CAR(1) model with observation noise fitted to the Extinctions series, as indicated in Table 8.2. These simulated series were sampled at the same time points as the Extinctions series and fitted by a CZAR(2,κ) model with observation noise, including a linear trend term. In a high proportion of cases, a line spectrum was again estimated. Figure 8.5(b) shows the corresponding cycle fitted to the simulated series in one of these cases. The deviance reduction on fitting this cycle again apparently provides strong evidence of its significance, yet we know that this must be spurious, because the series is simulated from a process with a simple smooth spectrum.

8.4   The spectrum of irregularly sampled series

As stated earlier, we now explain how we construct the sample spectrum of an irregularly sampled series. We show the results of this for the Extinctions series in Figure 8.6. There is a strong peak close to the frequency 0.007, which gives insight into why model I of the previous section converged to a line spectrum and why the fitted cycle of this frequency appeared to be highly significant. However, the spectrum peak does not appear to be so dominant that it could not have arisen through the sampling fluctuation of a smooth spectrum, and Figure 8.5(b) confirms this point using a simulation of a series with a smooth spectrum.

For a regularly sampled discrete series, the sample spectrum would provide direct and readily assessed evidence for a discrete sinusoidal component. For irregularly sampled series, our approach to constructing a sample spectrum

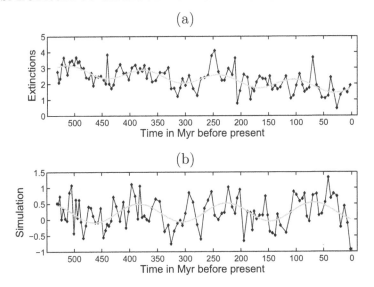

Figure 8.5 *(a) The Extinctions series with the fitted cycle implied by model I and (b) a series simulated from the CAR(1) model with observation noise fitted to the Extinctions series, with a fitted cycle derived in the same manner.*

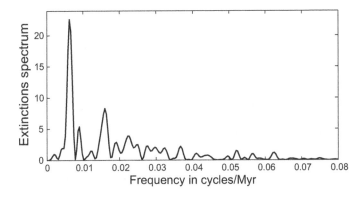

Figure 8.6 *The sample spectrum of the Extinctions series assuming the prior spectrum of a CAR(1) model fitted to the series with observation noise and trend terms.*

is based on the prior assumptions of some form of smooth *model spectrum* which constrains the spread of the sample spectrum over the frequency range. This is necessary because we impose no upper limit on this range such as is given by the Nyquist frequency for regularly sampled series. We will assume a simple form of this prior spectrum, the CAR(1) model with an observation noise term fitted to the series, as indicated in Table 8.2. The sample spectrum constructed in this way is shrunken in magnitude by an amount determined by the smooth form of this prior spectrum, but is still largely determined by the data. We show below how the sample spectrum so constructed is affected by changes to the form of this prior. Shrinkage is a property of estimation methods widely used in statistics; see Cox and Hinkley (1979).

Our method of constructing the sample spectrum is first presented in an intuitively appealing manner, based on a regression equation which approximates the observed series values in terms of the sum of cycles in (3.43). The uncorrelated increments δA_k and δB_k appearing there as coefficients of the cycles are, for convenience, rewritten as α_k and β_k. Including added observation noise, the regression is then

$$y_j = c + b\,t_j + \sum_k (\alpha_k \cos 2\pi f_k t_j + \beta_k \sin 2\pi f_k t_j) + w_j, \qquad (8.6)$$

where the frequencies f_k are on a sufficiently fine grid with spacing δ. The trend terms are also included. For our present example, we will take 1050 frequency points, which is ten times the number of series values, up to the frequency of 0.401, which is five times the Nyquist frequency appropriate if the data were equally spaced. This gives over 2000 regressors with only 105 responses. However, assuming that the series is represented by the smooth spectrum $S(f)$ shown in Figure 8.7(b), the coefficients α_k and β_k of the harmonic regressors are uncorrelated with known variances $S(f_k)\,\delta$. Thus (8.6) is a *mixed effects model* with the trend terms as fixed effects and the harmonic terms as random effects; see Robinson (1991). We write it in the usual form as

$$y = X\theta + Z\psi + w \qquad (8.7)$$

where $X\theta$ represents the trend and $Z\psi$ the harmonic regressors. The variance matrix D of ψ is diagonal with elements $S(f_k)\,\delta$, and the variance matrix of the error w is $\sigma_w^2 I$ where σ_w^2 is the variance of the white observation noise term in (8.5). The fitted coefficients are then determined by

$$\begin{pmatrix} X'X & X'Z \\ Z'X & Z'Z + \sigma_w^2\,D^{-1} \end{pmatrix} \begin{pmatrix} \hat{\theta} \\ \hat{\psi} \end{pmatrix} = \begin{pmatrix} X'y \\ Z'y \end{pmatrix}. \qquad (8.8)$$

Note that, although the coefficients α_k and β_k are uncorrelated, their fitted values are not. They are known as best linear unbiased predictors (BLUP), but we will refer to them simply as the estimates $\hat{\alpha}_k$ and $\hat{\beta}_k$. The regression terms in (8.6) for these coefficients are not orthogonal. The estimates for coefficients that are close together in frequency will be correlated, but, depending on the spacing of the time points at which the series is observed, those with a greater separation of frequency may also be correlated. The standard errors and correlations of these estimated coefficients may be determined from the matrix of the equations in (8.8).

The sample spectrum is found as

$$S^*(f_k) \propto \left(\hat{\alpha}_k^2 + \hat{\beta}_k^2\right) \qquad (8.9)$$

with the constant of proportionality chosen so that the (quadrature) integral of $S(f)$ is the sample variance of the observed series. This is how the sample spectrum shown in Figure 8.6 was obtained.

The fitted harmonic terms are also readily evaluated at all times intermediate as well as equal to those at which the series has been sampled. This provides interpolating estimates or best linear unbiased predictors of the unobserved CAR(1) component of the original series $x(t)$. These interpolations (and extrapolations) of the unobserved CAR(1) component $x(t)$ of the series are shown in Figure 8.7(a). Figure 8.7(b) shows the same spectrum estimates as in Figure 8.6 with the smooth spectrum of the CAR(1) model towards which they have been shrunk.

The regression (8.7) involved solving a set of over 2000 equations, and, although this is quite rapid for a modern computer, it is much more direct to reverse the order of the above procedure. First determine the BLUP interpolates of the series on a fine time grid, then construct the sample spectrum from this interpolated discrete series. This gives estimates of the interpolates and spectrum which are visually identical to those in Figures 8.7(a) and (b), though numerically slightly different due to reversing the discretization steps of time and frequency.

To illustrate the influence of the choice of prior spectrum in the resulting sample spectrum, Figures 8.7(c) and (d) show the series interpolated over a shorter time range and the sample spectrum obtained assuming a CAR(1) model with a much lower magnitude of $\xi = -0.01$ and with *no* observation noise, similar in structure to a random walk. The interpolates are very close to linear between the observation points. Compared with Figure 8.7(b), the sample spectrum in Figure 8.7(d) is enhanced because it now includes the variance of the noise. Figures 8.7(e) and (f) show the sample spectrum derived using again a CAR(1) model with no observation noise but with a much greater magnitude of $\xi = -2.0$, so that the CAR(1) itself is similar in structure to white noise. The interpolated series consists of sharp peaks located at, and rapidly decaying away from, each observation. The sample spectrum in Figure 8.7(f) is displayed to a much higher frequency and even then is not decaying away in the manner of that in Figure 8.7(d). It extends to frequencies much higher than the nominal Nyquist frequency of 0.1 that we first discussed on the supposition of equal spacing. Although, in consequence of the irregular sampling, precise aliasing will not occur, the variance that was concentrated in lower frequencies of the spectrum of Figure 8.7(d) will now be shared with the higher frequencies.

One lesson we draw from these figures is that, for irregularly sampled series, construction of the sample spectrum (and similarly the sample autocorrelation) will implicitly depend on some kind of interpolation of the data. In contrast, there is no such difficulty when constructing the sample spectrum of a discrete time series. No model was invoked when this was defined in Chapter 3, although it may also be constructed from the BLUPs of the harmonic coefficients based on the spectrum of some assumed time series model. It is,

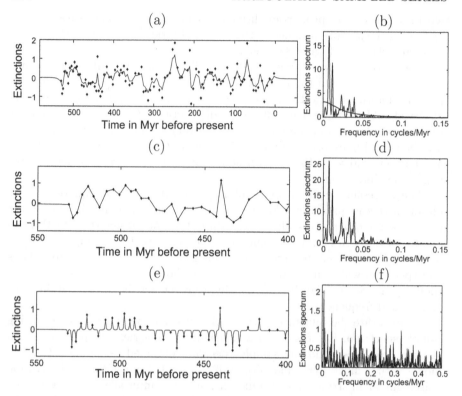

Figure 8.7 *(a)The trend corrected Extinctions series with the extracted and interpolated CAR(1) component of the model with observation noise and (b) the sample spectrum of this component with the spectrum of the CAR(1) model superimposed. (c) The early part of the same series interpolated on the assumption of a CAR(1) model with slow decay of autocorrelation and (d) the sample spectrum of this interpolated series. (e) The early part of the same series interpolated on the assumption of a CAR(1) model with rapid decay of autocorrelation and (f) the sample spectrum of this interpolated series.*

however, relatively insensitive to this assumed model. We can see this because the sample spectrum so constructed is the transform of the series *extrapolated* according to the assumed model. These extrapolations will typically not extend very far into the future or past and will have a limited effect on the sample spectrum unless the series is very short.

To complete this section, we explain how the vector $\hat{x}$ of interpolated values $\hat{x}(s_i)$ of the unobserved component series $x(t)$ at a grid of time points s_i is given in terms of the vector y of observed values for this example. We set

$$\hat{x} = G V^{-1} (y - X\hat{\theta}), \qquad (8.10)$$

where the elements of G and V are given in terms of the fitted CAR(1) model with observation noise, by $G_{i,j} = \mathrm{Cov}(x_i, y_j) = \sigma_x^2 \exp(\xi_1|s_i - t_j|)$ and $V_{i,j} = \mathrm{Cov}(y_i, y_j) = \sigma_x^2 \exp(\xi_1|t_i - t_j| + \sigma_w^2)$. The observations y in (8.10) are corrected for the fixed trend using the estimated coefficients $\hat{\theta} = (X'V^{-1}X)^{-1}(X'V^{-1}y)$. The link between the estimation equations (8.8) and (8.10) used in these two approaches is that $V \approx ZDZ' + \sigma_w^2 I$, and their solutions will be close if the frequency grid for the mixed effects model is chosen sufficiently fine and wide in extent. Note also that, in order to generate the same sample spectrum as given by the mixed effects model approach, the vector $\hat{x}$ needs to include, besides interpolates, the extrapolates extending beyond the time range of the observations up to the point where they decay effectively to zero on the trend corrected scale, as seen in Figure 8.7(a).

8.5   Recommendations on VCZAR model selection

We now return to the inference problem under consideration at the end of Section 8.3. The implication from the simulated series is that model I of order 2 for the univariate Extinctions series, which results in a line spectrum, is spurious. We must enquire into the reasons for this problem, and how it can be avoided, if possible. The line spectrum of model I, shown in Figure 8.4(c), is located within a band of low frequency spectral peaks containing most of the power in the sample spectrum, shown in Figure 8.6. The problem arises from using a second order model with observation noise fitted using a low value of κ. With such a low value of κ the model is particularly attuned to approximating low frequency features of the spectrum.

Figure 8.8(a) shows the sample spectrum of the Extinctions series on the warped frequency scale based on $\kappa = 0.08$. The line spectrum of model I is also shown on this scale and is seen to be located towards the center, nearly coincident with a prominent peak of the sample spectrum. Figure 8.8(b) is similar but using $\kappa = 0.5$ to define the frequency warp. Peaks in the sample spectrum at higher frequencies receive more emphasis in this figure, and the warped smooth spectrum of model II, also shown on this scale, extends across all these peaks. This illustrates how the respective choices of $\kappa = 0.08$ and $\kappa = 0.5$ influence the fitting of the line spectrum model I and continuous spectrum model II. Note, however, the higher value of κ does not necessarily ensure that a line spectrum fit is avoided. When the observation noise term is included, the line spectrum model may be fitted using the CZAR$(2, \kappa)$ model even with $\kappa = 0.5$ by carefully selecting the starting values of the coefficients when estimating the model.

The CZAR$(2, \kappa)$ model with observation noise does in fact encompass, in the limiting form of model I, the model which represents the series as a sinusoidal cycle with white noise errors. In this example, using a low value of κ, the estimation procedure has automatically located the frequency which maximizes the magnitude of this cycle. Fisher (1929) showed how the nominal level of significance had to be adjusted when testing for the presence of a cycle in white noise, the cycle having been located at the maximum of the

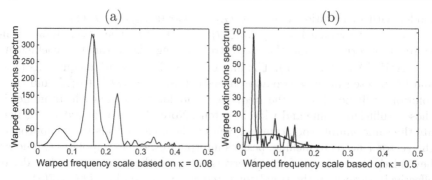

Figure 8.8 *The sample spectrum of the Extinctions series on the warped frequency scale (a) based on* $\kappa = 0.08$ *with the line spectrum of model I added, and (b) based on* $\kappa = 0.5$ *with the smooth spectrum of model II added.*

periodogram. Failure to make this adjustment leads to spurious detection of cycles. A very similar effect is present in this example; moreover, the probability of selecting such a model is increased here by the incorporation of trend terms. Fitting and removing these terms depletes the very lowest frequencies in the series, resulting in the appearance of a peak close to frequency zero in the sample spectrum. The presence of a strong trend in the series means we cannot discover what these low frequency components might have been, but we can illustrate this effect with the simulated series. Figure 8.9(a) shows the sample spectrum of the simulated series derived assuming the simulation model and estimated both with and without the presence of a trend term. There is a peak at frequency zero in the spectrum estimated without the trend term which vanishes when the trend term is estimated.

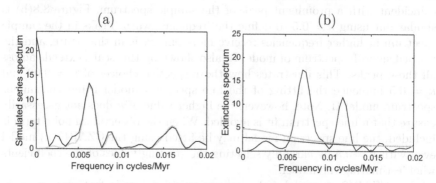

Figure 8.9 *(a) The effect of trend estimation on the sample spectrum of the simulated series at low frequencies. The black line shows the spectrum of the raw series and the gray line the spectrum of the trend corrected series. (b) The sample spectrum of the Extinctions series with the spectrum of a CZAR(3, κ) model with observation noise fitted using standard maximum likelihood, shown in black, and the spectrum of the same model fitted using restricted maximum likelihood, shown in gray.*

Estimation of a CZAR model for the series is affected by this distortion of the very low frequencies in the data. For processes similar in nature to our simulated series, the high natural variability of the sample spectrum often results in a prominent peak at some low frequency close to, but distinct from, the origin. This peak is open to mis-estimation by a parametric model as a deterministic cycle unless weight is given to the possibility that it is due to natural sampling fluctuation from a smooth model spectrum which rises to a maximum at the origin. A suitable choice of κ, as we have discussed, is one way to give weight to a smooth spectrum model. A further way is to estimate the model using the restricted (or residual) maximum likelihood (REML) criterion. The low frequency spectral information in the data that is removed by including trend terms cannot be restored, but the bias in model estimation that results from ignoring this distortion can be avoided by using the REML criterion. This was initially developed to overcome such bias in the estimation of variance component models (Harville, 1977); it estimates the process model using only that information in the data which is invariant to the procedure of trend correction. Its importance in time series modeling is also demonstrated by Tunnicliffe Wilson (1989), who uses the term marginal likelihood in place of restricted likelihood. The REML criterion may be evaluated quite simply for time series models by including the estimated coefficients of the trend terms in the state space model (Salla and Harville, 1988). These estimates are updated sequentially along with the other model states, starting with a large initial variance matrix representing diffuse information. The prediction error decomposition then yields the restricted likelihood. A recent exposition of this approach is Francke et al. (2010).

Figure 8.9(b) shows the effect on the fitted model spectrum of using the REML criterion when fitting the CZAR$(3, \kappa)$ model with observation noise term to the Extinctions series, setting $\kappa = 0.5$. We use this model because the spectrum has more flexibility than that of the CZAR$(1, \kappa)$ model. The REML model spectrum is seen to rise more rapidly towards the origin. In examples such as this, with relatively little data, the model fitted using the REML criterion will better capture the properties of the series. Our experience is also that when trend terms are included in the model, the REML criterion has less tendency to estimate a model with a low frequency line spectrum.

8.6   A model of regularly sampled bivariate series

We will first illustrate the fitting of a bivariate continuous time model to regularly observed discrete time series. This type of modeling has been advocated as being more appropriate than discrete modeling when the series truly arise from the sampling of a continuous process (Bergstrom, 1990). We will use the Gas furnace series for our example. First, we fit the natural form of model with structural error for orders up to six, in order to select the order p and parameter κ. We assume a sampling interval of one time unit since no other is provided. The Nyquist frequency is then 0.5, and an appropriate choice of κ should therefore be in the region of $0.5\pi \approx 1.5$. We fitted the models over a range of κ from 1.0 to 2.0 in steps of 0.1. Figure 8.10(a) shows the AIC for these models superimposed on one plot for the 11 distinct values of κ. The main point is that the plots are very similar for lags 4, 5 and 6, so that the model is robust to the choice of κ over the chosen range. The minimum envelope of AIC values is almost constant over these lags. The overall minimum AIC is at lag 6 with $\kappa = 1.9$, but this is only very slightly lower, by 0.28, than the AIC at lag 4 with $\kappa = 1.1$. The properties of these competing models are similar but the error limits are appreciably wider for the model of order 6. The SIC and HQC select lags 3 and 4, respectively, and the AIC is prone to overestimate the order, so we will fix on the model of order 4. The model was transformed to the predictive form which has exactly the same deviance. The innovation variance matrix of this form was close to diagonal, indicating that there is no structural dependence between the series. The model was re-fitted with this matrix constrained to be diagonal, resulting in a non-significant increase of 1.73 in the deviance.

Figure 8.10(b) shows the squared coherency between the Input and Output series derived from this model with two standard deviation limits. This is very similar to that shown in Figure 3.19(b) derived using direct spectral estimation. Figure 8.10(c) shows the phase between the series, which is linear up to frequency 0.15, indicating from its gradient a low frequency lag of 5 from the input to output. The VCZAR model does not allow for a simple time lag between the series, so this must be implicitly approximated by the model using a combination of powers of the CZAR operator Z. Figure 8.10(d) shows the step response from the input to the output derived from the fitted model. This is calculated by using the model to construct future states from specific initial conditions. All past values of input and output are taken to be zero and the input set as $x_1(t) = 1$ at $t = 0$. The noise variance is set to zero, as are the model coefficients which represent the (feedback) dependence of the Input on the Output. The future Output states then become the step response. This response is seen to be precisely estimated and becomes significantly positive just before lag 3. The impulse response is the derivative of the step response and is shown in Figure 8.10(e). This is very variable from lag 0 to 3, which we ascribe to the difficulties that the model has in approximating a step response which is close to zero up to lag 3 followed by a steep rise. This difficulty may be avoided by lagging the Input by two time units, which results in the impulse response shown in Figure 8.10(f).

8.7   A model of irregularly sampled bivariate series

Finding a good VCZAR model for the regularly sampled Gas furnace series proved to be quite straightforward, but important factors in this success were that, besides being regularly sampled, the series are strongly dependent, relatively long and suffer no discernible measurement error. We now turn to the bivariate Extinctions-Originations data, which enjoys none of these properties.

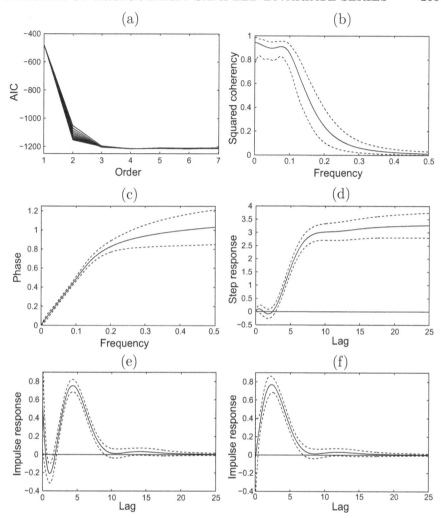

Figure 8.10 *Bivariate CZAR modeling of the Gas furnace series: (a) The AIC for increasing order of model with plots for $\kappa = 1, 1.1, \ldots, 2.0$ superimposed; (b) the squared coherency of the selected model of order 4 for $\kappa = 1.1$, with two standard error limits; (c) the phase of the selected model; (d) the step response of the selected model; (e) the impulse response of the selected model; (f) the impulse response of the model fitted with the Input series lagged by two time units.*

From our experience with univariate modeling of these data, we will include observation noise in both series. In order to avoid the degenerate form of fitted model representing a discrete spectral component, we will, initially, restrict the set of κ values that we explore. A value of $\kappa = 0.3$ was considered appropriate in (8.1), based on Nyquist frequency arguments, with the subsequent discussion suggesting slightly higher and lower values might be considered. For our bivariate model we initially investigated just three values, $\kappa = 0.25$, 0.5 and 1.0. The steps in model building were as follows.

1. For each value of κ we fitted the natural form of model up to order 4 using a structural form of the disturbance term which, with no loss of generality, represented the Originations as depending upon the Extinctions series. This form of model was chosen because of our general findings that this parametrization converged reliably.

2. We transformed the estimated parameters to the predictive form of model with the same structural form of the disturbance term and refitted using the transformed parameters as starting values. The refitted models converged immediately with the same deviance values.

3. For each of the three values of κ we noted that in the models of order 3 and 4 the disturbance term in the Originations series was estimated to have a variance very close to zero. If, when searching for the minimum deviance, a variance term in the model approaches zero, the search procedure may prematurely halt at the boundary of this constraint. It is then advisable to fix this variance parameter at, or close to, the zero boundary and re-start the search to find the minimum deviance over the remaining parameters. We did this, the only noticeable, but still small, reduction in deviance being for the model with $\kappa = 0.25$ at order 3.

A disturbance variance of zero in the presence of an observation noise term in a bivariate model is not necessarily spurious or to be avoided. In this example it can quite properly explain part of the Originations series as the output of a transfer function applied to the noise-free component of the Extinctions series.

Figure 8.11(a) shows the AIC for models up to order 4, superimposed for the three values of κ. The main conclusion is the clear selection of the model order $p = 2$ for each of the three values of κ, and for these models the disturbance variance of the Originations series was estimated to be distinct from zero. That the three lines are so close suggests that the selected model is relatively insensitive to the precise value of κ.

Being confident of the model order, we then extended the range of values of κ over which we searched for minimum deviance. Also, the estimated coefficient of the structural dependence of the disturbance term for Originations upon that for Extinctions had a t value of 0.1, so this term was fixed at zero, i.e., removed from the model. Furthermore, we used the REML likelihood so as to reduce the possibility of model degeneracy for lower values of κ and to reduce bias in the estimation of the low frequency spectrum properties of the model. Figure 8.11(b) shows the deviance plot with a clear minimum at $\kappa = 0.125$. Our final VCZAR model is therefore of order 2 fitted in this way for $\kappa = 0.125$. Various of its properties are presented in Figures 8.11 and 8.12 with two standard error limits for all except the univariate spectra, which are shown with one standard error limits.

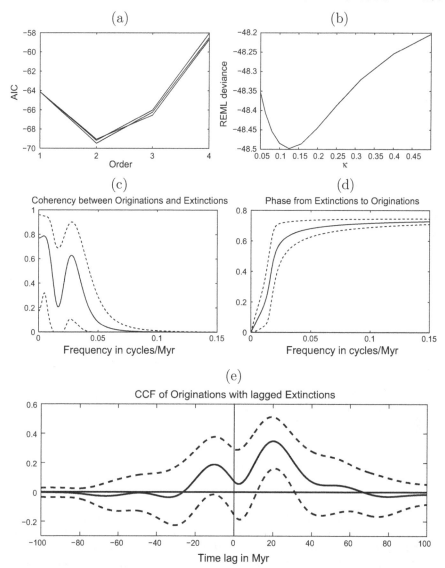

Figure 8.11 *Bivariate CZAR modeling of the Extinctions and Originations series: (a) the AIC for increasing order of model with plots for $\kappa = 0.25$, 0.5 and 1.0 superimposed; (b) the REML deviance of the order 2 model as a function of κ; (c) the estimated squared coherency between the series; (d) the estimated phase lag from Extinctions to Originations; (e) the estimated cross correlation between Originations and lagged Extinctions. Broken lines show 2 SE limits.*

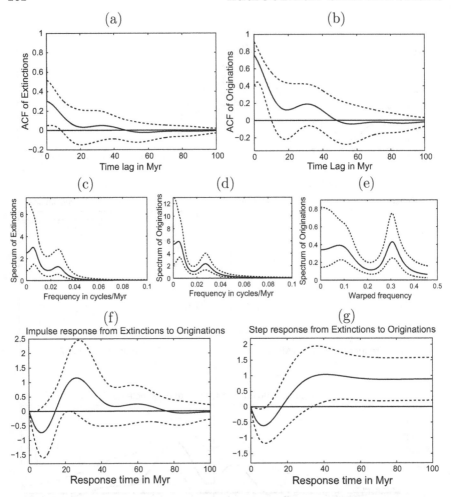

Figure 8.12 *Bivariate CZAR modeling of the Extinctions and Originations series: (a) and (b), respectively, the autocorrelations for the Extinctions and Originations series; (c) and (d) the univariate spectra of the Extinctions and Originations series; (e) the warped univariate spectra of the Originations series; (f) the impulse response from Extinctions to Originations; (g) the step response from Extinctions to Originations in the absence of feedback. Broken lines show 2 SE limits except for the spectra, for which they show 1 SE limits.*

The estimated squared coherency between the series is shown in Figure 8.11(c), indicating significant dependence between the series at low frequencies. This is computed taking into account the observation noise. The estimated phase in Figure 8.11(d) indicates that the Originations lags Extinctions at these frequencies. Figure 8.11(e) shows a clearly significant positive cross-correlation around a lag of 20 Myrs, supporting the conclusions of

Kirchner and Weil (2000). Figures 8.12(a) and (b) show the estimated univariate autocorrelations of the two series. Note in particular how small are those of the Extinctions series, though they are just significant at very low lags. These autocorrelations also take into account the observation noise, but the model spectra in Figures 8.12(c) and (d) are formed for the noise-free components of the series. These have a very similar appearance, with the power being predominantly associated with frequencies of period greater than 100 Myr. Figure 8.12(e) shows that with the selected value of $\kappa = 0.125$ the warped spectrum of the Originations series is well distributed over the range of warp frequencies and the same applies to the Extinctions series (not shown).

Figure 8.12(f) shows the response, predicted by the model, of the Originations series to an instantaneous unit impulse in Extinctions, in the absence of stochastic disturbances. After an initial reduction there is a surge of recovery followed by a return to equilibrium after about 70 Myr. In determining this response we did *not* suppress the feedback from Originations to Extinctions. The coefficients of the predictive form of model are shown in Table 8.3 with their t values in parentheses. There is significant dependence of Extinctions upon Originations through $\xi_{1,2,1}$ and of Originations upon Extinctions through $\xi_{2,1,1}$. Removing the two coefficients $\xi_{1,1,1}$ and $\xi_{1,2,2}$ with very low t values increased the significance of the remaining coefficients. Figure 8.12(g) shows the pattern of Originations in response to a step increase in Extinctions, determined by zeroing all (feedback) coefficients by which Extinctions are affected by Originations or by their own past values. The error limits are wide, but the stable value at higher lags indicates a significant long term response close to one. Because the series are modeled on a logarithmic scale, this implies that, say, a 10% increase in the rate of extinctions would eventually be compensated for by a 10% increase in the rate of originations.

Table 8.3 *Table of coefficients* $100 \times \xi_{i,j,k}$, *with t values in parentheses, for the predictive VCZAR(p, κ) model of the Extinctions and Originations series with $p = 2$ and $\kappa = 0.125$. The coefficients for $k = 1$ apply to $x(t)$ in the predictive model form (7.29) and the coefficients for $k = 2$ apply to $Zx(t)$. Note that the CZAR operator Z with this value of κ has a low frequency lag of 16 Myr.*

$k =$	1		2	
$j =$	1	2	1	2
$i = 1$	0.842 (0.2)	5.318 (2.9)	-10.045 (-3.6)	-0.693 (-0.5)
$i = 2$	-5.315 (-1.3)	-5.845 (-1.2)	14.690 (2.3)	-4.685 (-1.5)

For the model of Table 8.3 fitted by restricted maximum likelihood, the estimated trend coefficients (with t values) for the Extinctions and Originations are, respectively, -0.250 per 100 Myr (-4.71) and -0.128 per 100 Myr (-1.98). Fitting the model using standard maximum likelihood gave almost exactly the same estimates but respective t values somewhat more significant at -5.20 and -2.32. This is a consequence of the lower estimate of the spectrum at zero frequency when using standard maximum likelihood. For this model these spectrum values were, respectively, 1.7 and 3.6, whereas for the model fitted using restricted maximum likelihood they were 2.6 and 5.6, as seen in Figures 8.12(c) and (d).

In conclusion, constructing a VCZAR model for these two series provides significant and meaningful inferences regarding the dynamic dependence between these two series that may not be derived by other means, and we recommend this approach for similar forms of irregularly sampled multivariate time series.

Chapter 9

Linking graphical, spectral and VZAR methods

9.1 Outline of topics

In this chapter we bring together the different approaches presented in earlier chapters. First, we see how partial coherency graphs can be applied in the frequency domain to extend insight into the graphical modeling approach of Chapter 5. Second, we show how spectral analysis can be used directly to estimate the impulse response relationships between time series even in the context of feedback between the series. We also extend this approach to continuous, or high frequency sampled, time series by applying the frequency warp associated with VCZAR models. Finally, we show how partial correlation graphs can be extended to identify structural forms of the VZAR model.

9.2 Partial coherency graphs

We introduce this topic by returning to the example of the Flour price series, shown in Figure 1.2, for which we developed a structural VAR model, represented in Figure 5.17. Spectral analysis can also be a useful tool in the development of such models. We illustrate this for the Flour price series using Figure 9.1, which displays, in the form introduced by Dahlhaus (2000, p. 167), the squared coherencies between each pair of series and the partial coherencies of each pair given the third series. We call this a coherency array plot.

We note that the squared coherency between all pairs of series is highly significant over the whole frequency range, reflecting the obvious strong visual similarities between the series. However, the partial coherency between the first and the third series (Buffalo and Kansas City) is not at all significant, whereas that between the first and second is strongly significant and that between the second and third is significant over nearly all of the frequency range. Correspondingly, though not shown, the lagged responses estimated by spectral analysis are nowhere significant between the first and third series, whereas they are very strongly significant at lag zero between the other two pairs of series. Referring to the series in order, as X, Y and Z, i.e., X represents the set of x_t for all times t, we can represent the information in the partial

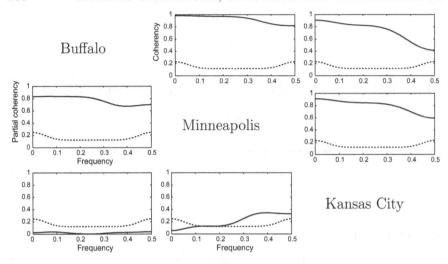

Figure 9.1 *An array of plots showing the squared coherencies between pairs of the Flour price series above the diagonal and, below the diagonal, the partial coherencies between pairs of the series given the third. These are all derived from the differenced series with bandwidth 0.25. The significance limits are shown by the broken lines.*

coherencies by the CIG in Figure 9.2, which we call a partial coherency graph (PCG). Two nodes are linked if there is significant partial coherency between the series.

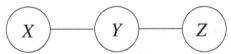

Figure 9.2 *The CIG between the three Flour price series implied by their partial coherencies.*

The partial coherencies and lagged responses provide a limited description of the causal relationships between the series. They can, however, be more directly related to causal structure, as described by Dahlhaus and Eichler (2003). They show, inter alia, how the Granger causality graph, Eichler (2007), is related to the partial coherency graph (which they term the partial correlation graph). The Granger causality graph combines characteristics of the conditional dependence of each series upon the past of others and dependence between current series values conditional upon both the past and the remaining current series values (Swanson and Granger, 1997). We introduce a very slightly different graph which we will call the structural dependence graph (SDG) and illustrate this in Figure 9.3.

This graph is *not* defined by conditional dependencies; it is derived from a proposed structural VAR such as that represented by the DAG of Figure 5.17, but permitting also the possible extension to relationships between current

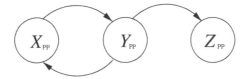

Figure 9.3 *A structural dependence graph representing a model for the three Flour price series.*

variables described by simultaneous equation models, as illustrated in Figure 5.27. The nodes are subscripted with PP to indicate that they now represent the set of all present and past values of the series at some time t. All the edges are directed; cycles naturally arise and are permitted, as are edges in both directions between a pair of nodes. The presence of an edge *to* one node *from* another indicates that in the SVAR there is *some* dependence of the *present* series value associated with that one node upon the present or past values associated with the other.

The value of this SDG is that upon moralization it generates the partial coherency graph (PCG). One may therefore postulate forms of the SDG that are consistent with the PCG estimated from the series values. This may be of further help in exploring the range of possible SVAR models for the series. For example, the SDG in Figure 9.3 is only one of two that are generally consistent with the PCG of Figure 9.2. The other is that with X and Z interchanged. In fact, one of the two links between X and Y in Figure 9.3 could be removed, but the direction of the link between Y and Z could not be reversed. The form shown is the one which is consistent with the SVAR identified in Figure 5.17.

The PCG may be of value in its own right for various scientific applications, without necessarily being a stepping stone to identification of an SVAR. We briefly illustrate this with three climate series, illustrated in Figure 9.4. These are mean annual values recorded from 1958 to 2009 of the concentration of atmospheric carbon dioxide (CO_2), the northern hemisphere temperature anomaly and the southern oscillation index (SOI), which measures the atmospheric pressure difference between Tahiti and Darwin.

Figure 9.5(a) shows the squared coherency between the CO_2 increase series and the temperature anomaly series, both having been trend corrected. This is very strong over much of the frequency range, reflecting the fact that the rate of absorption of CO_2 by the oceans, and possibly also the uptake and release of CO_2 by plants, is strongly affected by temperature.

Figure 9.5(b) shows that the partial coherency between the two series given the SOI is much lower than the unconditional squared coherency *at higher frequencies*. The SOI is known to be strongly associated with sea surface temperature variations in the southern Pacific which influence the rate of absorption of CO_2 into the ocean, but this result suggests that at higher

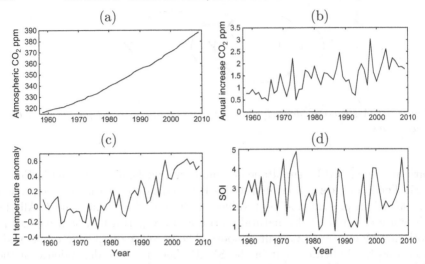

Figure 9.4 *Mean annual climate indicators from 1958 to 2009: (a) atmospheric CO_2 concentration in ppm, (b) annual increase in CO_2 concentration, (c) northern hemisphere (NH) temperature anomaly in degrees Celsius and (d) southern oscillation index.*

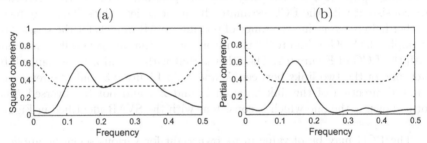

Figure 9.5 *(a) Squared coherency between the trend corrected CO_2 increase and temperature anomaly series and (b) partial coherency between the same series given the southern oscillation index. The bandwidth used was 0.15 with no tapering and the broken line indicates the 5% significance limit.*

frequencies it is also strongly related to temperatures in the northern hemisphere. Having made allowance for the SOI, the association of the CO_2 level increase with the northern hemisphere temperature anomaly vanishes at these frequencies. However, at lower frequencies this association remains significant. Because the PCG differs between the higher and lower frequency bands, there is no simple implication for SVAR modeling.

The two examples presented in this section involve the minimal number, three, of series to which the PCG can be usefully applied, but the PCG is readily constructed for a larger number of series and can be recommended as potentially providing further insight into the dependence between series. Figure 9.6 shows a further example, the coherency array plot of a section from the later part of the seven daily USA dollar term rates shown in Figure 1.7.

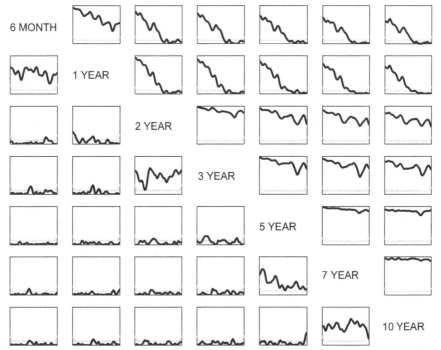

Figure 9.6 *The squared coherencies above the diagonal and partial coherencies below the diagonal of seven daily USA dollar term rates with terms indicated on the diagonal. The gray line is the 5% significance limit. The spectrum estimation was applied to the differenced series using a bandwidth of 0.05.*

The squared coherencies are all high, though less so at high frequencies for the shorter term rates. The most noticeable feature of the partial coherencies is that the only strongly significant values are between adjacent term rates, with all remaining partial coherencies being barely, if at all, significant. There are even two of the partial coherencies between adjacent term rates that are not very strongly significant. This pattern is useful in understanding and verifying the structural VZAR model for these series constructed in Section 9.4. The methods illustrated here are being developed further in the area of neuroscience using the concept of *partial directed coherence* (see Scheltera et al. (2009)) to quantify the causal dependence between series.

9.3 Spectral estimation of causal responses

We use the term causal here in the context where each series in a set depends on the past values of the other series, and that their dynamic behavior arises from this interdependence. In a control context, the output series depends on past values of the input through the causal mechanism of the system and the input depends on past values of the output through a designed causal feedback mechanism. It is this dependence on past values that we describe as causal.

In the general context, the series are not classed as input or output, but their interdependence is taken as being described in the same way, using the term feedback in a more general manner.

The estimation of feedback between mutually dependent time series has been an ongoing theme of previous chapters. It has long been a topic of interest in control engineering and econometrics; see, for example, Tsang and Bacon (1980) and Geweke (1982, 1984). It is of interest in neuroscience, as referred to in the previous section, and has potential application in environmental science and climatology. We now describe a spectral approach to this estimation problem which is similar to that which is being developed in the neuroscience field; see Dhamala et al. (2008) and Seth and Barnett (2014).

In the absence of feedback, the lagged response coefficients estimated by spectral analysis can be used to provide estimates of the causal impulse response at positive lags. More generally, and, in particular, in the presence of feedback, the direct estimation of the causal impulse response relationships between time series using the estimated spectral density is based upon the expression (4.22). On substituting for the smoothed spectral density estimate, we write this as

$$\widehat{S}_x(f) = \widehat{\Psi}(\exp-2\pi i f)\widehat{V}_e\widehat{\Psi}(\exp 2\pi i f)'. \tag{9.1}$$

Given $S_x(f)$, (4.22) can be uniquely solved for $\Psi(\exp-2\pi i f)$ and V_e under wide conditions; see Tunnicliffe Wilson (1978), which is also the basis of an efficient numerical procedure for constructing the solution. This procedure is called spectral factorization and $\Psi(\exp-2\pi i f)$ the spectral factor. Equivalently, given the autocovariances Γ_k, the equations (2.13) can be solved for the coefficients Ψ_k and V_e. Applying this method to $\widehat{S}_x(f)$ gives the coefficients $\widehat{\Psi}_k$ of $\widehat{\Psi}(\exp-2\pi i f)$ and therefore the estimated IRF or OLIRF defined in Chapter 2. Also furnished in this solution is $\widehat{V}_e$. It is, in fact, possible to derive $\log\det V_e$ directly from (4.22) by the formula

$$\log\det V_e = \int_{-0.5}^{0.5} \log\det S_x(f)df. \tag{9.2}$$

Substituting $\widehat{S}_x(f)$ gives the estimate $\log\det\widehat{V}_e$; see Mohanty and Pourahmadi (1997). This can be exploited to construct an information criterion for selecting the bandwidth of smoothing to be applied in the estimate $\widehat{S}_x(f)$. Figure 9.7(a) shows this criterion for the bivariate Gas furnace series, plotted against a function of the bandwidth that we call the *equivalent order*. The minimum of the criterion occurs at an equivalent order of 17.5 corresponding to a bandwidth of 0.028, which is somewhat less than that used for the spectral analysis of these series in Chapter 3. Figure 9.7(b) shows the estimated OLIRF from Input to Output, derived from the estimated coefficients $\widehat{\Psi}_k$ using this bandwidth. This is similar, for positive lags, to that in Figure 3.19(a). It is, however, a causal estimate; there are no negative lag response coefficients estimated. Note that the response value is constrained to be 0 at lag 0. This is reasonable because the correlation between the innovations of input and output is estimated to be only 0.05, i.e., there is no current dependence. If this correlation had been large, a structural representation for the dependence of the Output innovation upon the Input could have been used, which would have estimated a lag zero coefficient.

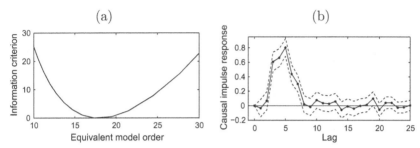

Figure 9.7 *Spectral estimates for the Gas furnace series: (a) the information criterion used to select the smoothing bandwidth and (b) the causal estimate of the OLIRF.*

We will use an information criterion NIC (N for non-parametric) to choose the bandwidth b. This is similar to the AIC defined in (4.31), using the form of deviance defined in (4.34) with the tapering efficiency factor γ in (4.35):

$$\text{NIC}(b) = \gamma n \log \det \widehat{V}_e + 2\, m^2\, p. \tag{9.3}$$

Here, however, p is the equivalent order:

$$p = \sum_{k \geq 1} w_k^2, \tag{9.4}$$

where w_k are the lag window coefficients used to smooth the spectrum and illustrated in Figure 3.12. For the smoothing window that we are using, p is given to an excellent approximation in terms of the bandwidth b as $p = 0.5(b^{-1} - 1)$. The derivation of the criterion (9.3) is based on the result that for a multivariate time series e_t of independent white noise the coefficients of the estimated spectral factor $\widehat{\Psi}_e(\exp -2\pi i f)$ are given in large samples by

$$\widehat{\Psi}_{e\,k} = w_k r_{e\,k}, \tag{9.5}$$

where $r_{e\,k}$ is the lagged sample correlation matrix of e_t and we are now using an additional subscript to indicate the associated series. The sampling properties of $\widehat{\Psi}_x(\exp -2\pi i f)$ can also be related to these white noise correlations in large samples as

$$\widehat{\Psi}_x(\exp -2\pi i f)\widehat{R}_e = \Psi_x(\exp -2\pi i f)\, R_e \widehat{\Psi}_e(\exp -2\pi i f), \tag{9.6}$$

where $\widehat{R}_e$ is the left Choleski factor of $\widehat{V}_e$ and $\Psi_x(\exp -2\pi i f)$, R_e are the true quantities. The error limits shown on the estimated OLIRF in Figure 9.7 are derived by applying to (9.6) the simulated delta method described in Section 4.4.

We now extend this method of impulse response estimation to high frequency sampled (effectively continuous) time series. The essential step is to apply spectrum smoothing *after* transforming the sample spectrum to a warped frequency space. In Chapter 7 we pointed out that almost all the power in the spectrum of the high frequency sampled Oxygen level series resided in the lower 5% of the frequency space and illustrated in Figure 7.2 how the warped spectrum covered the lower 50% of the warped frequency space. In that illustration the spectrum was smoothed *before* applying the warp. However, we now apply smoothing to the spectrum on the warped frequency space, which results in applying *less* smoothing at lower frequencies on the original space, where most of the power resides and it is more important to follow the features of the spectrum. Correspondingly, more smoothing is applied at the higher frequencies where there is little power. Spectral estimation on a warped frequency space is of recent interest in speech signal processing; see Wölfel (2006). We first demonstrate the effect of this procedure for the Gas furnace series using the warp function of the discrete Z operator with discount factor $\theta = 0.5$. Figure 9.8(a) shows the sample spectrum of the Gas furnace Output series over the lower part of the frequency range. The thick black line superimposed shows the spectrum smoothed on the original space with a band width of 0.028. The thick gray line shows the spectrum smoothed on the warped space with a bandwidth of 0.035 then transformed back to the original space. This last estimate more closely follows the pattern of the sample spectrum.

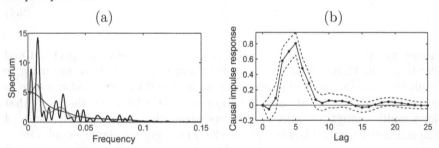

Figure 9.8 *Spectral estimates for the Gas furnace series: (a) sample spectrum (thin line), smoothed spectrum (thick black line) and smoothed spectrum of the Output using a frequency warp (thick gray line); (b) causal estimate of the OLIRF derived using the smoothed warped spectrum.*

The factorization of the smoothed spectrum can also be carried out on the warped frequency space, with the advantage of appreciable or even substantial improvement of the numerical conditioning of the procedure. See Bauer (1955) for an early application of such a transformation, equivalent to the discrete ZAR model operator Z, to factorization procedures. The spectrum factor can then be transformed back to the original space. Figure 9.8(b) shows the estimate of the OLIRF resulting from applying this approach. This is more similar to the estimate shown in Figure 4.3(a), derived from fitting a VAR(6) model.

We illustrate this method further, using the high frequency sampled series of the Infant monitoring system, and explain how the inference procedures are modified when frequency warping is employed. We emphasize the importance of using the frequency warp for this example by displaying the sample spectrum of the Respiration rate series and its logarithm in Figure 9.9.

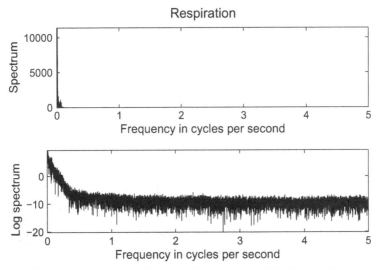

Figure 9.9 *Sample spectrum and log sample spectrum of the Respiration rate series.*

Although the upper plot of the sample spectrum shows little detail, it is deliberately included to emphasize that the power of the spectrum lies in a very small fraction of the frequency space. However, the frequencies over which the sample spectrum appears zero cannot be ignored. After a rapid initial fall, the log spectrum reduces very slowly through the upper 95% of the frequency space above the frequency of 0.5 cycles per second. But we note from (9.2) that the behavior of the spectrum on the logarithmic scale has a substantial influence on the innovation variance of the series. Frequency domain regression to estimate the gain and phase of an open loop relationship between the series, as presented in Section 3.7, can be carried out in each frequency band separately. However, when solving (9.1), the values of the spectral factor in one frequency band are influenced by, and can be sensitive to, the values of the spectrum over the *whole* frequency space. The factorization of the smoothed spectrum can become very ill conditioned numerically when applied directly to spectra similar to those illustrated in Figure 9.9. For a truly continuous process, there is no upper limit to the frequency space. The application of the frequency warp then provides a theoretical as well as a practical and robust response to these issues.

For the present example with 10 samples per second we will apply the discrete warp (6.9) using the discount factor $\theta = 0.92$. This is the value used in the high frequency sampled VZAR model, whose results are illustrated in Figure 7.11. Over the frequency space up to $f = 5$ cycles per second, this warp is almost identical to the continuous time warp (7.24) with rate parameter $\kappa = 0.833$. Figure 9.10(a) shows the sample spectrum on the warp scale with the smoothed warped spectrum superimposed. These now occupy the lower half of the frequency space. The smoothing bandwidth of $b = 0.0323$ was again selected by an information criterion. Figure 9.10(b) shows the same quantities on the logarithmic scale. The slowly reducing log spectrum within the upper 95% of the original frequency scale is transformed to the frequencies above 0.4 on the warped scale. The variations in the log spectrum appear comparable over the whole range of this scale, which confirms the suitability of the chosen value $\theta = 0.92$ and could be used as a criterion in the selection of this warp parameter.

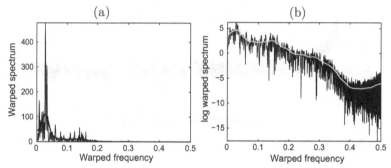

Figure 9.10 *(a) Sample and smoothed (thick line) warped spectrum of the Respiration rate series and (b) logarithms of the sample and smoothed warped spectrum.*

The smoothed multivariate warped spectrum of all three series is then factorized and the factor transformed back, by reversing the effects of the frequency warp, to give coefficients of the open loop impulse response and step response functions from Respiration rate to Oxygen level. These are formed with the Oxygen series constrained to the form of, respectively, an impulse or step, with the feedback to Oxygen from the other series suppressed, but with the dependence of the other two series upon Oxygen and upon themselves included in the model. They are shown in Figure 9.11.

These are very similar to those shown in Figure 7.11, derived by fitting a high frequency sampled VZAR model, though they are shown here up to a higher time lag at which the impulse response appears to be dying away. However, the time lag of this decay is constrained by the spectral smoothing. In this example the lag window used to smooth the warped spectrum decays to one half by lag 20 and effectively disappears by lag 60. Each lag of this corresponds to $2/\kappa = 2.4$ seconds on the un-warped time scale, so the impulse

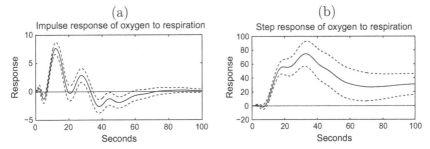

Figure 9.11 *(a) Impulse response and (b) step response of Oxygen level to a unit impulse in Respiration, estimated using spectral factorization of the smoothed warped spectrum.*

response estimated by this means can be expected to be decaying after 50 seconds and effectively to disappear after 140 seconds. The estimated impulse response from the VZAR model is not constrained in this manner, and though not shown to such high lags in Figure 7.11, it continues with minor oscillations to beyond 300 seconds. Given a sufficient length of series and careful model selection using the respective information criteria, one can, however, expect the properties of both the parametric VZAR model and the non-parametric spectral analysis to be similar and to capture the essential features of the lagged dependence between the series.

We conclude this section by explaining the inference procedures used when frequency warping is employed. We take the warped spectrum of the series, as illustrated by the thick line in Figure 9.10(a), to be an estimate $\widehat{S}_X(g)$ of the spectrum of the transformed process X_k defined by (7.25). This is obtained by first calculating the sample covariances of the warped sample spectrum shown by the thin line in Figure 9.10(a). These are, however, conveniently calculated directly from the sample spectrum of $x(t)$, rather than from the *warped* sample spectrum, as

$$\widehat{C}_{X\nu} = \int_{-\ell}^{\ell} Z^{-\nu} S_x^*(f) df. \qquad (9.7)$$

Here, in theory, the limits are given by taking $\ell = \infty$ for a truly continuous series $x(t)$ and $Z = \exp(-2\pi i g)$ is the continuous Z operator with $g = g(f)$ set to the continuous frequency warp function. In practice, for a high frequency sampled discrete series, $\ell = 0.5/\delta$, where δ is the sampling interval and Z and g are the discrete versions with the scaled frequency range up to $f = \ell$, as used in Figure 9.9 with $\ell = 5$. The estimated spectrum is then formed using the standard lag window as

$$\widehat{S}_X(g) = \sum_{\nu} w_{\nu} \widehat{C}_{X\nu} \exp(-2\pi i \nu g). \qquad (9.8)$$

The spectrum so formed is factorized in a similar manner to (9.1) as

$$\widehat{S}_X(g) = \widehat{\Psi}_X(\exp -2\pi i g)\widehat{V}_n\widehat{\Psi}_X(\exp 2\pi i g)', \tag{9.9}$$

where we use $\widehat{V}_n$ because it corresponds to the variance matrix of the estimated error $n(t)$ or n_t from the prediction of $Z^{-1}x(t)$ or $Z^{-1}x_t$ in the natural forms of, respectively, the continuous and discrete VZAR models (7.32) and (6.20). We can therefore, in the continuous case, represent

$$x(t) = \widehat{\Psi}_X(Z)n(t) \tag{9.10}$$

where $n(t)$ is the VCAR(1) process defined in (7.34) with scalar parameter κ and differential disturbance variance $V_\varepsilon = 2\kappa V_n$. Again, in practice, for the high frequency sampled discrete case we obtain

$$x_t = \widehat{\Psi}_X(Z)n_t \tag{9.11}$$

where n_t is the VAR(1) process defined in (6.23) with scalar parameter θ and disturbance variance $V_\varepsilon = (1 - \theta^2)V_n$. To obtain the estimated impulse response of x_t, we expand

$$\widehat{\Psi}_X(Z)/(1 - \theta B) = A(B) = \sum_{k=0}^{\infty} A_k B^k, \tag{9.12}$$

which can be achieved algebraically to generate A_k up to any desired maximum lag. The estimated coefficients in the Wold representation of x_t are then $\widehat{\Psi}_{x\,k} = A_k A_0^{-1}$ and the innovation variance is $\widehat{V}_e = A_0\widehat{V}_\varepsilon A_0'$. Note that on setting $B = 0$ in (9.12), $A_0 = \widehat{\Psi}_X(-\theta)$. These quantities can be used to provide an approximation of the estimated continuous time Wold representation:

$$x(t) = \int_{h=0}^{\infty} \widehat{\Psi}_x(h)dW(t - h) \tag{9.13}$$

with $\widehat{\Psi}_x(h) = \widehat{\Psi}_{x\,k}/\delta$ at $h = k\delta$ and $\text{Var}(dW) = dtV_e/\delta$. Alternatively, $\widehat{\Psi}_x(h)$ in (9.13) can be derived directly from the continuous representation (9.10) by summing the response functions of powers of Z illustrated in Figure 7.1. These can be calculated from the state transition form (7.20) using matrix exponentials.

The sampling properties of these estimates are derived as follows for the high frequency sampled series x_t. We will comment at the end on how these are modified for continuous $x(t)$. The derivation is based upon the approximation to the estimated factor similar to (9.6):

$$\widehat{\Psi}_X(\exp -2\pi i g)\widehat{R}_n = \Psi_X(\exp -2\pi i g)R_n\widehat{\Psi}_E(\exp -2\pi i g). \tag{9.14}$$

Here, $\widehat{R}_n$ is the left Choleski factor of $\widehat{V}_n$ and $\Psi_X(\exp -2\pi i g)$, R_n are the true quantities. Also, $\widehat{\Psi}_E(\exp -2\pi i g)$ is the spectral factor formed from the warped process $E_k = Z^{-k}n_t$ where now the components of n_t are indepen-dent AR(1) processes with parameter θ and unit variance. Let $\widehat{\Psi}_E(\exp - 2\pi ig)$ be formed by factorization of the smoothed spectrum $\widehat{S}_E(g)$ exactly as for $\widehat{\Psi}_X(\exp - 2\pi ig)$. We do not need $\widehat{\Psi}_E(\exp - 2\pi ig)$ explicitly, just the large sam-ple properties of its coefficients, which are given by

$$\widehat{\Psi}_{E\nu} = w_\nu r_{E\nu}. \tag{9.15}$$

Here, $r_{E\nu}$ are the lagged sample correlation matrices of E_k obtained from the covariances $\widehat{C}_{E\nu}$, formed exactly as for $\widehat{C}_{X\nu}$ in (9.7). The sampling properties we require depend, therefore, upon those of $r_{E\nu}$. These are, in large samples,

$$
\begin{aligned}
\mathrm{E}\left(r_{E\,i,j,\nu}\right) &= 0 \\
\mathrm{Var}\left(r_{E\,i,j,\nu}\right) &= \frac{1}{n}\frac{1+\theta^2}{1-\theta^2} \\
\mathrm{Corr}\left(r_{E\,i,j,\nu}, r_{E\,i,j,\nu+1}\right) &= \frac{\theta}{1+\theta^2}.
\end{aligned}
\tag{9.16}
$$

All other correlations between the elements of the lagged sample correlations are zero. The consequence of these properties is that selection of the bandwidth b of smoothing is now based on minimizing the criterion:

$$\mathrm{NIC}(b) = \gamma n \log \det \widehat{V}_n + 2\, m^2\, p, \tag{9.17}$$

where

$$\log \det \widehat{V}_n = \int_{-0.5}^{0.5} \log \det \widehat{S}_X(g) dg \tag{9.18}$$

and p is the equivalent order:

$$p = \left(\frac{1+\theta^2}{1-\theta^2}\right) \sum_{k\geq 1} w_k^2. \tag{9.19}$$

The confidence intervals for the estimated responses, such as those in Figure 9.11, are based on using (9.14) and (9.15) with the delta method and $r_{E\nu}$ sampled from the normal distribution with mean and covariance properties presented in (9.16).

For a continuous series $x(t)$ the main modification is that now $E_k = Z^{-k}n(t)$ where the components of $n(t)$ are independent CAR(1) processes with parameter κ and unit variance. The variances in (9.16) are now $1/(\kappa T)$ where T is the length of the observed series, and the correlations are 0.5. The information criterion becomes

$$\mathrm{NIC}(b) = \gamma T \log \det \widehat{V}_n + 2\, m^2\, p, \tag{9.20}$$

where

$$p = \left(\frac{1}{\kappa}\right) \sum_{k\geq 1} w_k^2. \tag{9.21}$$

9.4   The structural VZAR, SVZAR

In Chapter 6 we showed how the VZAR model, with the flexibility of the discount factor θ, could have an advantage over the standard VAR model for multivariate time series. In this section we demonstrate by example that a further modeling advantage may be gained by applying to the VZAR model the methods introduced in Chapter 5 to construct a sparse structural form of this model: the SVZAR model. Our example is that of the seven US dollar term interest rates illustrated in Figure 1.7. We use the sub-series over the period from 15 Dec. 1987 to 3 Dec. 1990 because we have used this in a previous analysis (Tunnicliffe Wilson et al. (2001)) of a structural VARMA(1,1) model. We will call these the Term rate series. The coherency array plot shown in Figure 9.6 was constructed using these series. Three forms of the VZAR model are presented in Chapter 6 and we will use the natural form in this exercise. This is appropriate because it extends the standard VAR model in a simple manner by replacing the operator B by Z, so that (5.1) becomes

$$\Phi_0 x_t = \Phi_1 Z x_t + \Phi_2 Z^2 x_t + \cdots + \Phi_p Z^p x_{t-p} + n_t, \qquad (9.22)$$

where the components of n_t follow the AR(1) model with coefficient θ, but are now uncorrelated with each other. We tentatively select a value for the factor θ by first plotting the AIC of the predictive form of model fitted for a range of values of θ and p. Figure 9.12(a) shows the AIC plotted against the model order for a range of values of θ, some plots having their minimum at order 2 and others at order 3. It is not possible to annotate each AIC plot in this figure with the value of θ with which it is associated, but some indication is given by the fact that for the models of order 2 the values of the AIC are decreasing as θ increases from 0 to 0.65. Figure 9.12(b) shows the minimum of the AIC over the order, plotted against θ. The minimum order is in fact 2 for θ between 0.2 and 0.8, and 3 outside this range, including for $\theta = 0$, which is a standard VAR model. The overall minimum AIC occurs for $p = 2$ and $\theta = 0.65$.

We then investigate whether this selection is robust by fitting the general form of model with $\rho = 0.65$, the value of θ found from the minimum AIC of the predictive model. Figure 9.12(c) shows the ZIC (6.59) for this model fitted for the same range of values of θ and p. In this plot, for the models of order 4 the values of the AIC are increasing as θ increases from 0 to 0.85. Again, the minima occur at either $p = 2$ or $p = 3$. Figure 9.12(d) shows the minimum of the ZIC over the order, but with the further adjustment using the full penalty term (6.61) for comparability of these minima. In this plot the minimum order is in fact 2 for θ between 0.3 and 0.8, and 3 outside this range. However, the overall minimum again occurs for $p = 2$ and $\theta = 0.65$, which reassures us that the model is robust. We will use these values for the natural form of model in our subsequent analysis. For this value of θ, the low frequency lag of the operator Z is approximately five days or one working week, which is a plausible lag for assessing dependence in the series.

The procedure for building the structural VZAR model parallels that for

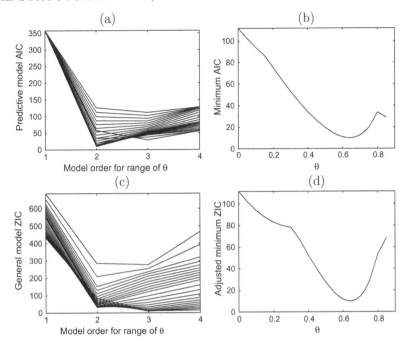

Figure 9.12 *Selection of order and discount parameter for a VZAR model of the Term rate series: (a) the AIC plotted against model order for a range of values of θ for the predictive model form, (b) the AIC minimum with respect to order plotted against θ, (c) the ZIC plotted against model order for a range of values of θ for the general model form with $\rho = 0.65$, (d) the ZIC minimum with respect to order, bias adjusted and plotted against θ.*

the structural VAR: we construct a partial correlation (or conditional independence) graph and use this to postulate a succession of DAG (or SVZAR) representations of the series. These are fitted by OLS regression, diagnostic cross-correlations are inspected and the model extended as appropriate to rectify any inadequacies in these. Model terms are removed where not significant and the overall model assessed by comparison with the saturated model of the same order using the ZIC. The final model is also assessed by comparison of its implied CIG structure with the sample CIG used in its identification.

The sample CIG is constructed from the series appearing in (9.22) and represented in our example for $p = 2$ using the notation of (6.38) as $X_0 = x_t$, $X_{-1} = Z\, x_t$ and $X_{-2} = Z^2\, x_t$. These can be derived from the estimation of the VZAR$(2,\theta)$ model for x_t, which derives initial states to reduce transient errors in their generation. Note that, apart from these errors, the states are *not* dependent on the model. In fact, we used the Kalman filter with the same model to estimate X_{-k} for $k = 0, \ldots, 4$, which are used later. This filter also indicates that the estimation error in these states is quite negligible beyond the first 40 values, which were therefore removed, leaving 600 values for model building. The sample cross-correlation between the 21 series comprising X_0, X_{-1} and X_{-2} were then evaluated. Figure 9.13 shows the consequent CIG with the significant links inserted. For clarity, two diagrams are shown, the upper one indicating links both within elements of X_0 and between elements of X_0 and X_{-1}. The lower indicates links between elements of X_0 and X_{-2}. Links shown with solid lines were all significant with an equivalent t value greater than 2.5; those shown with broken lines were significant with an equivalent t value greater than 2.0 but less than 2.5. The partial correlations are shown adjacent to these links.

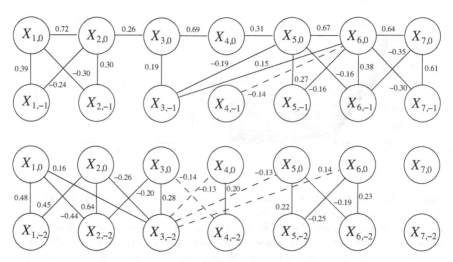

Figure 9.13 *The CIG for the ZAR model variables $X_0 = x_t$, $X_{-1} = Z x_t$ and $X_{-2} = Z^2 x_t$ of the Term rate series. The upper diagram indicates links both within elements of X_0 and between elements of X_0 and X_{-1}; the lower indicates links between elements of X_0 and X_{-2}. The partial correlations are shown adjacent to the links.*

To assess the significance of the partial correlations, we need to modify the test such as (5.13) introduced for SVAR modeling. The standard errors of the partial correlations are based on the same quantities used in (6.66) to evaluate the large sample variance matrix of the estimated coefficients of the general VZAR(p) model. However, for the natural form of model we set $\rho = 0$. Also, we will take the state vector S_t used to define these quantities to be the ordered elements of $X_0, \ldots, X_{-p}$ at time t, because we are considering structural forms of the VZAR(p) model in which there may be dependence between any pair of variables in this vector. The state transition matrix T of the state representation is then constructed from the predictive form of the VZAR($p+1$) obtained by extending the VZAR(p) model with coefficients set to zero. With these modifications, the required matrices are $V_s = \text{Var} S_t$ and

$$\Omega = \sum_{k=-\infty}^{\infty} \theta^{|k|} \text{Cov}(S_t, S_{t-k}) = V_S + \theta (I - \theta T)^{-1} V_S + \theta V_S (I - \theta T')^{-1}. \quad (9.23)$$

The modifications to the partial correlation test are given by the vector of factors f with elements

$$f_k = U_{k,k} / W_{k,k} \quad (9.24)$$

where

$$W = V_S^{-1} \quad \text{and} \quad U = W \Omega W. \quad (9.25)$$

The elements of f are in the corresponding order to the elements of X_0, ..., X_{-p} in S_t and the large sample variance of the partial correlation ρ_k between any *single* element of X_0 and any element of S_t *other than itself* is f_k/n. For all partial correlations of variables within X_0 and between X_0 and X_{-p}, the value of f_k is $(1 + \theta^2)/(1 - \theta^2)$. To apply the test we construct f from the estimated coefficients of the fitted natural VZAR(p, θ) model.

To fit a sparse SVZAR model we use OLS to regress each of the m elements $x_{i,t}$ of X_0 upon a selection of explanatory variables (other than itself) from X_0, ..., X_{-p}. The selection of variables must be acyclic and the initial selection is made by consideration of the CIG. This selection is then assessed and developed using the various criteria previously mentioned. Among these are the t values of the estimated coefficients and the information criterion ZIC, both of which require modification in a similar manner to the partial correlation test. Let $\overline{V}_S$ and $\overline{\Omega}$ be sub-matrices of V_S and Ω with rows and columns corresponding to the selection of explanatory variables used to fit the component $x_{i,t}$ of X_0. Then the large sample variance matrix of the estimated coefficients of these variables is

$$\frac{1}{n} \overline{V}_S^{-1} \overline{\Omega} \, \overline{V}_S^{-1} \hat{\sigma}_i^2, \quad (9.26)$$

where σ_i^2 is the error variance in the OLS prediction of $x_{i,t}$.

To compute the information criterion used to compare models, we require for each of the m OLS regressions a quantity that we refer to as the equivalent degrees of freedom (EDF). For the regression of $x_{i,t}$ this is given by

$$\text{EDF}_i = \text{trace}\left(\overline{V}_S^{-1} \overline{\Omega} \right). \quad (9.27)$$

In the OLS fit to $x_{i,t}$, let s_i be the sample variance of the residuals *not* corrected for degrees of freedom. Then an estimate of the error variance that is unbiased in large samples is

$$\hat{\sigma}_i^2 = \frac{n}{n - \text{EDF}_i} s_i. \quad (9.28)$$

Following (6.60), we introduce a generalized form of the deviance as a function of the sums of squares minimized to fit the model:

$$\text{ZDev} = \sum_i n \log(s_i). \quad (9.29)$$

The determinant form in (6.60) reduces to this because the errors in the separate OLS regressions are assumed to be uncorrelated. Also, following (6.59), we call the related information criterion for comparison of SVZAR models:

$$\text{ZIC} = \text{ZDev} + 2 \sum_i \text{EDF}_i. \tag{9.30}$$

The last term in this corresponds to the penalty term in (6.61), to which it would reduce for a saturated model of order p.

As a guide to the acceptability of a sparse SVZAR model, we will monitor the difference in ZDev between that and the saturated model of the same order. This will be compared with the corresponding change in total equivalent degrees of freedom, $\sum_i \text{EDF}_i$. For a predictive VZAR model form the difference in deviance has a chi-squared distribution with degrees of freedom equal to the change in degrees of freedom, under the hypothesis that each parameter removed in the sparse model has a true value of zero. This distributional property is no longer valid for differences in ZDev for the natural form of SVZAR model fitted by OLS, as described above. However, under the same hypothesis, the expected value of the difference in ZDev is equal to the change in the total equivalent degrees of freedom, so it is useful to compare the two. We will also monitor the difference in ZIC between the sparse SVZAR model under consideration and the saturated model of the same order, and will require that our final model has a lower ZIC if it is to be acceptable.

Cross-correlations of the residuals are an important diagnostic tool. The residuals n_t, derived from the model (9.22) are simply the OLS residuals of the regressions used to estimate the sparse model coefficients. But for our model diagnostics we need to form the cross-correlations of this series, which we now refer to as N_0, with $N_{-1} = Zn_t$, ..., $N_{-p} = Z^p n_t$. For our example we form, for $k = -1, -2$,

$$N_k = \hat{\Phi}_0 X_k - \hat{\Phi}_1 X_{k-1} - \hat{\Phi}_2 X_{k-2}, \tag{9.31}$$

which is the reason why we previously suggested forming $X_1, \ldots, X_{-4}$. We use the sample cross-correlations between the components of N_0 and between the components of N_0 and those of both N_{-1} and N_{-2} for diagnostic checks. These are formed from the sample of values of N_k generated over the span of time $t = 1, \ldots, n$. We will refer to the matrices of these cross-correlations as $r_{N,k}$ for $k = 0, 1, 2$ and their elements as $r_{N,ijk}$. If formed using the true model parameters in (9.31), these sample correlations will have mean zero and variance $(1/n)(1 + \theta^2)/(1 - \theta^2)$, and this will only be affected to a small order by model estimation.

We will not describe the model building steps in complete detail for our example, but before presenting the final selected model we will list some of the considerations that influenced these steps.

1. The length of the series is $n = 600$, which is sufficiently large that coefficients of small magnitude and not therefore of great importance in applications of the model may nevertheless prove to be statistically significant.

The coefficients and their t values are displayed in the diagram representing the final model, and broken lines are used to represent the links corresponding to those considered to be of small magnitude. All the series and their generalized lagged values are also highly positively correlated, so that collinearity problems can be expected in determining the dependence between these variables.

2. We note that the elements of X_0 in the upper diagram of Figure 9.13 are simply linked in sequence. If this is correct, their dependence can be represented by a DAG in which one element is selected as a *pivot* and the links in the CIG are replaced by arrows pointing away from this pivot. We are told that the 2-year rate is dominant in the evolution of term rates, so taking $X_{3,0}$ as the pivot, this diagram would have the appearance

$$X_{1,0} \leftarrow X_{2,0} \leftarrow X_{3,0} \rightarrow X_{4,0} \rightarrow X_{5,0} \rightarrow X_{6,0} \rightarrow X_{7,0}.$$

3. We can check the assumption that such a pivotal representation is adequate by first fitting a DAG in which *all* links to the past are included. All pivotal representations of the current links are then likelihood equivalent, i.e., they have the same (quasi-) deviance. If such a pivotal representation is valid, the diagnostic cross-correlations $r_{N,0}$ should all be acceptable. This was done, using $X_{3,0}$ as the pivot, but a value of $r_{4,6,0} = -0.22$ was found, much larger than the two standard deviation limit of 0.124. The pivotal representation does not then appear to be adequate.

4. We would hope that consideration of links of X_0 to X_{-1} and X_{-2} in Figure 9.13 would help to indicate which pivotal, or other, representation to use. For example, the lack of a moral link between $X_{3,-1}$ and either of $X_{2,0}$ and $X_{4,0}$ suggests the DAG structure $X_{2,0} \leftarrow X_{3,0} \rightarrow X_{4,0}$, which supports taking $X_{3,0}$ as the pivot. However, the lack of a moral link between $X_{5,-1}$ and $X_{4,0}$ also supports a link in the direction $X_{4,0} \leftarrow X_{5,0}$, which is not consistent with this pivotal representation.

5. To resolve the unacceptable diagnostic in $r_{N,0}$, the model, including all past links, was again fitted, retaining $X_{3,0}$ as the pivot, but adding the dependence of *each* variable in the sequence $X_{3,0}, X_{4,0}, X_{5,0}, X_{6,0}$ upon *all the previous* variables in that sequence. The elements of $r_{N,0}$ were then acceptable, as was the ZIC value of -14.92. The link $X_{4,0} \rightarrow X_{6,0}$ was *not* significant, and neither was $X_{3,0} \rightarrow X_{5,0}$. These were removed to give a ZIC of -17.02 and we took this structure as one option for further model development. However, this strategy could have been applied to any sequence of the same variables, giving the same deviance before removal of links that were not significant. Following step 4 above we also considered the sequence $X_{3,0}, X_{5,0}, X_{4,0}, X_{6,0}$ implied by the direction of links

$$X_{3,0} \rightarrow X_{4,0} \leftarrow X_{5,0} \rightarrow X_{6,0}.$$

The link $X_{4,0} \rightarrow X_{6,0}$ was again not significant and removing it gave a

ZIC of -15.65. We took this structure as an alternative option for further model development.

6. Taking each of these two options for the structure of current links, we then restricted the directed links from the past to present variables to the locations of the corresponding links in the CIG of Figure 9.13. These DAG models were assessed according to the ZIC, coefficients that were not significant were removed and diagnostic cross-correlations improved where necessary by insertion of further corresponding links.

The resulting model is represented by the DAG in Figure 9.14. Compared with the saturated model it has 36 rather than 119 coefficients and a ZIC of -101.48. The largest residual cross-correlation was $r_{5,3,2} = -0.19$ but only 5% of the 119 values were outside the 2 standard error limits of ± 0.124 about zero.

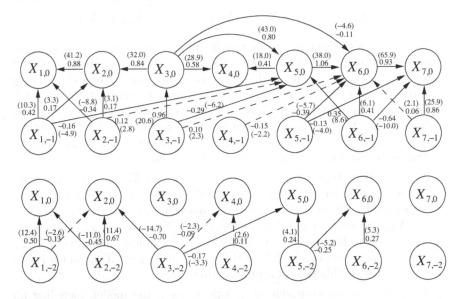

Figure 9.14 *The DAG fitted to the ZAR model variables* $X_0 = x_t$, $X_{-1} = Z\,x_t$ *and* $X_{-2} = Z^2\,x_t$ *of the Term rate series. The upper diagram indicates links both within elements of* X_0 *and between elements of* X_0 *and* X_{-1}; *the lower indicates links between elements of* X_0 *and* X_{-2}. *The regression coefficients and their t values (in parentheses) are shown adjacent to the links.*

Figure 9.15(a) shows a plot of the elements of the CIG implied by the model against the whole set of 119 elements of the sample CIG. The central horizontal cluster of points corresponds to all 66 implied zero CIG values of the model, for which the sample values lie between ± 0.11 apart from two at 0.14 and 0.16. The diagnostic checks on the model are therefore quite satisfactory and the model appears to have captured well the significant partial correlation properties of the series with a sparse structural parameterization.

The natural form of the structural model estimated in this way by OLS

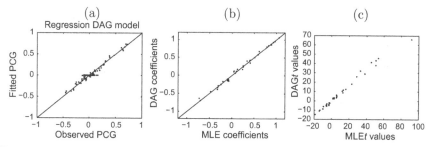

Figure 9.15 *Properties of the SVZAR model fitted to the Term rate series: (a) a plot of the partial correlations of the fitted model against the sample values; (b) a plot of the OLS estimates of the fitted model coefficients against the maximum likelihood estimates; (c) a plot of the t values of the OLS estimates against those of the maximum likelihood estimates.*

regression was also fitted by maximum likelihood. In this context the sparse structural natural VZAR model is considered simply as a parameterization of the standard predictive VZAR model of the same order. If the model closely represents the true structure of the series, then the OLS and ML estimates should be similar. It is when there are, possibly subtle, failures to represent the true structure that the robust properties of fitting by minimizing the prediction errors of $Z_\rho^{-1} x_t$ are advantageous for higher lead time prediction. Figure 9.15(b) shows a plot of the OLS estimates against the maximum likelihood estimates of the model coefficients. These are very close to each other, which is further evidence that the model does well represent the structure of the series. Figure 9.15(c) shows a plot of the t values of the OLS estimates against those of the maximum likelihood estimates. Those of the MLEs are typically some 20% greater, indicating the increased efficiency of these estimates.

We comment briefly on the fitted model represented in Figure 9.14. First, we note that the series $X_{3,0}$ depends only on its first generalized lag value $X_{3,-1}$. But a first order ZAR model of any form is also, simply, an AR(1), which therefore describes the series $x_{3,t}$, the 2-year term rate. The coefficient of this AR(1) can be determined from that of the ZAR(1) to be 0.991, indicating that it is very similar to a random walk. This is just one of the roots of the transition matrix for the state space form of the VZAR(2,θ) model of the whole set of seven series. There are five other roots (out of a total of $7p = 14$) greater than 0.9 and another greater than 0.8. By diagonalizing the transition matrix (see, for example, Belcher et al., 1994), the state vector can be transformed to a set of 14 correlated univariate AR(1) processes with coefficients equal to these roots. Thus, although $x_{3,t}$ appears to be the main driver of the whole set, it is only one of seven other highly autoregressive components. Note also that no dependence is shown of $X_{4,0}$ and $X_{5,0}$ upon their respective generalized lag values $X_{4,-1}$ and $X_{5,-1}$. Our practice is always, on prior grounds, to include the immediately preceding values of the series, even if no link appears in the CIG, as is the case between $X_{4,0}$ and $X_{4,-1}$. But on testing, these links in the DAG proved not to be significant.

A final check on the model is to construct the implied coherency array plot shown in Figure 9.16. That both squared coherencies and partial coherencies correspond well to the graph in Figure 9.6 provides further confirmation that the constructed SVZAR is a good representation of the series structure.

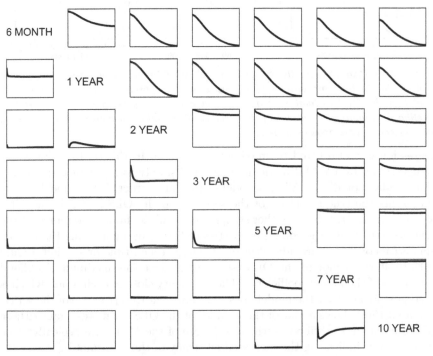

Figure 9.16 *The coherency array plots of the SVZAR model constructed for the seven US dollar Term rate series.*

We conclude by describing an alternative procedure for developing a sparse SVZAR representation given a sample CIG such as that of Figure 9.13. We have, in both Chapter 5 and this section, advocated that DAG models are postulated to explain the observed structure of the CIG. The main challenge of this procedure is to select a limited set of possible links, and their directions, for the current series values. For each model in this set the links from the past can be selected as an appropriate subset of the links in the CIG. Some of these may be removed as being accounted for by moralization, but if not, their importance can be tested by the OLS regression along with the remaining links.

We have, however, experimented with a more mechanistic procedure of selecting the directions of the links between the current variables, which is a development of the simultaneous equation model fitting described in Section 5.9 and illustrated in Figure 5.27. In the context of the present example, we start by fitting by maximum likelihood the natural form parameterization of the SVZAR model, including coefficients corresponding to *all* the links in the CIG. In particular, we included coefficients for links in *both* directions for those between current variables. The model is therefore not acyclic, but of the simultaneous equation form. There is no guarantee that the ML estimates can be uniquely determined without reducing the number of coefficients; this would, however, be detected by the estimation procedure. In this example there was no such difficulty because of the restrictions arising from the sparsity of the links from past values. Significant lag-zero residual cross-correlations were $r_{N,530}$ and $r_{N,540}$, which motivated the addition of a further bidirectional link between $X_{3,0}$ and $X_{5,0}$. Two further links from past values were introduced to reduce residual correlations: $X_{4,-1} \rightarrow X_{4,0}$ and $X_{4,-2} \rightarrow X_{5,0}$. Figure 9.17 shows just the links between current variables in this model, with the estimated coefficients and t values in parentheses. Current and lagged residual correlations were considered acceptable in this model. The links indicated with broken lines in Figure 9.17 correspond to the lower t values,

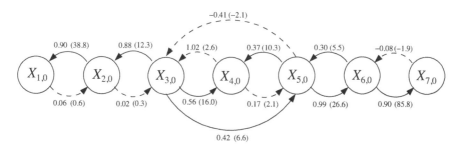

Figure 9.17 *The links between the current variables in the initial SVZAR model fitted to variables $X_0 = x_t$, $X_{-1} = Z\,x_t$ and $X_{-2} = Z^2\,x_t$ of the Term rate series by maximum likelihood, allowing a simultaneous equation form of structure with bidirectional dependence. The regression coefficients and their t values (in parentheses) are shown adjacent to the links.*

the remainder having very much larger t values. Only one bidirectional link, $X_{5,0} \leftrightarrow X_{6,0}$, remains if the broken lines are removed. The remaining model development consisted of sequentially removing those links with low t values between current and lagged variables and within current variables. This led to the removal of 12 links, resulting in the model represented in Figure 9.18. The corresponding increase in deviance was 18.25, giving a p value of 0.11 when referred to the chi-squared distribution on 12 degrees of freedom. The contribution to the increase in deviance of the four current links with the largest t values in Figure 9.17 was 10.07, giving a p value of 0.04 when referred to the chi-squared distribution on 4 degrees of freedom. Although this increase is considered to be just significant statistically, we do not consider it to be significant in a practical sense, because of the length of the series. Even so, we have retained five links from past to current series variables which have t values between 2.0 and 3.0.

The model presented in Figure 9.18 has a likelihood very close to that

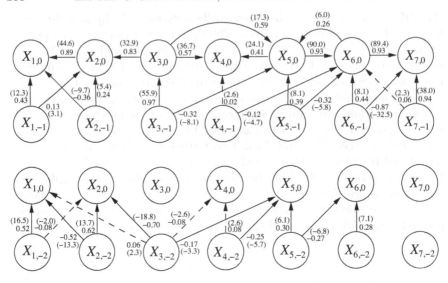

Figure 9.18 *The SVZAR model between variables* $X_0 = x_t$, $X_{-1} = Z x_t$ *and* $X_{-2} = Z^2 x_t$ *of the Term rate series. The upper diagram indicates links both within elements of* X_0 *and between elements of* X_0 *and* X_{-1}; *the lower indicates links between elements of* X_0 *and* X_{-2}. *The regression coefficients and their t values (in parentheses) are shown adjacent to the links.*

obtained by ML estimation of the previous model in Figure 9.14, but required only 33 coefficients, of which 5 are of marginal importance. Its structure is also simpler with the pivotal role of the 2-year term rate being more evident. The simplification does appear to be made possible by allowing the bidirectional dependence between $X_{5,0}$ and $X_{6,0}$. Without this, the earlier model requires dependence of these two variables on several other lag 1 variables. However, the CIGs implied by this and the previous model are very similar—they have many features in common. From this example we can recommend trying the approach of fitting models which allow bidirectionality of some, if not all, the current links as possibly assisting model development. Of course, OLS regression could no longer then be used. We have used ML estimation, which automatically provides the t values (valid in large samples) needed to assess the coefficients. This method still requires extension to the robust fitting of the natural form of model with bidirectional and other acyclic diagrams of the simultaneous equation type.

Our aim in constructing these models has been to understand the lagged dependence of these series upon one another. One practical application is to simulate long records of the series which can be used to assess the performance of financial products related to them. Model based forecasts of the series may also be of value, and they are also of interest as a further means of assessing and comparing different models. We therefore complete this section by presenting graphs of the relative forecasting performance of three models.

The first, which is taken as the baseline, is the *persistence* forecast, constructed for all future values as the latest observed value of the series. This is optimal for a series which follows a random walk that is not dependent upon any other series. The second model we take is the saturated VAR(3) model, the order 3 being selected by minimum AIC. The third model we take is the SVZAR model above constructed using OLS applied to the DAG in Figure 9.14. These three models were fitted to the first 500 values of the series and then used to predict a further 130 values, at lead times from 1 to 10 days. Figure 9.19 shows the percentage deviation of the mean square error of the VAR(3) model and SVZAR model predictions from the mean square error of the persistence predictions.

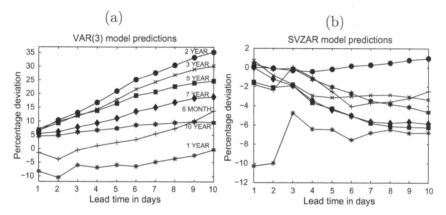

Figure 9.19 *Percentage deviation of the mean square error of predictions using (a) the saturated VAR(3) model and (b) the SVZAR model, from the mean square error of the persistence predictions for the Term rate series. The term periods for the lines in (b) are the same as for those in (a) with the corresponding marker symbols.*

The accuracy of prediction of the VAR(3) model compares poorly with the persistence predictor except for the 6-month and 1-year rate series. Even for the 6-month rate series, the predictions are worse beyond the 4-day lead time. The model is estimated to minimize the 1-day-ahead prediction error, but even for that, there is a component of the prediction error deviation of about 8% that is due to estimation of the large number of coefficients in the saturated model. This results in a worse prediction error deviation for the higher term rate series. As Figure 9.12(b) shows, the order 3 is selected by the AIC for the VAR model, but a much lower AIC is found for the VZAR(2,0.65) model. This implies that the VAR(3) is *not* an optimal model for the series, and though estimated to minimize the 1-day-ahead prediction errors, it can produce much poorer predictions at higher lead times than the much simpler random walk model. In contrast, Figure 9.19(b) shows that the SVZAR model generally outperforms the persistence predictor. The main exception is that the RW model is very slightly better for higher lead times than the SVZAR model for the 2-year rate series and is almost the same at lower lead times.

But this is the series that follows an AR(1) very close to a random walk and is not dependent upon any other series. On the 7-year and 3-year rate series the persistence predictor also very slightly outperforms the SVZAR model at the 1-day lead time, but this is also accounted for by the component of prediction error deviation due to estimation—though of a relatively small number of parameters. Large reductions in the prediction error variance, compared with the persistence predictor, are not expected for financial time series, but Figure 9.19(b) suggests that the SVZAR model may deliver some useful, if modest, reduction for several of these series.

9.5   Further possible developments

This last example, of the Term rates, involved just seven series. The methods we have described might be applied to much larger numbers of series, such as in the field of neuroscience. Then, for construction of the partial correlation or partial coherency graphs using the methods we have presented, correspondingly longer series are required to ensure a sufficient number of degrees of freedom for estimation of the many coefficients. However, methods have now been developed (see Khare et al. (2014) and references therein), for identifying sparse partial correlation graphs in examples where the number of degrees of freedom is insufficient for our methods. Although developed for very high dimensional data, these could also be applied to smaller numbers of series for which the series length was insufficient.

The ZAR models may also be generalized by replacing the states $Z^k x_t$, $k = 0, 1, \ldots, p$ by a set of states, $Z_{\theta_k} x_t$, defined using a spread of discount factors with distinct θ_k, $k = 0, 1, \ldots, p$, and omitting higher powers of the operators. The weights applied to past values by these operators may be orthogonalized; see Partington (1997, p. 123). For predicting a series described as a long-memory process (Beran, 1994), a selection of θ_k may be specified with a wide range of corresponding low-frequency lags. Models of this form are presented in the context of control systems theory by Wahlberg (1994), who further considers the case of equal but complex valued decay constants. The corresponding states of ZAR models would then be given by $Z_\theta^k x_t$ and $Z_\omega^k x_t$, where θ is complex with conjugate $\omega = \bar{\theta}$. Such an approach would concentrate the flexibility of the models for approximating the series spectrum, on an intermediate, rather than the lower, range of frequencies, depending on the specification of θ.

The class of ARARMA models proposed by Parzen (1982) may also be naturally generalized using ZAR operators. These models seek to capture the longer term dependence in a series using one AR or ARMA model, and then to represent any remaining short term dependence in the residuals by a further AR or ARMA model. An AR-ZAR or ZAR-ZAR model achieving the same objective by generalizing (6.14) would be of the form

$$[I - B\xi_1(Z_{\theta_1})] [I - B\xi_2(Z_{\theta_2})] x_t = e_t, \tag{9.32}$$

with θ_1 and θ_2 corresponding, respectively, to small and large low frequency lags, possibly with $\theta_1 = 0$.

The ZAR states have also been investigated by Ibañez (2005) as predictor variables in non-linear models for high lead time forecasting. Non-linearity is modeled by Gaussian regression, also known as support vector methods, and the results from simulation studies and applications to standard real time series show the methods to be competitive with a wide range of other non-linear modeling approaches, including artificial neural nets. A further exposition is given in Ibañez and Tunnicliffe Wilson (2007).

The novel methods presented in this book have been tested on a fair range of real time series as presented in the examples herein. We believe that this demonstrates their potential as a valuable supplement to existing time series methodology. We encourage our readers to explore further applications and to investigate some of the possible extensions outlined in this final section.

References

H. Akaike. Power spectrum estimation through autoregressive model fitting. *Annals of the Institute of Statistical Mathematics*, 21:407–419, 1969.

H. Akaike. A new look at statistical model identification. *IEEE Transactions on Automatic Control*, AC–19(2):716–723, 1973.

H. Akaike and T. Nakagawa. *Statistical analysis and control of dynamical systems*. Tokyo: KTK Scientific Publishers, 1988.

B. D. O. Anderson and J. B. Moore. *Optimal filtering*. Englewood Cliffs, NJ: Prentice-Hall, 1979.

T. W. Anderson. *An introduction to multivariate statistical analysis*. Wiley, 2003.

D. F. Andrews and A. M. Herzberg. *A collection of problems from many fields for the student and research worker*. New York: Springer-Verlag, 1985.

G. Athanasopoulos and F. Vahid. A complete VARMA modelling methodology based on scalar components. *Journal of Time Series Analysis*, 29: 533–554, 2008.

T. O. Awokuse and D. A. Bessler. Vector autoregressions, policy analysis, and directed acyclic graphs: An application to the U.S. economy. *Journal of Applied Economics*, 6:1–24, 2003.

M. S. Bartlett. *An introduction to stochastic processes*. Cambridge: Cambridge University Press, 1955.

F. L. Bauer. Ein direktes iterationsverfahren zur Hurwitz-zerlegung eines polynoms. *Archiv für Elektronik und Übertragungstechnik*, 9:285–290, 1955.

J. Belcher, J. S. Hampton, and G. Tunnicliffe Wilson. Parameterisation of continuous time autoregressive models for irregularly sampled time series data. *Journal of the Royal Statistical Society: Series B*, 56:141–155, 1994.

J. Beran. *Statistics for long-memory processes*. Boca Raton, FL: Chapman and Hall, 1994.

A. R. Bergstrom. *Continuous Time Econometric Modelling*. Oxford: Oxford University Press, 1990.

K. N. Berk. Consistent autoregressive spectral estimates. *The Annals of Statistics*, 2(3):489–502, 1974.

F. J. Beutler. Alias-free randomly timed sampling of stochastic processes. *IEEE Transactions on Information Theory*, 16:147–152, 1970.

R. J. Bhansali. Parameter estimation and model selection for multistep prediction: A review. In S. Ghosh, editor, *Asymptotics, nonparametrics and time series*, pages 201–225. New York: Marcel Dekker, 2007.

R. B. Blackman and J. W. Tukey. *The measurement of power spectra*. New York: Dover, 1958.

P. Bloomfield. *Fourier analysis of time series*. New York: Wiley, 1976.

M. K. Bodwick. *Multivariate time series: The search for structure*. PhD thesis, Department of Mathematics and Statistics, Lancaster University, 1988.

G. E. P. Box and G. M. Jenkins. *Time series analysis: Forecasting and control*. San Francisco: Holden-Day, 1970.

D. R. Brillinger. *Time series: data analysis and theory*. San Francisco: Holden-Day, 1981.

P. J. Brockwell and R. A. Davis. *Time series: theory and methods*. New York: Springer-Verlag, 1987.

R. G. Brown. *Smoothing, forecasting and prediction of discrete time series*. Englewood Cliffs, NJ: Prentice-Hall, 1962.

P. Burman and R. H. Shumway. Generalized exponential predictors for time series forecasting. *Journal of the American Statistical Association*, 101: 1598–1606, 2006.

W. W. Chen and R. S. Deo. The restricted likelihood ratio test at the boundary in autoregressive series. *Journal of Time Series Analysis*, 30: 618–630, 2009.

W. W. Chen and R. S. Deo. The restricted likelihood ratio test for autoregressive processes. *Journal of Time Series Analysis*, 34:325–339, 2012.

G. Chevillon. Direct multi-step estimation and forecasting. *Journal of Economic Surveys*, 21:746–785, 2007.

W. S. Cleveland. The inverse autocorrelations of a time series and their applications. *Technometrics*, 14(2):277–293, 1972.

K. O. Cogger. The optimality of general-order exponential smoothing. *Operations Research*, 22:858–867, 1974.

D. M. Cooper and R. Thompson. A note on the estimation of parameters of the autoregressive-moving average process. *Biometrika*, 64:625–628, 1987.

M. Corduas. La funzione di verosimiglianza marginale in un test di correlazione seriale. *Statistica*, 47:49–65, 1987.

D. R. Cox. Unbiased estimating equations derived from statistics that are functions of a parameter. *Biometrika*, 80(4):905–909, 1993.

D. R. Cox and D. V. Hinkley. *Theoretical statistics*. London: Chapman and Hall, 1979.

R. Dahlhaus. Asymptotic statistical inference for nonstationary processes with evolutionary spectra. In P. M. Robinson and M. Rosenblatt, editors, *Athens conference on applied probability and time series, Volume II*. New York: Springer, 1996.

R. Dahlhaus. Graphical interaction models for multivariate time series. *Metrika*, 51:157–172, 2000.

R. Dahlhaus and M. Eichler. Causality and graphical models in time series analysis. In P. J. Green, N. L. Hjort, and S. Richardson, editors, *Highly structured stochastic systems*. Oxford: Oxford Universiity Press, 2003.

R. A. Davis, T. C. M. Lee, and G. A. Rodriguez-Yam. Structural break estimation for nonstationary time series models. *Journal of the American Statistical Association*, 101:223–239, 2006.

S. Dégerine. Canonical partial autocorrelation function of a multivariate time series. *The Annals of Statistics*, 18:961–971, 1990.

M. Dhamala, G. Rangarajan, and M. Ding. Analyzing information flow in brain networks with nonparametric Granger causality. *NeuroImage*, 41: 354–362, 2008.

J. L. Doob. *Stochastic processes*. New York: Wiley, 1953.

M. Drton and M. D. Perlman. A SINful approach to Gaussian graphical model selection. *Journal of Statistical Planning and Inference*, 138:1179–1200, 2008.

H. Dym and N. Young. A Schur-Cohn theorem for matrix polynomials. *Proceedings of the Edinburgh Mathematical Society*, 33:337–366, 1990.

D. Edwards. *Introduction to graphical modelling*. New York: Springer-Verlag, 2000.

M. Eichler. Granger causality and path diagrams for multivariate time series. *Journal of Econometrics*, 137:334–353, 2007.

R. F. Engle and C. W. J. Granger. Co-integration and error correction: representation, estimation and testing. *Econometrica*, 55:251–276, 1987.

D. F. Findley. Model selection for multi-step ahead forecasting. In H. A. Barker and P. C. Young, editors, *Proceedings of the 7th Symposium on Identification and System Parameter Estimation*, pages 1039–1044, Oxford, 1985. Pergamon.

R. A. Fisher. Tests of significance in harmonic analysis. *Proceedings of the Royal Society of London*, 125:54–59, 1929.

M. K. Francke, S. J. Koopman, and A. F. de Vos. Likelihood functions for state space models with diffuse initial conditions. *Journal of Time Series Analysis*, 31:407–414, 2010.

J. F. Geweke. Measurement of linear dependence and feedback between multiple time series. *Journal of the American Statistical Association*, 77: 304–313, 1982.

J. F. Geweke. Measures of conditional linear dependence and feedback between time series. *Journal of the American Statistical Association*, 79: 907–915, 1984.

C. W. J. Granger. Investigating causal relations by econometric models and cross-spectral methods. *Econometrica*, 37:424–438, 1969.

C. W. J. Granger and M. Hatanaka. *Spectral analysis of economic time series*. Princeton: Princeton University Press, 1964.

C. W. J. Granger and A. O. Hughes. A new look at some old data: The Beveridge wheat price series. *Journal of the Royal Statistical Society Series A*, 134:413–428, 1971.

C. W. J. Granger and P. Newbold. Spurious regressions in econometrics. *Journal of Econometrics*, 2:111–120, 1974.

H. Grubb. A multivariate time series analysis of some flour price data. *Applied Statistics*, 41:95–107, 1992.

J. D. Hamilton. *Time series analysis*. Princeton: Princeton University Press, 1994.

S. Hammerling. Numerical solution of the discrete-time, convergent, nonnegative definite Lyapunov equation. *Systems and Control Letters*, 17: 137–139, 1990.

E. J. Hannan. *Multiple time series*. New York: Wiley, 1970.

E. J. Hannan and M. Deistler. *The statistical theory of linear systems*. New York: Wiley, 1988.

E. J. Hannan and B. G. Quinn. The determination of the order of an autoregression. *Journal of the Royal Statistical Society Series B*, 41:190–195, 1979.

L. P. Hansen and T. J. Sargent. A note on the Wiener-Kolmogorov prediction formulas for rational expectations models. *Economics Letters*, 8:255–260, 1981.

A. C. Harvey. *Forecasting, structural time series models and the Kalman filter*. Cambridge: Cambridge University Press, 1989.

A. C. Harvey and C. Chung. Estimating the underlying change in unemployment in the UK (with discussion). *Journal of the Royal Statistical Society Series A*, 163(3):303–339, 2000.

D. A. Harville. Maximum likelihood approaches to variance component estimation and to related problems. *Journal of the American Statistical Association*, 72:320–340, 1977.

J. Haywood and G. Tunnicliffe Wilson. Fitting time series models by minimizing multistep-ahead errors: A frequency domain approach. *Journal of the Royal Statistical Society Series B*, 59:237–254, 1997.

J. Haywood and G. Tunnicliffe Wilson. A test for improved multi-step forecasting. *Journal of Time Series Analysis*, 30:682–707, 2009.

J. R. M. Hosking. The multivariate portmanteau statistic. *Journal of the American Statistical Association*, 75:602–608, 1980.

C. M. Hurvich and C.-L. Tsai. Regression and time series model selection in small samples. *Biometrika*, 76(2):292–307, 1989.

L. Hurwitz. On the structural form of interdependent systems. In E. Nagel, editor, *Logic, methodology and the philosophy of science*. Palo Alto: Stanford University Press, 1962.

R. J. Hyndman. Yule–Walker estimates for continuous-time autoregressive models. *Journal of Time Series Analysis*, 14:281–296, 1993.

J.-C. Ibañez. *New Tools for Multi-Step Forecasting of Non-linear Time Series*. PhD thesis, Department of Mathematics and Statistics, Lancaster University, 2005.

J.-C. Ibañez and G. Tunnicliffe Wilson. Multi-stage time series forecasting using Gaussian regression and generalised lags. In A. Lendasse, editor, *European Symposium on Time Series Prediction, ESTSP'07*, pages 49–58. Multiprint Oy: Otamedia, 2007.

G. M. Jenkins. General considerations in the analysis of spectra. *Technometrics*, 3:133–166, 1961.

G. M. Jenkins and D. G. Watts. *Spectral analysis and its applications*. San Francisco: Holden-Day, 1968.

R. H. Jones. Fitting a continuous time autoregression to discrete data. In D. F. Findley, editor, *Applied time series analysis III*, pages 651–682. New York: Academic Press, 1981.

R. E. Kalman. A new approach to linear filtering and prediction problems. *Journal of Basic Engineering, Transactions ASME, Series D*, 82:35–45, 1960.

K. Khare, S.-Y. Oh, and B. Rajaratnam. A convex pseudo-likelihood framework for high dimensional partial correlation estimation with convergence guarantees. *Journal of the Royal Statistical Society Series B*, 2014.

J. W. Kirchner and A. Weil. Delayed biological recovery from extinctions throughout the fossil record. *Nature*, 404:177–180, 2000.

R. E. Kromer. *Asymptotic properties of the autoregressive spectral estimator*. PhD thesis, Stanford University Department of Statistics, 1969.

S. L. Lauritzen. *Graphical models*. Oxford: Oxford University Press, 1996.

S. L. Lauritzen and D. J. Spiegelhalter. Local computations with probabilities on graphical structures and their application to expert systems. *Journal of the Royal Statistical Society Series B*, 50:157–224, 1988.

D. S. Lee, M. Zahari, G. Russell, B. A. Darlow, C. J. Scarrott, and M. Reale. An exploratory investigation of some statistical summaries of oximeter oxygen saturation data from preterm babies. *ISRN Pediatrics*, 2011. doi: 10.5402/2011/296418.

W. Little, H. W. Fowler, and J. Coulson. *The shorter Oxford English dictionary*. Oxford: Clarendon, third edition, 1965.

M. T. Lo. *Improvement in multi-step prediction by a class of extended autoregressive models*. PhD thesis, Department of Mathematics and Statistics, Lancaster University, 2008.

H. Lütkepohl. Discounted polynomials for multiple time series model building. *Biometrika*, 69:107–115, 1982.

H. Lütkepohl. *Introduction to multiple time series analysis*. New York: Springer-Verlag, 1993.

M. Marcellino, J. H. Stock, and M. W. Watson. A comparison of direct and iterated multistep AR methods for forecasting macroeconomic time series. *Journal of Econometrics*, 135:499–526, 2006.

S. L. Marple Jr. and A. H. Nuttall. Experimental comparison of three multichannel linear prediction spectral estimators. *IEE Proceedings, Part F*, 130:218–229, 1983.

E. Masry. Alias-free sampling: An alternative conceptualization and its applications. *IEEE Transactions on Information Theory*, 24:317–324, 1978.

E. Masry. Non-parametric covariance estimation from irregularly-spaced data. *Advances in Applied Probability*, 15:113–132, 1983.

R. Mohanty and M Pourahmadi. Estimation of the generalized prediction error variance of a multiple time series. *Journal of the American Statistical Association*, 91:294–299, 1997.

A. S. Morton. *Spectral analysis of irregularly sampled time series data using continuous-time autoregressions*. PhD thesis, Department of Mathematics and Statistics, Lancaster University, 2000.

A. S. Morton and G. Tunnicliffe Wilson. Extracting economic cycles using modified autoregressions. *The Manchester School*, 69:574–585, 2001.

G. Nason. A test for second-order stationarity and approximate confidence intervals for localized autocovariances for locally stationary time series. *Journal of the Royal Statistical Society Series B*, 75:879–904, 2013.

Y. Ogata and H. Akaike. On linear intensity models for mixed doubly stochastic Poisson and self-exciting point processes. *Journal of the Royal Statistical Society: Series B*, 44:102–107, 1982.

B. Øksendal. *Stochastic differential equations*. Heidelberg: Springer-Verlag, third edition, 1992.

L. Oxley, M. Reale, and G. Tunnicliffe Wilson. Graphical models for structural VARMA representations. In R. S. Anderssen, R. D. Braddock, and L. T. H. Newham, editors, *18th World IMACS Congress and MODSIM 2009 International Congress on Modeling and Simulation*, pages 1175–1180. Modelling and Simulation Society of Australia and New Zealand, 2009.

J. R. Partington. *Interpolation, identification and sampling*. Oxford: Clarendon Press, 1997.

E. A. Parzen. Mathematical considerations in the estimation of spectra. *Technometrics*, 3:167–190, 1961.

E. A. Parzen. ARARMA models for time series analysis and forecasting. *Journal of Forecasting*, 1:67–82, 1982.

E. A. Parzen. *Time series analysis of irregularly observed data*. New York: Springer-Verlag, 1984.

K. Pearson. Mathematical contributions to the theory of evolution. On a form of spurious correlation which may arise when indices are used in the measurement of organs. *Proceedings of the Royal Society of London*, 60: 489–498, 1896.

D. B. Percival and A. T. Walden. *Spectral analysis for physical applications*. Cambridge: Cambridge University Press, 1993.

D. B. Percival and A. T. Walden. *Wavelet methods for time series analysis*. Cambridge: Cambridge University Press, 2006.

P. C. B. Phillips and P. Perron. Testing for a unit root in time series regression. *Biometrika*, 75:335–346, 1988.

P. C. B. Phillips and J. Yu. Maximum likelihood and Gaussian estimation of continuous time models in finance. In T. G. Andersen, J.-P. Kreiß, R. A. Davis, and T. Mikosch, editors, *Handbook of financial time series*, pages 497–530. Berlin: Springer-Verlag, 2009.

M. B. Priestley. Evolutionary spectra and non-stationary processes. *Journal of the Royal Statistical Society Series B*, 27:204–237, 1965.

M. B. Priestley. *Spectral analysis and time series*. London: Academic Press, 1981.

M. H. Quenouille. *The analysis of multiple time series*. London: Griffin, 1957.

B. G. Quinn. Order determination for a multivariate autoregression. *Journal of the Royal Statistical Society Series B*, 42:182–185, 1980.

M. Reale and G. Tunnicliffe Wilson. The sampling properties of conditional independence graphs for structural vector autoregressions. *Biometrika*, 89:457–461, 2002.

G. C. Reinsel. *Elements of multivariate time series analysis*. New York: Springer-Verlag, 1993.

G. K. Robinson. That BLUP is a good thing: The estimation of random effects. *Statistical Science*, 6:15–32, 1991.

P. M. Robinson. Multiple local Whittle estimation in stationary systems. *The Annals of Statistics*, 36:2508–2530, 2008.

S. E. Said and D. A. Dickey. Testing for unit roots in autoregressive-moving average models of unknown order. *Biometrika*, 71:599–607, 1984.

W. M. Salla and D. A. Harville. Noninformative priors and restricted maximum likelihood estimation in the Kalman filter. In J. C. Spall, editor, *Bayesian analysis of time series and dynamic models*. New York: Marcel Dekker, 1988.

J. D. Scargle. Studies in astronomical time series. II. Statistical aspects of spectral analysis of unevenly spaced data. *The Astrophysical Journal*, 263: 835–853, 1982.

B. Scheltera, J. Timmera, and M. Eichler. Assessing the strength of directed influences among neural signals using renormalized partial directed coherence. *Journal of Neuroscience Methods*, 179:121–130, 2009.

A. Schuster. On the investigation of hidden periodicities with application to a supposed 26 day period of meteorological phenomena. *Terrestial Magnetism*, 3:13–41, 1898.

G. Schwarz. Estimating the dimension of a model. *The Annals of Statistics*, 6(2):461–464, 1978.

A. K. Seth and L. C. Barnett. The MVGC multivariate Granger causality toolbox: A new approach to Granger-causal inference. *Journal of Neuroscience Methods*, 223:50–68, 2014.

H. S. Shapiro and R. A. Silverman. Alias-free sampling of random noise. *Journal of the Society for Industrial and Applied Mathematics*, 8:225–248, 1960.

R. Shibata. Selection of the order of an autoregressive model by Akaike's information criterion. *Biometrika*, 63:117–126, 1976.

R. Shibata. Asymptotically efficient selection of the order of the model for estimating parameters of a linear process. *The Annals of Statistics*, 8: 147–164, 1980.

R. Shibata. An optimal autoregressive spectral estimate. *The Annals of Statistics*, 8:300–306, 1981.

R. H. Shumway and D. S. Stoffer. *Time series analysis and its applications*. New York: Springer-Verlag, 2000.

D. Slepian and H. O. Pollak. Prolate spheroidal wave functions, Fourier analysis and uncertainty. I. *The Bell Sys. Tech. J.*, 40(1):43–64, January 1961.

V. Solo and A. Pasha. A test for independence between a point process and an analogue signal. *Journal of Time Series Analysis*, 33:824–840, 2012.

N. R. Swanson and C. W. J. Granger. Impulse response functions based on a causal approach to residual orthogonalization in vector autoregression. *Journal of the American Statistical Association*, 92:357–367, 1997.

D. J. Thomson. Spectrum estimation and harmonic analysis. *Proc. of the IEEE.*, 70(9):1055–1096, 1982.

G. C. Tiao and R. S. Tsay. Model specification in multivariate time series. *Journal of the Royal Statistical Society Series B*, 51:157–213, 1989.

H. Y. Tsang and D. W. Bacon. Detection of unsuspected feedback in linear dynamic systems. *Technometrics*, 22:509–516, 1980.

G. Tunnicliffe Wilson. A convergence theorem for spectral factorization. *Journal of Multivariate Analysis*, 8:222–232, 1978.

G. Tunnicliffe Wilson. On the use of marginal likelihood in time series model estimation. *Journal of the Royal Statistical Society Series B*, 51:15–27, 1989.

G. Tunnicliffe Wilson and J. Haywood. Models for high lead time prediction. In W. R. Bell, S. H. Holan, and T. S. McElroy, editors, *Economic time series, modeling and seasonality*, pages 499–524. Boca Raton, FL: CRC Press, 2012.

G. Tunnicliffe Wilson and A. S. Morton. Modelling multiple time series: Achieving the aims. In J. Antoch, editor, *Proceedings in computational statistics, 2004*, pages 527–538, Heidelberg: Physica Verlag, 2004.

G. Tunnicliffe Wilson and M. Reale. The sampling properties of conditional independence graphs for I(1) structural var models. *Journal of Time Series Analysis*, 29:802–810, 2008.

G. Tunnicliffe Wilson, M. Reale, and A. S. Morton. Developments in multivariate time series modelling. *Estadistica*, 53:353–395, 2001.

B. Wahlberg. System identification using Laguerre filters. *IEEE Transactions on Automatic Control*, 36:551–562, 1991.

B. Wahlberg. System identification using Kautz models. *IEEE Transactions on Automatic Control*, 39:1276–1282, 1994.

B. Wahlberg and E. J. Hannan. Parametric signal modelling using Laguerre filters. *The Annals of Applied Probability*, 3:467–496, 1993.

M. P. Wand and M. C. Jones. *Kernel smoothing*. Boca Raton, FL: CRC Press, 1994.

J. C. Whittaker. *Graphical models in applied multivariate statistics*. Chichester: Wiley, 1990.

P. Whittle. The analysis of multiple stationary time series. *Journal of the Royal Statistical Society Series B*, 15:125–139, 1953.

P. Whittle. On the fitting of multivariate autoregression and the approximate canonical factorisation of a spectral density matrix. *Biometrika*, 50:129–134, 1963.

N. Wiener. *Extrapolation, interpolation and smoothing of stationary time series*. Cambridge, New York, 1949.

M. Wölfel. Warped and warped-twice mvdr spectral estimation with and without filterbanks. In S. Renals, S. Bengio, and J. G. Fiscus, editors,

Machine Learning for Multimodal Interaction, volume 4299 of *Lecture Notes in Computer Science*, pages 265–274. Berlin: Springer, 2006.

P. C. Young. *Recursive estimation and time-series analysis.* Heidelberg: Springer, second edition, 2011.

G. U. Yule. Why do we sometimes get nonsense-correlations between time-series?—A study in sampling and the nature of time-series. *Journal of the Royal Statistical Society*, 89:1–63, 1926.

A. Zellner. An efficient method of estimating seemingly unrelated regressions and tests for aggregation bias. *Journal of the American Statistical Association*, 57:348–368, 1962.

A. Zellner and H. Theil. Three-stage least squares: Simultaneous estimation of simultaneous equations. *Econometrica*, 30:54–78, 1962.

H.-C. Zhang. Reduction of asymptotic bias of AR and spectral estimators by tapering. *Journal of Time Series Analysis*, 13:451–469, 1992.

Subject Index

Author Index